INSIDE**OUT**

INSIDEOUT

DESIGN PROCEDURES FOR PASSIVE ENVIRONMENTAL TECHNOLOGIES

G. Z. BROWN
University of Oregon

JOHN S. REYNOLDS
University of Oregon

M. SUSAN UBBELOHDE
Tulane University

JOHN WILEY & SONS

New York Chichester Brisbane Toronto Singapore

The authors will be pleased to provide further information upon request sent to G. Z. Brown or John S. Reynolds at the Department of Architecture, University of Oregon, Eugene, Oregon 97403; or to M. Susan Ubbelohde at the School of Architecture, Tulane University, New Orleans, Louisiana 70118.

ACKNOWLEDGMENTS

This approach to integrating design and calculation procedures has developed over a four-year period. Much of this book was planned and written under grants from the United States Department of Energy and the University of Pennsylvania. Several people have helped in its preparation, and we would like to thank them. They are Jeffrey Cook, Tisha Egashira, Harrison Fraker, John Goldman, Iris Gould, Bruce Haglund, Don Harton, Leslie Holmes, Barbara Keller, Ron Kellett, John Jennings, Matt Miller, Marietta Millet, Murray Milne, Barbara Jo Novitski, Patti Pritikin, Don Prowler, John Welch, Mary Williams, and especially the students in Architecture 321 and 322, Environmental Control Systems, at the University of Oregon in 1980-81, who tested an earlier draft. We are also grateful to the following reviewers of the manuscript in its revised and expanded form: M. David Egan, P.E. FASA, Anderson, South Carolina; Professor Murray Milne, University of California at Los Angeles; Professor Charles C. Benton, Georgia Institute of Technology; Professor Roger G. Whitmer, AIA, University of Illinois at Chicago Circle; Professor Patricia O'Leary, University of Arkansas at Fayetteville; and Professor Joseph Olivieri, Lawrence Institute of Technology, Southfield, Michigan.

CONTENTS

This book is about design processes for the thermal, luminous, sonic and water and waste environments of buildings. Our intent is not to suggest that building design should be generated solely by response to these factors. Yet all design decisions, by intention or not, affect these environments, which in turn affect the natural environment to which we are bound. Because these decisions, which can seem secondary in the design process, can have potentially devastating impacts on the natural environment, we feel that it is important to stress their fundamental importance in the design process.

Robert Venturi noted in Complexity and Contradition in Architecture that "Architecture occurs at the meeting of interior and exterior forces of use and space. These interior and environmental forces are both general and particular, generic and circumstantial. Architecture as the wall between the inside and the outside becomes the spatial record of this resolution and its drama." We feel that it is important to understand the attitudes which lead to the definition of inside from outside by the intervention of the wall. The ideas of what is appropriately in and out and what inside and outside are like play primary roles in defining how buildings respond to their thermal and luminous environments.

You will note in the course of using this book that it doesn't attempt to treat each subject comprehensively, but concentrates on providing a set of procedures or a process which connects them. We have made an effort to provide access to the subjects by referencing well-known texts in each area. Specifically, this workbook uses as its companion reference text Mechanical and Electrical Equipment for Buildings, 6th Edition, by McGuinness, Stein, Reynolds (Wiley, 1980). In the exercises, it is referred to as "MEEB".

The subjects with which we are going to deal have conventionally fallen within the province of a survey course in mechanical and electrical systems in buildings, and there are a number of other good texts which cover that material.

However, they tend to treat their subject narrowly without stressing connections to contextual issues. The values which underlie the approach taken by these texts are not made explicit. As a result, designers schooled in these subjects, while gaining competence in specific techniques, are unaware of the implicit values which accompany the techniques and may be incapable of responding to changing contexts.

The power of the underlying values in the design and use of technical systems (which include most of the parts of a building) has become apparent in the last decade as a result of the obvious degradation of the natural environment.[1] There has been and continues to be considerable debate about the seriousness of environmental problems and the extent to which technical solutions are appropriate for these problems.[2]

We believe that environmental problems are substantial, and that basic changes in both the technical systems used in the building and the design philosophies which govern design methodologies are required to ameliorate these problems.

An example of technical systems which intensify environmental problems is energy use in building. A number of writers have made the connection between energy use and environmental damage.[3] Energy use in buildings has increased dramatically in the 1960-68 period. In New York, energy use per square foot of building doubled between 1950 and 1970.[4] The reasons for this increase are complex and are deeply rooted in the values of our society.[5]

There is a spectrum of opinion concerning development of these values, which places the origin in everything from Judeo-Christian religious belief to our capitalistic/materialistic culture.

One of the characteristics of these values is a profound trust in technological solutions to problems. This has been evidenced in building design by a reliance onmechanical systems to solve climate-related heating and cooling problems, which have been created and/or accentuated by the building design.

Frequently this is a result of the building design being completed with little concern for the building's mechanical systems, which are treated as a later add-on that rely on the brute power of concentrated energy forms.[6]

In other words, the problem boundary (the area in which the aspects of the problem are considered changeable) of heating buildings, as an example, has been defined narrowly as being inside the building's skin. A constrained problem boundary is useful because there are fewer variables to consider and they are easier to measure. However, a narrowly defined problem boundary can lead to neglect of the universe in which the problem exists. For example, it is quite possible to design a building for a hot climate whose features make it more difficult to cool. Therefore, no matter how efficiently the cooling system is designed, it exists in a building which is inherently inefficient.

The process which we suggest is one which attempts to define the problem boundaries broadly, but tries to simplify the variables to the point that they can be measured and their connection to the context understood.

These problems are:

1. How heat flows within and around buildings.
2. How light articulates surfaces.
3. How water is taken into buildings and how waste is expelled.
4. How sound moves through buildings.

The process we will use to approach the resolution of these problems follows a sequence of progressively reduced scales. Take energy use in buildings as an example. If our goal was to reduce energy use in a given building, considerations at various scales might be: Off-Site -- could the materials used to construct the building be selected on the basis of how little energy was used in their manufacture, thereby reducing the buildings' embodied energy? Site-Scale -- can the building be located on the site such that it takes advantage of the climate-modulating characteristics of topography and vegetation? Building-Scale -- can the shape and orientation of the building be designed so that it can utilize solar energy for its heating and cooling? Component-Scale -- can heating and cooling machines be used, the basic process of which makes them highly efficient converters of fossil or natural fuels to heat or cool?

You will note this scaled sequence allows us to consider problems with progressively reduced problem boundaries. In addition, we also favor problem solutions which rely on natural solutions rather than artificial ones. This is based on the assumption that natural solutions tend to be the least energy-intensive and have the most potential for making a positive contribution the natural environment.

For example, in most cases you will find that we favor utilizing solar heating first and fossil fuel heating second, or natural lighting first and electric lighting second.

To reiterate the assumptions which govern this approach to building design:

1. There has been and will continue to be a serious degradation of the natural environment, partly as a result of the way buildings have been and are being designed.

2. Because of the unbreakable bond to natural systems, in the long run environmental cost equals social cost, and conversely environmental benefit equals social benefit.

3. The relationship between the environment and buildings can be understood if the problem boundaries are broadly defined.

4. If the problems are approached in a progressively reduced scale, it reduces the possibility that efficient small scale solutions will exist within inefficient large scale solutions.

5. Designs which utilize natural before artificial solutions are more likely to reduce environmental degradation.

GZB, JSR, MSU
June 1982

1.1 STRUCTURE OF THE WORKBOOK

This workbook is organized in such a way that it is useful in both a lecture and a studio or office situation. The lecture organization is strongly related to the subject of the exercises, and is geared to the initial learning of the material. The division of exercises into steps of the design process allows the tools associated with each step to be combined across many subject areas by people who are already familiar with the subjects, as in studio or office. Example: rules of thumb for the schematic design stage in heating, cooling and daylighting can be compared simultaneously.

Pedagogically, we feel that it is important that calculation exercises be presented in the consistent context of a particular building/site. Traditionally, the teaching of calculation procedures has been presented in a series of often unre-lated contexts to illustrate the particular applicability of a given procedure.

For example, reverberation time is calculated for a large auditorium, water supply sizing for an office building, and heat loss for a residence. We prefer to show calculation procedures' interrelationships to one another in a single building/site, rather than having them done for their own sake, isolated from the larger architectural problem.

We have defined calculation procedures broadly to include ideas of "reckoning" as well as numerical manipulation. Therefore, we have included organization attitudes, rules of thumb, graphic techniques, and numerical calculation. This broader definition of calculation, including intuition and estimation, is important if these procedures are to be used in the beginning stages of the design process.

The basic structure of the exercises is illustrated below.

	HEATING				COOLING				LIGHTING				WATER/WASTE				ACOUSTICS			
	SITE	CLUSTER	BUILDING	COMPONENT	SITE	CLUSTER	BUILDING	COMPONENT	SITE	CLUSTER	BUILDING	COMPONENT	SITE	CLUSTER	BUILDING	COMPONENT	SITE	CLUSTER	BUILDING	COMPONENT
Conceptualizing																				
Scheming																				
Developing																				
Finalizing																				

Fig. 1-1 Structure of the Exercises

The horizontal axis deals with the major subject areas: heating, cooling, light-ing, water and waste, and acoustics. Each of the major subject categories is sub-divided into site, cluster, building, and component scales of consideration. The inter-mixing of subject categories like heating with scales of consideration reinforces the idea that in the design process, the same problem may be addressed at varying levels of specificity with a wide spectrum of design responses. This helps to widen a narrow view of a particular subject area; for example, heating becomes more than only HVAC.

The vertical axis identifies steps in the design process. The specific number of steps is less important than the idea that different kinds of information are needed in each stage in the process. In some cases the same information is needed at each stage but in progressively more detail. In other cases information that is extremely important at the scheming phase is much less important at the finalization stage.

Probably the most important aspect of matching calculation procedures with steps in the design process is to understand the nature of the decisions at each step. Conceptualizing concerns itself with ways of thinking about buildings. Decisions at this point are frequently independent of (or very selective in terms of) program or context. Calculation procedures associated with this design stage should concentrate on providing ways of thinking about the subject areas which are compatible with a range of starting points and are powerful in their organization and crystalizations of ideas. Scheming calculation procedures should concentrate on revealing major context and form relationships in the project (for example, orientation, shape and size). These calculation procedures should be presented in a form which is generative, that is, one which aids in the development of design concepts. Developing procedures should be presented in a way which encourages the comparison of whole systems (such as M. Millet's graphic method of daylight analysis, which allows comparison of daylight factor contours that result from

different window sizes and location). Finalizing techniques should allow detailed comparison of system performance and sizing.

Another important aspect of matching calculation procedures with design stages is ease of use. This is especially important in the beginning stages. Procedures must be perceived as being simple and fast to apply with substantial rewards in the form of aiding decision making. In general, calculation procedures can be divided into three types related to the beginning, middle, and end of the design process: rules of thumb, whole piece (like graphic or sun peg methods), and numerical sizing.

A major tenet of the workbook is that many Environmental Control Systems (ECS) decisions are strongly climate related. Therefore, we have treated climate as the context within which the heating, cooling, and lighting problem is defined. This approach has the advantage of being conceptually similar to most design problems, which tend to occur in a single climate where heating, cooling and lighting take on different degrees of importance depending on that climate.

There seem to be substantial differences between how calculation procedures can be most efficiently taught and how they are used in the design process. In a lecture format course it is easiest to relate the course work to a vertical subject area like passive heating and deal with it from general to specific, and then move on to the next subject area. In the design process the subject areas are most likely linked together and divided horizontally according to design phase. In order to accomodate this difference the workbook is organized in such a way that the information is usable whether access is horizontal or vertical.

CLIMATE				
HEATING	COOLING	LIGHTING	WATER/WASTE	ACOUSTICS
Conceptualizing				
Scheming				
Developing				
Finalizing				

Fig. 1-2 Organization for Lecture Format

The vertical organization by subject, used for the lecture format, allows a broad area of learning to be subdivided into more bite-sized chunks, each of which have some internal consistency and unique features. This organization easily accomodates lecturers with different areas of expertise related to specific subject areas. The tendency of designers to see the subject areas as unrelated is obviously at odds with the notion of design as a synthesis process. The subject areas will be strongly linked by the study building/ site. At the end of all the exercises the designer will emerge with a design which has been considered from the perspective of each area and is a product of their combination.

CLIMATE				
HEATING	COOLING	LIGHTING	WATER/WASTE	ACOUSTICS
Conceptualizing				
Scheming				
Developing				
Finalizing				

Fig. 1-3 Organization for Studio or Office Format

The horizontal division of the studio or office format will allow the designer to consider heating, cooling, daylighting, etc. at varying levels of specificity. At each stage in the design process all the subject areas can be seen as interdependent.

We have divided calculation procedures into three major sections: design strategies, analysis tools, and evaluation tools. The design strategies concentrate on changing the building design at any of the various scales of consideration to meet ECS objectives. Because general building design objectives are much broader than ECS objectives it is important to present redundant design strategies. Any of several avenues then may be used to approach an ECS design objective. The more ways there are to manipulate the building design to meet the ECS objectives, the better chance there is that ECS objectives will be included in a building design.

Analysis tools concentrate on establishing the context in which design strategies and evaluation tools are used. The analysis tools characterize important variables within particular climates and set the stage for understanding the relative importance of each subject such as heating. It is important to note it is not until the building design is envisioned that the magnitude of the environmental control problem can be understood.

Evaluation tools are used to assess a design to see if the ECS objectives have been met. These tools are used in conjunction with design strategies in a re-iterative pattern of design-evaluate-redesign-reevaluate-redesign-reevaluate and so on, which is constant throughout the design process. The nature of the design process demands that evaluation tools for preliminary design stages be extremely easy and fast to use. There are increasing degrees of both time consumption and accuracy as one moves through the design phases.

1.2 ADVICE TO STUDENTS:

Our observations of students using this book indicate that some approaches are more efficient and instructive than others.

Probably the single most important piece of advice we can give is: Figure out the purpose of each procedure before you begin. Understand the objective - don't just fill in the blanks. Since the exercises don't have one correct answer (they have several which depend on your design), it's possible to follow all the right procedures and still end up with an incorrect answer. Time spent at the beginning planning your approach and thinking about basic assumptions will save a lot of time in the long run.

People who are most successful in complet-

ing the exercises are the ones who have a good understanding of what is going on in the exercise before they start it. Before you start, read the entire exercise through carefully to determine the logic and intent. Formulate a plan of attack so that you can complete the exercise in the least amount of time and learn the most.

This book is divided into three main sections: (1) site and building; (2) exercises; and (3) a worked example.

A. Site and Building
This section describes the buildings and the site you will use for all the exercises. There are two building types; a factory and an office. Beyond their obvious use differences, they are also very different in the amount of internal heat that they generate. This difference is important because it can largely determine the relative importance of heating, cooling, and, indirectly, daylighting.

In this section we request that you design the buildings using a grid with specified opening sizes. Even though it may seem inhibiting from a design standpoint, DO NOT deviate from the grid or vary the window size. Small deviations will result in increased time spent in each exercise. The grid allows rapid, accurate calculation of the areas of walls, roofs, etc., because each piece is 100 square feet. Once free of the tediousness of area calculations, the designer can concentrate on more important aspects of the process.

One final piece of advice. Your building/site will be redesigned several times in the course of the term, so the first design should be SIMPLE.

B. Exercises
The exercise section is the heart of this workbook. This is where you will spend the most time and learn the most informative detail.

Information is provided for four climatically different locations: Phoenix, Arizona; Dodge City, Kansas; Charleston, South Carolina; and Madison, Wisconsin.

These cities represent typical climate regions in the U.S.: hot-arid, temperate, hot-humid, and cool. Climate is a major determinant of the relative importance of heating, cooling, and daylighting.

If you do all the exercises for one climate, don't neglect the others. A valuable learning experience is to compare your design to someone's in a different climate.

NOTE: It is easy to also adapt this workbook to your particular climate. To do so, you will need to assemble the following information for your location:

1. Annual summary for your NOAA Local Climatological Data Station (See Samples, pages 3-28 through 3-37).

2. From J. D. Balcomb, et al "Passive Solar Design Handbook" (listed in Bibliography) you will need:

Solar Rules of Thumb, from Table D.1 (see pages 20-26)
Approximate Solar Savings Fraction, from Appendix F
Average Solar Data, from Appendix A

3. Soil type for water/waste design, available from your county sanitation office.

4. Solar radiation at various tilt angles for "best" and "worst" months, from NOAA's SOLMET data. Inquire at your local climatological station. (These are also shown in the User's Manual for PASCALC II, published by TEA Incorporated, Harrisville, N.H.)

The exercises are presented in this order: climate in general, heating, cooling, daylighting, water and waste, and acoustics.

At the end of the climate exercise you will have a good idea about which of the remaining are most important, but you will perhaps not work on the most important problem first. For example, if you are designing for Charleston, a hot-humid climate, you may do the heating exercise first even though you know that cooling is the most important problem. If you find yourself in this situation, read ahead in the cooling exercise to learn about important building design considerations, so that when you design for heating you don't make cooling impossible. Since most climates in the U.S. require heating and cooling don't be afraid to look ahead even if you are designing for frosty Madison.

Each exercise is divided into four steps: conceptualizing, scheming, developing and finalizing. Each step is increasingly more detailed and time-consuming. At each step in the process you adopt a design strategy, propose a design, and evaluate it against given criteria. If your design doesn't meet the criteria you need to redesign and reevaluate before moving on to the next design phase.

These steps match a typical design process. In any design problem you address all aspects of the problem in each step. The fairly straightforward heating, cooling and daylighting goals herein become much more complex when they are combined with other architectural design issues.
You will find that the analysis and evaluation tools tend to be specifically designed to answer a particular question.
The design strategies, on the other hand, are deliberately redundant; several ways of solving the problem are offered. The designer must address many questions in addition to environmental controls in a site/building design, and some strategies will work better than others to resolve all questions.

In addition, these subject areas may be addressed at several different scales: site, cluster, building and component.

These scales are not meant to imply sizes, but scale relationships. Problems may be solved at one scale or at several. For example, in the summer, the building cluster might provide shade for an outside court in the morning and afternoon when the sun is low, and the court could have an awning (component scale) to shade itself from the high sun at mid day.

C. Example

The third section of the book shows the example we worked for Salem, Oregon. Salem's climate is different from the ones you are working with, so the building/site design will probably be different. You should read the example before you do each exercise. A good way to use the example is to imagine how a building/site designed for the Salem climate would be different from yours, and then check your design against the Salem solution. Was it as different as what you expected? If not, why not?

The procedures suggested should be useful to you in your professional work and in taking the registration exam. We have left margins for your notes. We believe that if you take the time to understand the exercises and annotate them, this book will be a valuable reference.

1.3 ADVICE TO TEACHERS:

We have designed this workbook so that it can be added to and rearranged to meet your specific needs. For example, although it addresses passive heating and cooling and daylighting specifically, it leads into conventional HVAC and electric lighting, and these exercises could be easily added.

We have been developing and testing this workbook for several years at the University of Oregon. Its current form was tested in 1980-81. Although this edition has been revised in response to student feedback, it is essentially the same as the one tested.

Our Environmental Control Systems (ECS) class is a large lecture course (enrollment 250-275 students) taught as a two-term sequence. The ten week first term covers climate, heating, and cooling, and the second term lighting, water/waste, and acoustics. In a third term, approximately 70 students do a large project stressing the integration of design and ECS systems. This three term sequence is offered in alternate years, with advanced seminars in climate, solar heating and cooling, daylight, electric lighting, acoustics and HVAC offered the other years. ECS is not a required course, but the first two terms are taken by over 90% of the students in the architecture program. Students can elect to take the course sequence on either a graded or a pass/no pass basis. The students in the course are required to have a minimum of four terms of design or to be of graduate standing in the school. The faculty responsible for the ECS course are G. Z. Brown and John Reynolds. Five Graduate Teaching Fellows (GTF's) were assigned to the course 1980-81. All had taken this course and had a range of experience and background in the field, which included advanced seminars, or a prior GTF appointment in the course, or involvement with the University of Oregon Solar Energy Center.

Class sessions are of three types: lectures, help sessions, and discussion sections. Brown and Reynolds conduct the two two-hour lectures per week. Three types of reading assignments are associated with the lectures: introductory, which present non-technical introductions to the material; required; and supplementary, optional materials which enlarge upon the regular readings, and are more technical, detailed, or specialized than the regular readings. Required books are: Mechanical and Electrical Equipment for Buildings (MEEB), 6th Edition, by McGuinness, Stein, and Reynolds; this workbook; and the Libbey-Owens-Ford Sun Angle Calculator. Recommended texts are Design with Climate by Olgyay and Architectural Interior Systems by Flynn and Segil.

The discussion sections and help sessions, conducted by the GTF's, focus on the exercises in this workbook. With few exceptions, the students work in teams of two on the exercises. There are four exercises per term, two weeks per exercise. For each exercise, in the first week we hold two large help sessions (attendance not required) where questions about how to do the exercise are answered. In the second week, the completed exercise is due at the one hour discussion section. In 1980-81 there were 15 sections with approximately17 students each. Attendance at these sessions was required and the projects were discussed.

The workbook exercises are graded by the GTF's and account for 50% of the final grade. A multiple-choice, machine graded mid-term and final exam each contribute 25% to the final grade.

There is of course a broad range of student work in the course. The two projects discussed below are typical of how student teams approach the exercises.

The first team designed for the climate of Charleston, South Carolina. These students were new to ECS and passive design, and knew nothing about Charleston. Generally, they took formal cues from the Salem, Oregon example, followed the analysis steps and rules of thumb very carefully, and understood what they were doing conceptually. Their design changes in response to evaluation steps were typically relocating the building at the site scale, and manipulating openings and building materials.

In the climate exercise they determined that they would need the sun for heat at some times during the year, that they couldn't shade at the site scale when they need shade, and that winds from each direction would need to be admitted and blocked at different times of the year. Their site design located and oriented the factory differently than the office building in response to the combined analysis of climate and building program.

The cooling exercise was the most difficult for them. They chose cross ventilation as their cooling strategy. Speculating on the factory ventilation flows showed some hot areas, so they added ventilators in the roof. The office sections redesigned in the heating exercise gave them some problems, so they redesigned those sections again. Use patterns over 24 hours clearly illustrated that they could reduce the cooling loads by 22% in the office building by shifting schedules, but the new schedules didn't make sense socially. In the factory, on the other hand, even though changing theschedules made sense, the cooling load reduction achieved was only 4%. This difference between the two buildings clearly emphasized the role of internal gains. Plotting the overheated times of the year indicated a need for an adjustable shading device.

The second team took a different, but also fairly typical approach to the exercise. These students had some ECS background and a lot of interest in passive design and their climate was Madison, Wisconsin. The winds in Madison pose an opportunity for site and building scale seasonal response, unlike those in Charleston, because of the winter-summer directional shift. These students put a lot of work into design at the end of the climate exercise, doing building designs which considered heating, cooling and daylighting. These designs explored the potential in variations in site location and orientation, quantity of building skin and volume, and amount of aperture. With this background, they proceeded with a design that worked through all the exercises, meeting all criteria with a minimum of redesign. In the factory, cooling and daylighting was emphasized, while in the office building heating was the major design goal.

This team did no redesign in the heating exercise, and only minimal redesign in the cooling exercise.

The greatest success in the course came from comparing the two buildings and four climates. There was also a marked improvement in the quality of questions asked in lectures and discussions over the previous years, much of which we attribute to the revised exercises.

The exercises have been arranged so that they can be graded easily by course graders, GTF's, or instructors. Grading does require knowledge of the field. We concentrate our grading effort on the design documentation and the evaluation procedure.

---

[1] See the following references for an introduction to the effect of technical development on the natural environment.
Carson, R. -- Silent Spring
Leopold, A. -- Sand Country Almanac
Commoner, B. -- The Closing Circle
These are classic in the awakening and formation of the environmental movement in the U.S.

[2] Meadows, et al - The Limits to Growth
Ford Foundation -- A Time to Choose
Kahn, H. -- The Next 200 Years

[3] Commoner, B. -- The Poverty of Power
See especially Chapters one and two.

[4] Stein, Richard -- Architecture and Energy
See Chapter 1 for a description of how, and how much, energy is used in buildings.

[5] White, Lynn -- "Historical Roots of the Ecological Crises" in Science, 10 March, 1978.

Nash, R. -- Wilderness and the American Mind
Leiss, W. -- The Domination of Nature
McHarg -- Design with Nature
See especially: "The Plight," "The Case and the Capsule," and "On Values".

[6] Banham, R. The Architecture of the Well Tempered Environment.

2.0 SITE AND BUILDING DESCRIPTION

Throughout the exercises you will be designing a small factory, an associated office/conference building, and outside spaces for a site situated near a freeway and railroad line. (See the site context map.) Inherent in the structure of the exercises is an emphasis on redesign and quick manipulation of the building form. To facilitate this, your designs will be relatively abstract and must conform to a 10' grid as specified below. We have provided background grids for design presentations in the exercises. Remember, it is extremely important not to become too attached to your design, model or drawings at any point because they should be changing.

A. General Requirements

1. In order that all calculations and comparisons might be more clear and concise, your design must conform to the 10 foot grid at all stages. This means that walls, roofs and floors must fall on a plane of the 10' grid or on a 45° cut through the grid. This is not a real building; it will not be built. It is a matrix on which to test various design ideas.

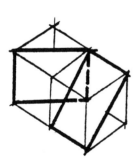

2. At all stages you must have two thermally separate buildings, although they may be connected by circulation space.
3. At all stages the apertures must be in 50 or 100 s.f. increments.
4. Initially you must have:

a. window openings in both the walls and roof of each building
b. at least one window or skylight in each room
c. at least one window facing each orientation (N,E,S,W)
d. at least two entrances, on different sides of the building, in each building

B. Exterior Program Requirements

1. Outside space (300 s.f. minimum) associated with building use
2. Parking for staff and visitors at 325 s.f. per vehicle
3. Vehicular access to the factory

C. Factory Building Requirements

1. 72,000 cu. ft. total
2. 16 machine/work stations
   a. 1 person working per station
   b. average 2.5 horsepower per machine
   c. 500 watts lighting per station
   d. 150 s.f. per station
3. 100 s.f. of exterior door (minimum)
4. Two 8 hour shifts = 16 hours occupancy

D. Office/Conference Building Requirements

1. Multi-purpose space with mezzanine (used for display, conference, lunch room) 10,000 cu.ft.
2. Two office space - 2,000 cu.ft. each
3. Two toilet spaces which are accessible from the factory when the offices are closed - 2,000 cu.ft. each. Minimum floor area is 150 s.f. each. Interior portions need not fall on the grid.
4. Circulation area connecting all spaces 10' wide
5. Tower - 3,000 cu.ft.
6. One 8 hour shift = 8 hours occupancy

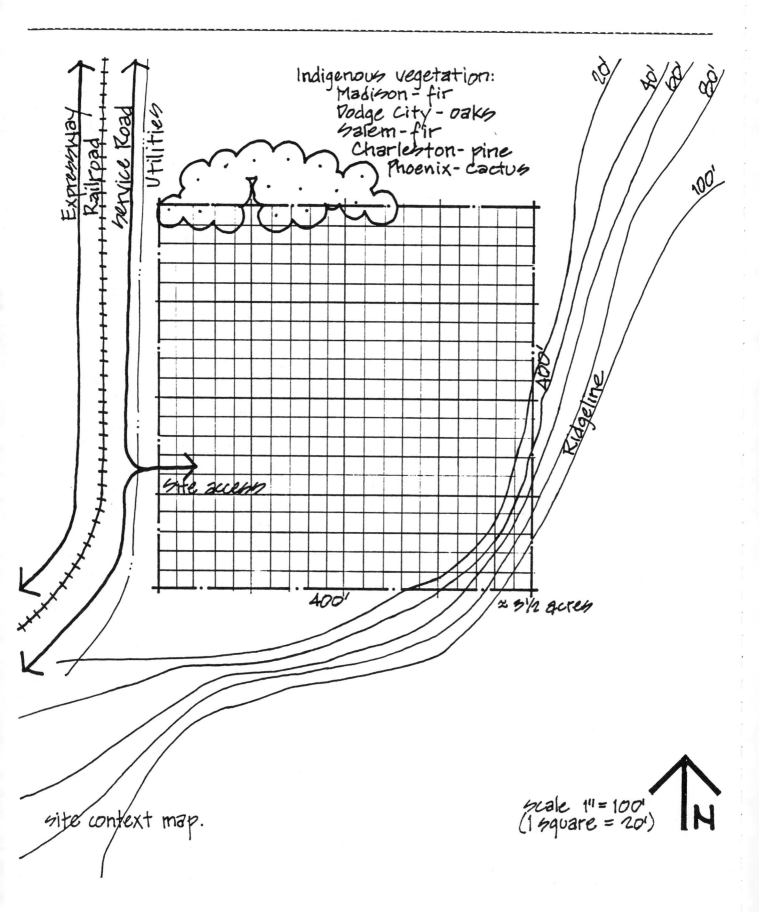

Indigenous vegetation:
Madison - fir
Dodge City - oaks
Salem - fir
Charleston - pine
Phoenix - cactus

Expressway
Railroad
Service Road
Utilities

20' 40' 60' 80'

100'

400'

Ridgeline

site access

400'

≈ 3½ acres

site context map.

scale 1" = 100'
(1 square = 20')

N

3.1 INTRODUCTION

A beginning step in many design projects is analyzing the context of the problem; identifying what you have to work with, what is demanded in the solution, and where the potential assets and liabilities lie in the interaction of these. This exercise is structured to explore the climate as context, using a range of analytical tools to answer questions.

You need to determine when your climate is comfortable for people, potential thermal benefits and problems of the sun and wind on the site, what potential benefits and problems that daylighting offers on the site, and the implications of building program and configuration for choosing a design strategy.

The exercise is structured as a series of goals which are reached by analyzing available data and information about the site and program. Summary questions help you establish design strategies for your climate that you will use throughout the exercises.

One evaluation tool involves building a model of the site and observing sun shadow patterns for different sun positions (summer morning, winter noon, etc.). These observations require direct sun. Given the unpredictability of the weather, we recommend that you build your model <u>now</u> and have it ready to test when the sunny weather occurs.

Remember that a sample of how a climate is analyzed with these tools is found in "Climate; the Salem Oregon Example".

A good way to begin is to read the narrative summaries of the five cities to get a comparative sense of your climate.

3.0 CLIMATE

3.1 INTRODUCTION

Contents

3.2 THERMAL COMFORT

GOAL

Determine how people stay thermally comfortable in your climate.

A. Analysis tool - "Climate Plotted on Bioclimatic Chart"
Use the NOAA climatological data for your city (found at the end of this exercise section 3.7), to <u>plot</u> the average maximum and minimum temperatures for each month against the relative humidity (R.H.) that occurs approximately the same time (generally minimum temperature occurs approximately 3-4 a.m., maximum temperature 3-4 p.m.). <u>Connect</u> the 2 points for each month to get a picture of how the temperature/RH condition changes over a day.

For summer temperatures above the average maximum, find the summer dry bulb (DB) from MEEB 6th edition (<u>Mechanical and Electrical Equipment for Buildings</u>), Appendix C table C.2a. Then <u>extend</u> your hottest month line until it crosses this temperature on the Bioclimatic Chart. This will represent the worst case design condition.

(A full description of the bioclimatic chart and how to use it will be found in <u>Design with Climate</u>, Victor Olgyay, pp. 19-23.)

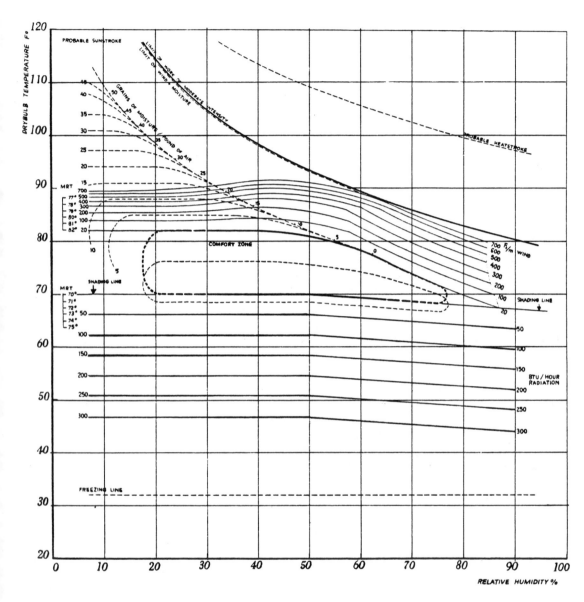

Figure 3-1   Bioclimatic Chart.   From Victor Olgyay, _Design With Climate: Bioclimatic Approach to Architectural Regionalism_ (copyright (c) 1963 by Princeton University Press) Fig. 45, p. 22.   Reprinted by permission of Princeton University Press.

QUESTIONS

Use what you have plotted to answer the following:
A. In your climate, most of the time people will be (too hot, too cold, comfortable, too dry, etc.)?

B. In what months does the difference between high and low temperatures seem significant? (For example, on opposite sides of the comfort zone?) Such diurnal changes can be used to advantage by storing heat or coolth in periods when they are not needed for use in periods when they are.

C. At what times of the year will you want to use the wind for cooling?

D. At what times of the year will you want to use the wind for cooling only if you also add moisture?

E. At what times of the year will you want to block wind?

F. At what time of year will you need to shade yourself by blocking the sun?

G. At what time of year will you need to use the sun to stay warm?

H. What will be the role of the building in keeping people comfortable
  - in summer?

  - in fall?

  - in winter?

  - in spring?

3.3 SITE THERMAL POTENTIAL

GOAL

Determine how you can use the natural energy provided by sun and wind to keep people comfortable.

A. Analysis Tool - "Wind Roses and Site Wind Flows"
Wind "roses" (found in 3.8) show two important characteristics for each city in each month: a) what directions are most prevalent each month, and b) what % of the time is calm. Another important characteristic which must be added is the average speed of the wind, which may be obtained from the yearly climate summaries in 3.7. (For those months in which data is not given for Madison, use the month immediately following or preceeding.)
1. From your answers in 3.2
Months to admit wind:
Months to block wind:
2. On one of the following site plans, show the approximate wind flows in the months where wind is to be admitted. Show directionality, approximate flow around obstacles, and indicate monthly % calm and average wind speeds for these "admit" months.
3. On the other site plan, show the approximate wind flows in the months where wind is to be blocked. Show directionality and flow around obstacles.
4. In months in which wind is to be admitted, how reliable a resource is it? (Consider % calm, consistency of direction, and average speed; higher speeds mean more thorough and rapid "flushing" of heat from a building.)

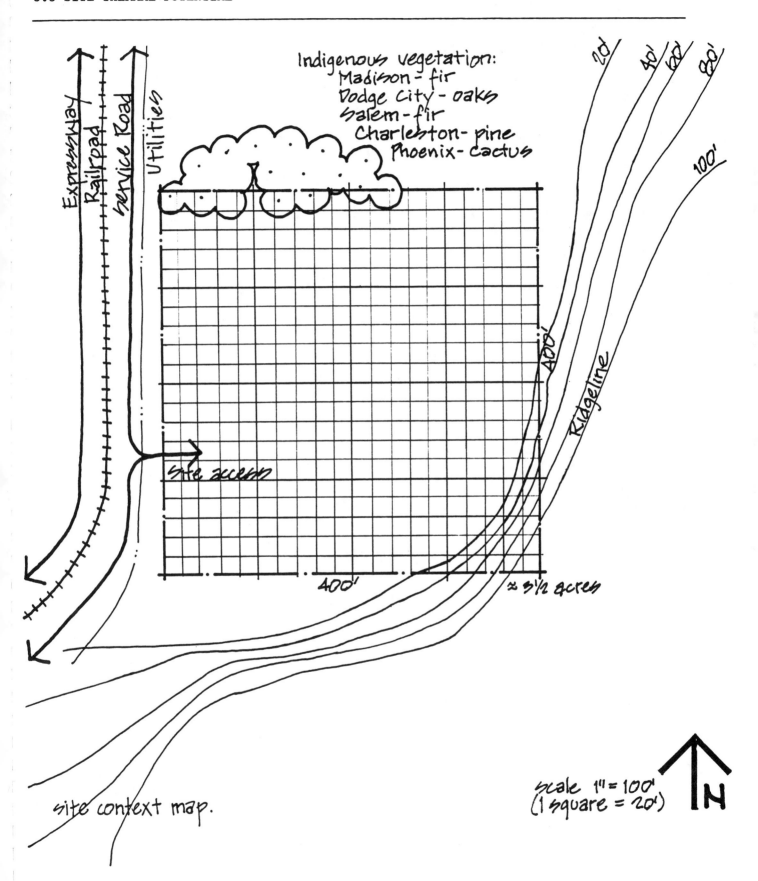

Indigenous vegetation:
 Madison - fir
 Dodge City - oaks
 Salem - fir
 Charleston - pine
  Phoenix - cactus

Expressway
Railroad
Service Road
Utilities

20' 40' 60' 80'

100'

400'

Ridgeline

Site access

400'

≈ 5½ acres

site context map.

scale 1" = 100'
(1 square = 20')

N

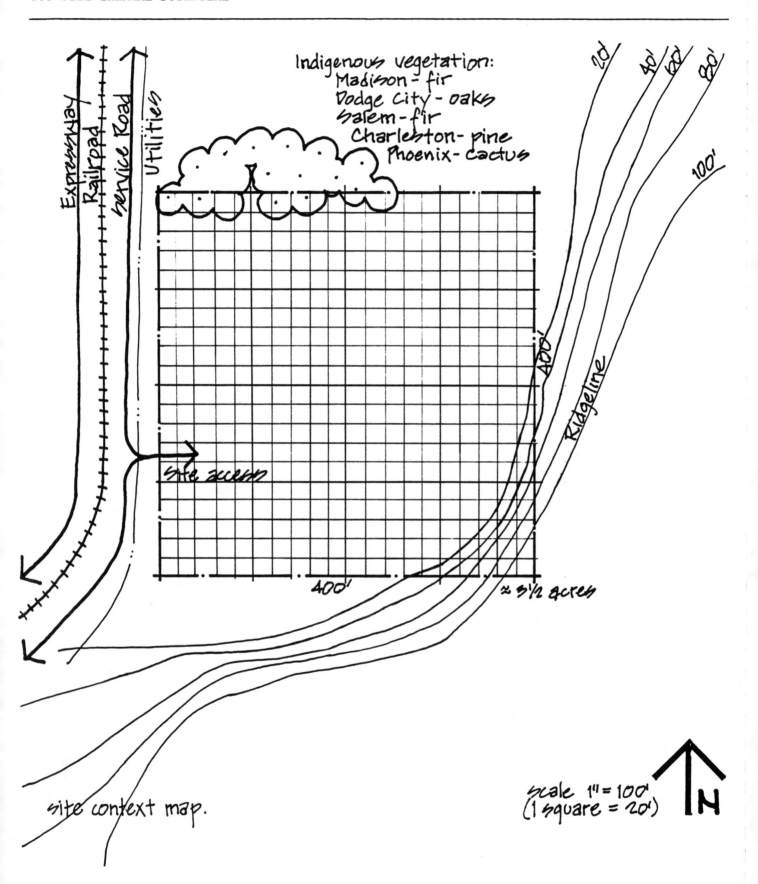

Indigenous vegetation:
Madison - fir
Dodge City - oaks
Salem - fir
Charleston - pine
Phoenix - cactus

site context map.

scale 1" = 100'
(1 square = 20')

B. Analysis Tool - "Sun Peg Shadow Plot"
Using the sun peg will give you information on how the sun and your site conditions will interact daily and seasonally.
1. Make a site contour model, scale 1"=40' from the site context map in the building/site program, section 2.2. Mark the surface into a 10' grid and include all vertical obstructions such as vegetation and land forms. Leave approximately 9" of horizontal space across the south end of your model base for mounting the chart.

Use the "peg shadow plot chart" for your latitude in section 3.9.
Mount the chart perfectly <u>horizontally</u> on your model base, south of the site. The north arrow at the center of the chart must face due north on the model.
Mount a peg exactly the height shown on the shadow plot where indicated, so that the peg remains perfectly <u>vertical</u>.
2. Find a sunny day and take the model and mounted chart out under full sun (indoor lamps do not work; they will give you divergent shadows). By tilting your model in the sun, you can make the <u>end of the peg's shadow</u> fall on any intersection <u>on the shadow plot.</u> Each intersection represents a given time on a given day. When the peg shadow falls on this intersection, the shadows and sun penetration in your model are the actual condition for that time of day and year.
3. <u>Explore</u> how the sun and shading on the site changes over the course of a day and seasonally over the year. On your model record the shadows cast at 9 a.m., noon, and 3 p.m. for the month you most need to block the sun and the month you most need to admit the sun. Transfer this information onto the two site plans which follow, one for each month you looked at.
4. Indicate (on the admit sun site plan) in which areas you could locate a 30' high building without blocking someone else's access to the sun during these hours. A more thorough discussion of solar rights (and the resulting "solar envelope" within which buildings can exist without shading their neighbors) can be found in Ralph Knowles <u>Energy and Form</u>.

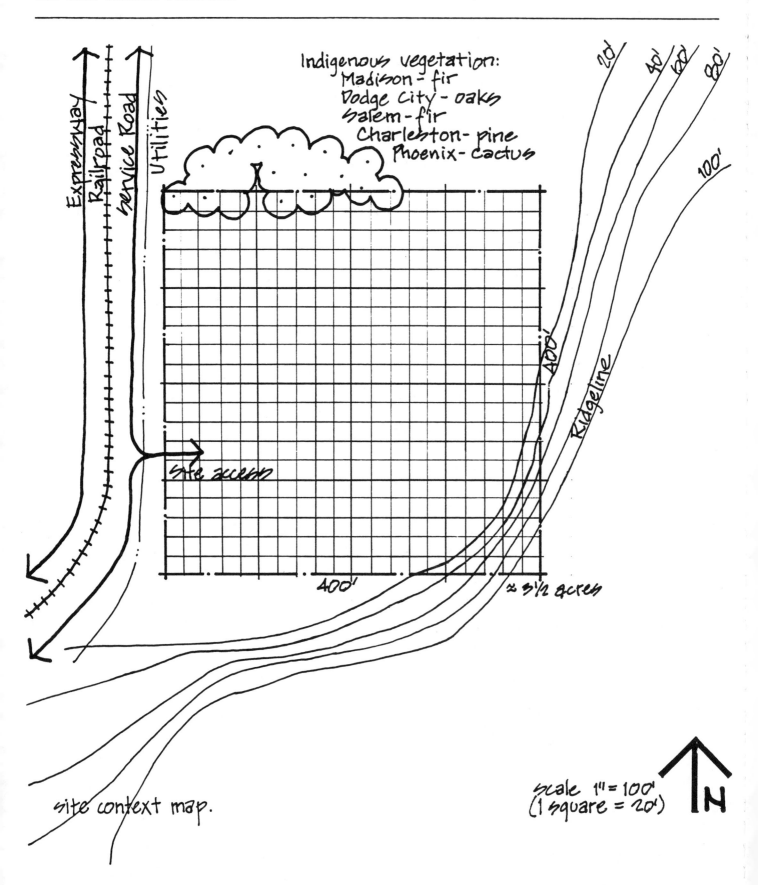

Indigenous vegetation:
Madison - fir
Dodge City - oaks
Salem - fir
Charleston - pine
Phoenix - cactus

site access

400'

≈ 3½ acres

Ridgeline

20' 40' 60' 80'

100'

400'

site context map.

scale 1" = 100'
(1 square = 20')

N

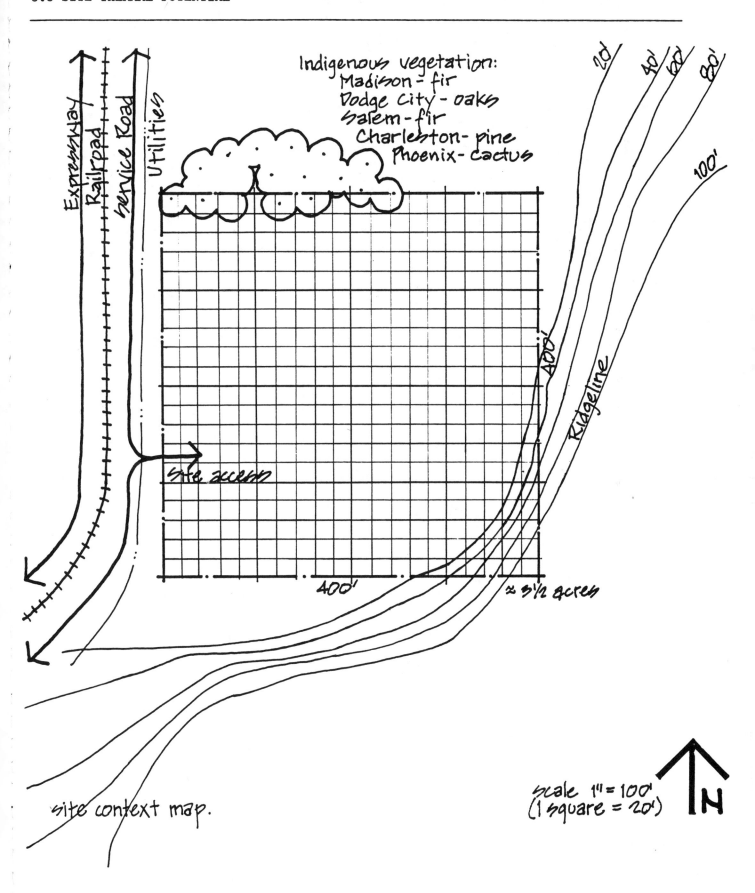

Indigenous vegetation:
Madison - fir
Dodge City - oaks
Salem - fir
Charleston - pine
Phoenix - cactus

site context map.

scale 1" = 100'
(1 square = 20')

C. Analysis Tool - "Climatic Response Matrix"
From the Bioclimatic Chart, you have found how to extend
the "comfort zone", by admitting or blocking sun or wind
at appropriate temperature/R.H. combinations. These conditions can be represented by a matrix:

	ADMT SUN	BLOCK SUN
ADMT WIND	A	B
BLOCK WIND	C	D

Diagram 2

As a result of sun position, wind direction, topography
and vegetation, these conditions may occur on your site,
resulting in "microclimates" that may be favorable building sites.
1. On the first of the following site plans, for your
hottest month, at noon, label each 100' x 100' square on
the site with the letter that describes conditions
occuring naturally.
2. On the other site plan repeat this procedure for noon
of your coldest month.

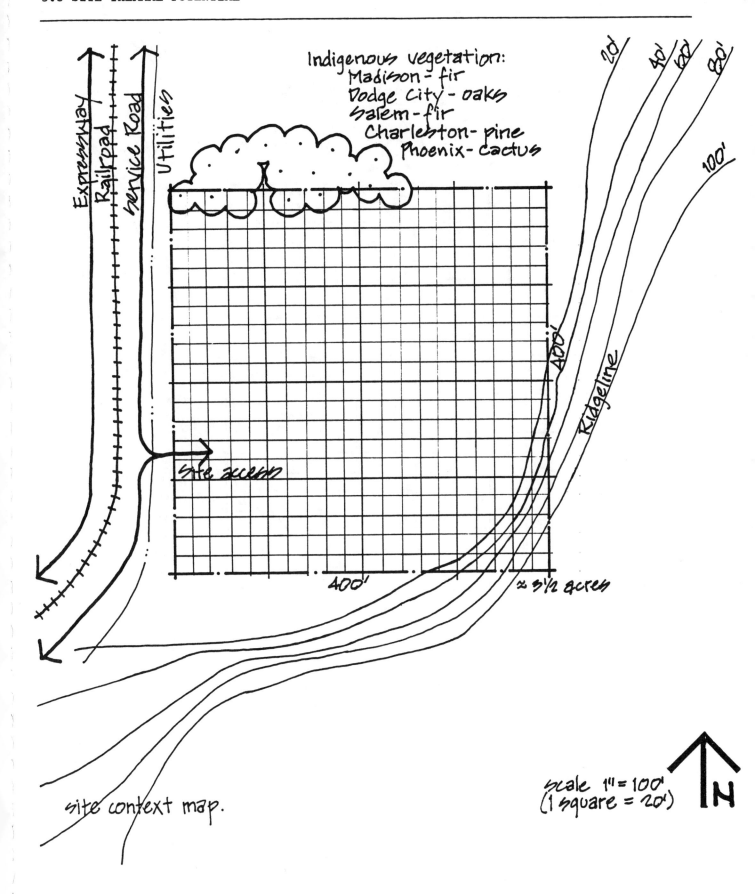

Indigenous vegetation:
  Madison - fir
  Dodge City - oaks
  Salem - fir
  Charleston - pine
  Phoenix - cactus

Expressway
Railroad
Service Road
Utilities

site access

400'

Ridgeline

≈ 3½ acres

20' 40' 60' 80'

100'

400'

site context map.

scale 1" = 100'
(1 square = 20')

N

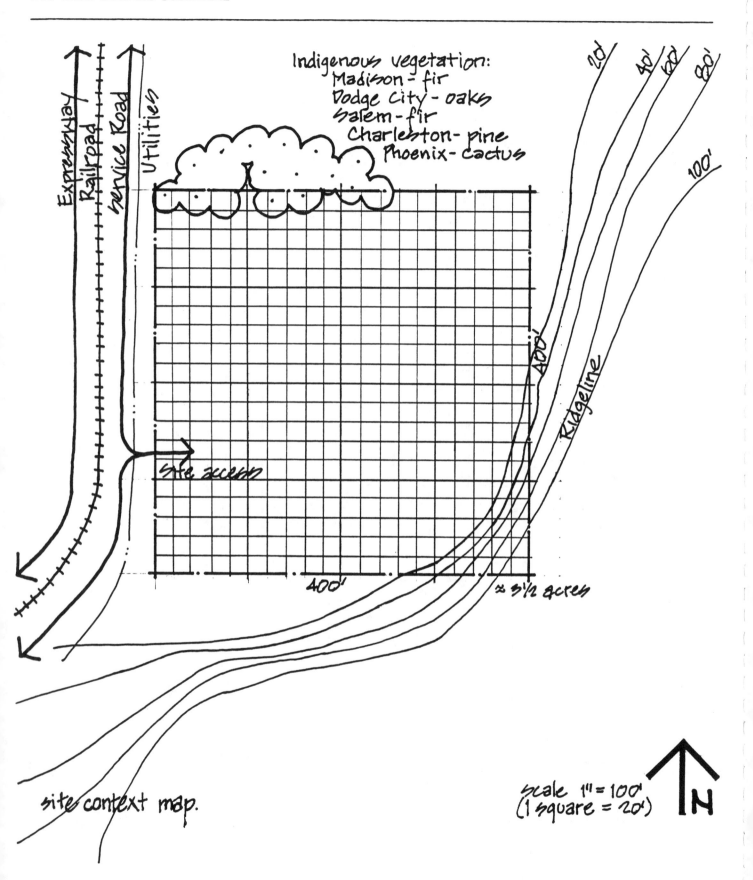

Indigenous vegetation:
Madison - fir
Dodge City - oaks
Salem - fir
Charleston - pine
Phoenix - cactus

Expressway
Railroad
Service Road
Utilities

Site access

400'

≈ 3½ acres

Ridgeline

20' 40' 60' 80'

100'

400'

site context map.

scale 1" = 100'
(1 square = 20')

N

QUESTIONS

A. Which block/admit condition (A, B, C or D from the matrix) is the most favorable for heating?

B. Which is most favorable for cooling?

C. Considering the results of the last three analysis tools, which site location is optimum for heating?

D. Which site location is optimum for cooling?

3.4 VISUAL COMFORT

GOAL

Determine how much light people need to do their work.

A. Analysis Tool - "Recommended Footcandle Levels"
Refer to the energy conservation lighting level recommendations, such as those in MEEB (p. 736) to complete this table:

Task	Recommended Footcandles
Circulation, as in hallways	fc
Conference Areas	fc
Display areas	fc
Office Work	fc
Toilet rooms	fc
Work Stations and their immediate surroundings	fc

3.5 ILLUMINATION POTENTIAL

GOAL

Determine if you will need lots of big windows, or a few
little ones, or somewhere between to provide adequate
daylighting in your climate.

A. Analysis Tool - "Sky Condition"
Use the NOAA climate data in 3.7 to graph your climate in
terms of sky conditions:

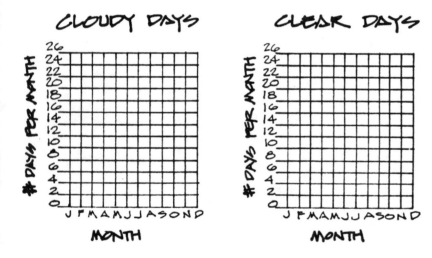

Diagram 3

B. Analysis Tool - "Available Illumination"
The previous analysis tool indicates whether clear or
overcast skies are prevalent in your climate throughout
the seasons.
1. From tables of available daylight at various latitudes
(MEEB Table 19.2a for overcast skies, Table 19.2c for
clear skies), graph the following information on diagram
4:

Available exterior footcandles over a typical day (in
each case, specify sky condition--overcast or clear?)
for:
December 21
March or September 21
June 21

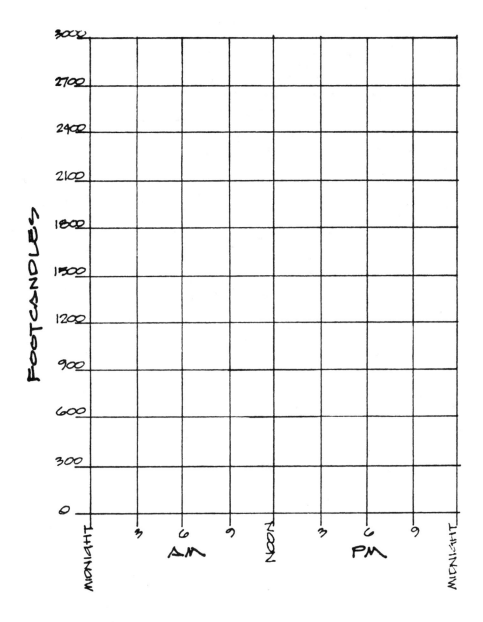

Diagram 4

Note: For seasons where clear sky predominates, graph
both north and south windows.

_____

C. Analysis Tool - "Required Exterior Illumination"
Roughly, "lots of big windows" (which amount to a total
window area of about 1/5 of the total floor area) will
produce <u>indoor</u> daylight levels of about 10% of outdoor
levels. "A few little windows" (total window area about
1/50 of total floor area) will produce indoor daylight
levels of about 1% of outdoor levels.

1. Extend the table you began in the previous analysis
tool:

Task	Recommended Ft.C. (Interior)	Min. Required Exterior Illumination	
		If interior=10% exterior level	If interior=1% exterior level
Circulation	_____	_____	_____
Conference	_____	_____	_____
Display	_____	_____	_____
Office	_____	_____	_____
Toilets	_____	_____	_____
Work Stations	_____	_____	_____

2. Graph these minimum required exterior illumination
levels, according to the daily hours of operation of your
factory and office building, on diagram 4 in Analysis
Tool B.

_____

QUESTIONS

A. What percentage of exterior illumination would you attempt to provide <u>indoors</u> to meet the minimum task light requirements on your darkest days?

What time of year is this?

What is the sky condition?

B. What percentage of exterior illumination would you attempt to provide indoors to meet the minimum task light requirements on your brightest days?

What time of year is this?

What is the sky condition?

3.6 BUILDING LOAD _____

GOAL

Determine whether the buildings have an internally dominated load (IDL) or a skin dominated load (SDL)?

"Load" refers to an imbalance of heat between inside and outside that makes a building skin necessary. If a building has a "heating load" that means you need to provide heat. Conversely a building with a "cooling load" will need to be cooled.

IDL - means that the major thermal problem over the year in the building is caused by what's going on <u>inside</u> (internal), <u>not</u> by what's going on outside the building skin. What's going on inside the building is that people, lights, and machines are giving off large quantities of heat compared to how fast it is dissipated. Therefore the major problem for an internally dominated load building over the year will be removing heat, or cooling.

SDL - means that the major thermal problem over the year will be caused by forces acting <u>on the skin</u> of building. There may still be internal heat generated; however, the skin effects will be relatively more important.

Depending upon the weather, the forces acting on skin make it lose heat (cold environment) or gain heat (hot environment). The building's thermal problem will be either heating or cooling or both depending on the season. The distinction between an SDL and IDL building depends on how the building interacts with the climate and therefore on many factors: how hot and cold the climate is, how much heat is generated out of the building through the walls and with the air flow, and the overall building configuration.

A. Analysis Tool - "Internal Heat Gain Estimate"
1. Heat from people:

Factory: the number of people is determined by the number of work stations, _____people, total.

Office/conference building: In this case, the number of people is determined by total floor area. When the spaces defined in the program are added together, along with circulation, the total floor area will probably range from 1,000 sq.ft. to 2,000 sq.ft. Only the actual areas occupied by the functions of office and conference need to counted when determining occupancy.

Use	sq.ft. floor area per occupant
1. Aircraft Hangars (no repair)	500
2. Auction Rooms	7
3. Assembly Areas, Concentrated Use (without fixed seats) Auditoriums Bowling Alleys (Assembly areas) Churches and Chapels Dance Floors Lodge Rooms Reviewing Stands Stadiums	7
4. Assembly Areas, Less-concentrated Use Conference Rooms Dining Rooms Drinking Establishments Exhibit Rooms Gymnasiums Lounges	15
5. Children's Homes and Homes for the Aged	80
6. Classrooms	20
7. Dormitories	50
8. Dwellings	300
9. Garage, Parking	200
10. Hospitals and Sanitariums - Nursing Homes	80
11. Hotels and Apartments	200
12. Kitchen - Commercial	200
13. Library Reading Room	50
14. Locker Rooms	50
15. Mechanical Equipment Room	300
16. Nurseries for Children (Day-care)	50
17. Offices	100
18. School Shops and Vocational Rooms	50
19. Stores - Retail Sales Rooms Basement Ground Floor Upper Floors	20 30 50
20. Warehouses	300
21. All others	100

Figure 3-2 Floor Area Per Occupant.

3.0 CLIMATE

3.6 BUILDING LOAD

Use the above table from the <u>Uniform Building Code</u> to figure

office area occupancy: _____ people

conference area occupancy: _____ people

Total number of people: factory + office + conference = people.

Estimate the heat generated by these people from MEEB Table 4.30 (p. 175). Use <u>sensible</u> heat gains, assuming moderately active office work (_____ Btuh per person) and light bench work in the factory (_____ Btuh per person).
2. Heat from electric lighting. Compare recommended footcandles, section 3.4, to MEEB Figure 20.2 (p. 867). With reasonably efficient sources such as fluorescents, the work and surrounding circulation area footcandle levels can be obtained by installing _____ watts/sq.ft. of electric lighting. To convert to heat (col. 5) 1 watt = 3.41 Btuh.)
3. Business machinery in the office building may be assumed at 1 watt/sq.ft. (half that for which electric circuits are designed). See MEEB Table 17.1, p. 645) Again for col. 5, 1 watt = 3.41 Btuh.
4. Factory machinery heat generation for col. 5 may be figured by the standard conversion factor of 1 hp = 746 watts = 2544 Btuh.

---

5. Summarize your internal heat gains as follows:

Col.1 Building	Col. 2 Heat Source	Col. 3 Area(sq.ft.) x	Col. 4 Unit per sq.ft. x	Col. 5 Btuh per unit =	Col. 6 Btuh total
Office:	• People	•(number of people=___)		•	
	• Lights	•	•	•	
	• Machines	•	•	•	
				•Total Btuh/sq.ft. office	
Factory:	• People	•(number of people=___)		•	
	• Lights	•	•	•	
	• Machines	•	•	•	
				•Total Btuh/sq.ft. factory	

B. Analysis Tool - "Balance Point Charts"
The external temperature at which internal loads are enough to adequately heat a building is called the "balance point" for that building. So, if the balance point of a given building is 40°: the building must be cooled when the external temperature is above 40°, and the building must be heated when the external temperature is below 40°.

The balance point depends on the climate, the internal gains, and the skin/volume characteristics. Once the internal heat gain has been estimated, the skin area must be examined to see how it interacts with the climate outside and the heat gains inside. The volume is also important, because it so often directly relates to the amount of internal heat produced. A long, thin cylinder has relatively more skin, and less volume, than a sphere; the cylinder's skin is likely to dominate its thermal behavior, the sphere's volume is likely to dominate its thermal behavior.

3.6 BUILDING LOAD

A building with relatively little skin area and a lot of
internal gain will be an "internal-dominated load" (IDL)
building. The hotter the climate, the more severely the
internal loads will dominate, and the more difficult it
will be to get rid of excess internal heat in summer.

A building with relatively great skin area and not much
internal gain will be a "skin-dominated load" (SDL)
building. The colder the climate, the more severely the
skin loads will dominate and the more difficult it will
be to keep the interior warm enough in winter.

1. Skin/volume characteristics. Shown here are several
ratios of skin area to total enclosed volume for buildings similar to the office and factory:

Total volume 20,000-27,000 cuft (similar to office
building)
Low skin/volume: (30' high, 30' wide, 30' long): .167
High skin/volume: (20' high, 10' wide, 100' long): .27

Total volume 72,000-75,000 cuft (similar to factory)
Low skin/volume: (30' high, 50' long, 50' wide): .113
High skin/volume: (30' high, 20' wide, 120' long): .15

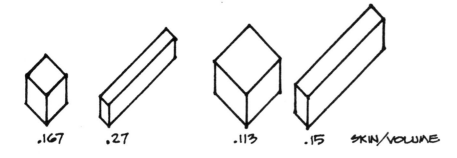

Diagram 5

2. Use these skin/volume ratios for each building and the
internal heat gains from the previous analysis tool to
plot the buildings on the charts below. The charts will
give you the estimated balance point temperature. Plot
both high and low skin/volume ratios and the occupied
hourly gains. Then, plot the balance point(s) of your
buildings on the Chart of Average Temperatures in the
next analysis tool, Diagram 7, and on the Bioclimatic
Chart, back in section 3.2.

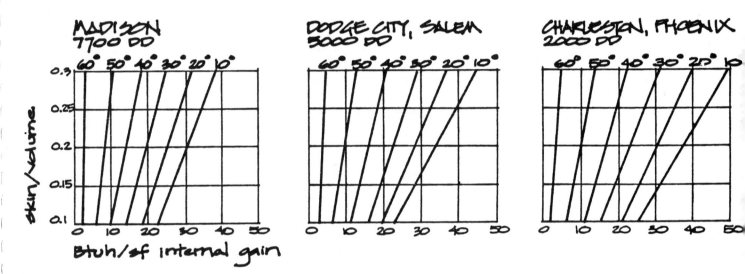

Diagram 6

Figure 3-3 Balance Point Charts

Technical note: To generate these charts, the following
assumptions were made:
Skin heat loss: roof and wall U$_o$ were half those allowed
for "type B" buildings, by ASHRAE 90-75 (see
Heating/Cooling Finalization Exercise), in keeping with
recent more strict standards. Balance points vary with
climate because ASHRAE 90-75 recommendations for Uo vary
with climate. Further, roof area assumed to be 1/3,
walls 2/3, of total skin area.
Infiltration (lung heat loss): assumed at 1 ACH (air
change per hour)
The floor area was kept constant as the skin/volume
ratios changed.
The internal temperature was held constant at 65°F.

C. Analysis Tool - "Graph Average Temperatures"
On the graph below, plot the average daily maximum and minimum temperatures, by month, from the NOAA data you used earlier.

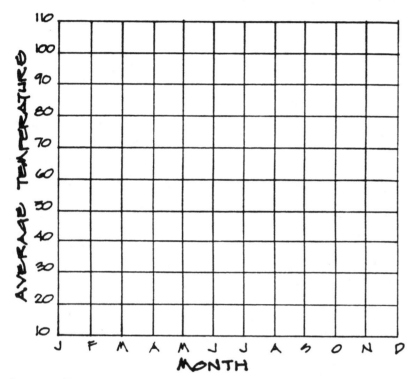

Diagram 7

QUESTIONS

A. Will the skin/volume ratio affect the Balance Point of the office significantly?

Of the factory?

B. On which months will the balance point for each building be greater than the temperature outside (heat required)?
office:

factory:

C. In which months will the balance point for each building be <u>less than</u> the temperature outside (cooling required)?
office:

factory:

D. Which is the dominant problem, heating or cooling, for each building?
office:

factory:

E. Daylighting is particularly important for buildings with cooling problems. Does your daylight analysis indicate enough exterior daylight to adequately daylight tasks inside?

F. Would you design to achieve a high or low skin/volume for the factory?
Why?

For the office?
Why?

G. What % of internal gains is due to lighting in the office, factory? What % is due to people and machines in the office, factory?

H. Will considerations of heating, cooling or lighting have the greatest influence on the form of your office, factory?

CHARLESTON, SOUTH CAROLINA

Narrative Climatological Summary

The city of Charleston, prior to expansion that began in 1960, was limited to the peninsula bounded on the west and south by the Ashley River, on the east by the Cooper River, and on the southeast by a spacious harbor. Weather records for the City are from observation sites on the lower portion of the peninsula, while airport records are from sites some 10 miles inland. The terrain is generally level, ranging in elevation from sea level to 20 feet on the peninsula, with gradual increases in elevation toward inland areas. The soil is sandy to sandy loam with lesser amounts of loam. The drainage varies from good to poor. Because of the very low elevation, a considerable portion of this community and the nearby coastal islands are vulnerable to tidal flooding, though only a few tides have exceeded 8 feet above mean low water.

The climate is temperate, modified considerably by the nearness to the ocean. The marine influence is noticeable during winter when the minimum temperatures are sometimes 10° to 15° higher on the peninsula than at the airport. By the same token, maximum temperatures are dampened 3° lower on the peninsula. The wind direction is vital to life and work along the coast. The prevailing winds are northerly in the fall and winter, southerly in the spring and summer.

Summer is warm and humid. Temperatures of 100° or higher are infrequent. Maximum temperatures are generally several degrees lower along the coast than inland due to the cooling effect of the sea breeze. Summer is the rainiest season with 41 percent of the annual fall. The rain, aside from occasional tropical storms, is generally of a shower or thundershower nature, producing variable amounts over scattered areas.

The fall season passes through the warm "Indian Summer" period to the prewinter cold spells which begin late in November. From late September to early November the weather is mostly sunny and extremes of temperature are rare. Late summer and early fall, however, is the period of maximum threat to the South Carolina coast from hurricanes. Some memorable hurricanes that have affected the Charleston area occurred in August 1885, August 1893, August 1911, July 1916, September 1928, August 1940, August 1952, and September 1959. The highest storm tide of record for which accurate heights were obtained was 11.2 feet above mean low water in the August 1893 storm.

The winter months, December through February, are mild with rainfall averaging 18 percent of the annual total. The winter rainfall is generally of a more uniform type, although a few thundershowers do occur. There is some chance of a snow flurry, with the best probability of its occurrence in January, but a significant amount is rarely measured. An average winter would experience less than one cold wave and severe freeze. Temperatures of 20° or less on the peninsula and along the coast are very unusual.

The most spectacular time of the year, weatherwise, is spring with its rapid changes from windy and cold in March to warm and pleasant in May. The spring rainfall represents about 20 percent of the total annual rain. Severe local storms are more likely to occur in the spring than in other seasons; however, some of the most destructive local storms of the 20th century were a series of severe tornadoes on September 29, 1938, and a single small tornado accompanying a hurricane on September 11, 1960.

The average date of the first freeze in the fall is December 10, and the average date of the last freeze before spring is February 19, giving an average growing season of 294 days. Freeze has been reported in the immediate inland areas as late as April 16 (1962) and as early as October 24 (1937).

noaa NATIONAL OCEANIC AND ATMOSPHERIC ADMINISTRATION / ENVIRONMENTAL DATA AND INFORMATION SERVICE / NATIONAL CLIMATIC CENTER ASHEVILLE, N.C.

Normals, Means, And Extremes

Charleston, South Carolina (Municipal Airport); Eastern Standard Time.
Latitude: 32°54'N; Longitude: 80°02'W; Elevation (ground): 40 feet.

Month	Temperatures °F — Normal			Temperatures °F — Extremes				Normal Degree days Base 65°F		Precipitation — Water equivalent							Precipitation — Snow, Ice pellets			
	Daily maximum	Daily minimum	Monthly	Record highest	Year	Record lowest	Year	Heating	Cooling	Normal	Maximum monthly	Year	Minimum monthly	Year	Maximum in 24 hrs.	Year	Maximum monthly	Year	Maximum in 24 hrs.	Year
(a)				37		37					37		37		37		37		37	
J	59.8	37.3	48.6	83	1950	11	1970	521	12	2.90	6.68	1966	0.63	1950	2.26	1946	1.0	1977	0.8	1966
F	61.9	39.0	50.5	86	1962	12	1973	419	13	3.27	6.32	1964	0.33	1947	3.28	1944	7.1	1973	5.9	1973
M	67.8	45.1	56.5	90	1974	21	1960	300	36	4.75	11.11	1959	0.99	1963	6.63	1959	2.0	1969	2.0	1969
A	76.2	53.0	64.6	93	1968	29	1944	69	57	2.95	9.50	1958	0.01	1972	4.10	1958	0.0		0.0	
M	83.1	61.1	72.1	98	1953	36	1963	5	225	3.81	9.28	1957	0.68	1944	6.23	1967	0.0		0.0	
J	87.7	68.1	77.9	103	1944	50	1972	0	387	6.30	27.24	1973	0.96	1970	10.10	1973	0.0		0.0	
J	89.1	71.2	80.2	101	1977	58	1952	0	471	8.21	18.46	1964	1.76	1972	5.81	1960	0.0		0.0	
A	88.6	70.6	79.6	102	1954	56	1979	0	453	6.44	16.99	1974	0.88	1979	5.77	1964	0.0		0.0	
S	84.5	65.9	75.2	99	1944	42	1967	0	306	5.17	17.31	1945	0.53	1971	8.84	1945	0.0		0.0	
O	77.1	55.1	66.1	94	1954	27	1976	74	108	3.05	9.12	1959	0.08	1943	5.77	1944	0.0		0.0	
N	68.4	44.1	56.3	88	1961	15	1950	271	10	2.13	7.35	1972	0.48	1966	5.24	1969	T	1950	T	1950
D	60.8	37.7	49.3	83	1972	8	1962	487	0	3.14	7.09	1953	0.82	1944	3.40	1944	2.1	1943	2.1	1943
YR	75.4	54.0	64.7	103	JUN 1944	8	DEC 1962	2146	2078	52.12	27.24	JUN 1973	0.01	APR 1972	10.10	JUN 1973	7.1	FEB 1973	5.9	FEB 1973

Month	Relative humidity pct.				Wind — Mean speed m.p.h.	Wind — Prevailing direction	Peak Gust # — Speed m.p.h.	Peak Gust # — Direction	Peak Gust # — Year	Pct. of possible sunshine	Mean sky cover, tenths, sunrise to sunset	Clear	Partly cloudy	Cloudy	Precipitation .01 inch or more	Snow, Ice pellets 1.0 inch or more	Thunderstorms	Heavy fog, visibility ¼ mile or less	Max. 90° and above (b)	Max. 32° and below	Min. 32° and below	Min. 0° and below	Average station pressure mb. Elev. 48 feet m.s.l.
	Hour 01	Hour 07	Hour 13	Hour 19 (Local time)																			
	37	37	37	37	30	14	15	15		21	30	31	31	31	37	37	37	30	37	37	37	37	7
JAN	81	83	55	72	9.3	SW	54	W	1964	58	6.3	9	6	16	10	0	1	4	0	*	11	0	1018.6
FEB	79	81	51	67	10.1	NNE	62	SW	1963	61	5.9	9	7	12	9	*	1	2	0	*	8	0	1017.8
MAR	81	83	50	66	10.2	SSW	51	SW	1973	67	5.9	9	9	13	11	*	2	2	*	0	3	0	1016.6
APR	83	84	49	66	9.9	SSW	71	NW	1971	71	5.3	11	8	11	7	0	3	2	4	0	0	0	1016.4
MAY	88	85	54	72	8.9	S	60	N	1967	71	6.0	8	11	12	9	0	7	2	11	0	0	0	1015.1
JUN	90	86	59	75	8.5	S	65	NW	1968	67	6.3	6	11	13	11	0	10	2	11	0	0	0	1015.1
JUL	90	88	63	78	8.0	SW	58	N	1971	68	6.6	5	12	14	14	0	13	1	15	0	0	0	1015.8
AUG	91	90	63	80	7.4	SW	69	N	1968	66	6.2	6	13	12	13	0	11	1	14	0	0	0	1016.9
SEP	91	91	63	81	8.0	NNE	46	SE	1964	62	6.3	6	11	13	9	0	6	3	5	0	0	0	1015.2
OCT	88	89	55	79	8.1	NNE	54	W	1967	67	5.1	12	9	10	6	0	1	3	*	0	4	0	1017.3
NOV	85	86	52	77	8.1	N	55	SW	1962	65	5.1	13	6	11	7	0	1	4	0	*	10	0	1018.7
DEC	82	83	54	73	8.7	NNE	54	S	1968	59	5.9	9	8	14	8	*	*	4	0	*	10	0	1018.7
YEAR	86	86	56	74	8.8	NNE	71	NW	APR 1971	66	5.9	103	114	151	114	*	56	28	49	*	37	0	1016.8

Means and extremes above are from existing and comparable exposures. Annual extremes have been exceeded at other sites in the locality as follows: Highest temperature 104 in June 1944; lowest temperature 7 in February 1899; maximum precipitation in 24 hours 10.57 in September 1933.

(a) Length of record, years, through the current year unless otherwise noted, based on January data.
(b) 70° and above at Alaskan stations.
* Less than one half.
T Trace.

NORMALS - Based on record for the 1941-1970 period.
DATE OF AN EXTREME - The most recent in cases of multiple occurrence.
PREVAILING WIND DIRECTION - Record through 1963.
WIND DIRECTION - Numerals indicate tens of degrees clockwise from true north. 00 indicates calm.
FASTEST MILE WIND - Speed is fastest observed 1-minute value when the direction is in tens of degrees.

Peak Gust through July 1975

DODGE CITY, KANSAS

Narrative Climatological Summary

The climate of Dodge City and southwestern Kansas is classified as semiarid. Dodge City is nearly 300 miles east of the Rocky Mountains, but the weather reflects the influence of the mountains. The mountains form a barricade against all except high level moisture from the southwest, west, and northwest. Chinook winds occur occasionally but with less frequency and effect than at stations farther to the west. Relatively dry air predominating with an abundance of sunshine contribute to broad diurnal temperature ranges.

The average annual precipitation accumulates to near 20 inches. Thunderstorms during the growing season contribute most of the moisture. In general, the thunderstorms are widely scattered, occurring during the late afternoons and evenings. They are occasionally accompanied by hail and strong winds, but due to the local nature of the storms, damage to crops and buildings is spotted and variable. Winter is the dry season; however, the moisture accumulated during the winter months is important for the hard winter wheat. Accumulated snowfall averages near 19 inches for the winter, but duration of snow cover is generally brief due to mild temperatures and an abundance of sunshine. The exception results from the occasional blizzard that spreads across the flat treeless prairies of the high plains.

The extreme temperatures recorded for Dodge City range from 109° to -26°. Afternoon temperatures in the nineties prevail during the summer months; temperatures above 100° are the exception. Due to low humidity and a continual breeze, these temperatures are effectively moderated. Temperatures drop sharply after sunset, allowing cool comfortable nights. During the winter months, large temperature changes are frequent, but duration of extreme cold spells is brief.

The visibility at Dodge City is generally unrestricted as the terrain is favorable for unrestricted movement of air and air masses.

Western Kansas is noted for clear skies and an abundance of sunshine.

The climate is conducive to the growing of hard winter wheat and maize. The winter wheat is pastured by cattle during the winter. The semiarid climate favors production of wheat high in protein.

noaa   NATIONAL OCEANIC AND   / ENVIRONMENTAL DATA AND / NATIONAL CLIMATIC CENTER
       ATMOSPHERIC ADMINISTRATION / INFORMATION SERVICE   / ASHEVILLE, N.C.

INSIDEOUT

Normals, Means, And Extremes

Dodge City, Kansas (Municipal Airport); Central Standard Time.
Latitude: 37°46'N; Longitude: 99°58'W; Elevation(ground): 2582 feet.

Month	Temperatures °F — Normal Daily maximum	Normal Daily minimum	Normal Monthly	Extremes Record highest	Year	Extremes Record lowest	Year	Normal Degree days Base 65°F Heating	Cooling	Precip. Water equivalent Normal	Maximum monthly	Year	Minimum monthly	Year	Maximum in 24 hrs	Year	Snow, Ice pellets Maximum monthly	Year	Maximum in 24 hrs	Year
(a)				37		37					37		37		37		37		37	
J	42.6	19.0	30.8	78	1967	-12	1979	1060	0	0.50	1.96	1949	T	1961	1.12	1945	10.3	1979	8.8	1971
F	47.1	23.2	35.2	85	1972	-15	1951	834	0	0.63	2.04	1971	0.01	1970	1.18	1948	16.5	1978	10.3	1971
M	53.9	28.4	41.2	93	1946	-15	1948	738	0	1.13	8.84	1973	0.02	1966	2.54	1973	24.0	1970	12.8	1970
A	66.9	41.1	54.0	94	1956	15	1975	344	14	1.71	6.26	1976	0.07	1963	4.64	1978	5.6	1956	5.9	1970
M	76.2	51.7	64.0	102	1967	26	1967	115	84	3.13	8.69	1951	0.12	1952	5.57	1978	0.9	1978	0.9	1978
J	86.0	61.4	73.7	108	1978	41	1954	21	282	3.34	7.95	1951	0.17	1946	4.17	1944	0.0		0.0	
J	91.4	66.9	79.2	109	1978	47	1952	0	440	3.08	9.13	1962	0.68	1953	3.18	1974	0.0		0.0	
A	90.4	65.7	78.1	107	1964	47	1950	0	406	2.64	7.44	1977	0.01	1947	3.27	1959	0.0		0.0	
S	81.4	56.3	68.9	106	1947	31	1942	41	158	1.67	6.80	1973	T	1952	4.55	1968	T	1945	T	1945
O	70.7	45.0	57.9	96	1968	20	1957	247	27	1.65	4.88	1968	T	1952	2.42	1971	3.8	1976	3.8	1976
N	55.2	30.4	42.8	83	1973	0	1958	666	0	0.59	3.75	1971	T	1966	1.35	1973	13.2	1948	6.7	1948
D	44.6	22.2	33.4	86	1955	-7	1968	980	0	0.51	2.01	1965	T	1957	1.35	1973	10.8	1942	6.6	1958
YR	67.2	42.6	54.9	109 JUL 1978		-15 FEB 1951		5046	1411	20.58	9.13	JUL 1962	T	NOV 1966	5.57	MAY 1978	24.0	MAR 1970	12.8	MAR 1970

Month	Relative humidity pct. Hour 00	Hour 06	Hour 12	Hour 18	Wind Mean speed m.p.h.	Prevailing direction	Fastest mile Speed m.p.h.	Direction	Year	Pct. of possible sunshine	Mean sky cover, tenths, sunrise to sunset	Mean no. of days Clear	Partly cloudy	Cloudy	Precipitation .01 inch or more	Snow, Ice pellets 1.0 inch or more	Thunderstorms	Heavy fog, visibility ¼ mile or less	Max. 90° and above	Max. 32° and below	Min. 32° and below	Min. 0° and below	Avg. station pressure mb Elev. 2592 feet m.s.l.
	16	16	16	16	37	14	36	36		37	37	37	37	37	37	37	37	37	16	16	16	16	7
JAN	72	76	58	59	13.6	S	56	NW	1950	67	5.6	11	7	13	5	1	*	3	0	10	29	3	926.3
FEB	70	75	54	52	14.1	N	72	NW	1953	65	5.7	9	7	12	5	1	*	3	0	5	25	1	925.1
MAR	67	74	49	44	15.8	N	74	NW	1952	66	5.9	9	9	13	6	2	1	3	*	2	19	*	921.2
APR	67	74	46	42	15.7	SSE	70	N	1955	68	5.8	9	9	12	7	*	4	2	*	*	5	0	921.7
MAY	70	78	49	46	14.7	S	71	W	1959	69	5.7	9	10	12	10	0	9	1	3	0	*	0	923.1
JUN	67	77	45	41	14.4	S	73	S	1947	76	4.8	12	11	7	8	0	11	1	14	0	0	0	925.0
JUL	63	75	44	38	12.9	S	78	SW	1951	78	4.7	12	12	7	9	0	11	1	22	0	0	0	925.2
AUG	66	78	46	42	12.7	S	72	NW	1944	77	4.5	14	10	7	9	0	9	1	18	0	0	0	925.8
SEP	71	80	50	46	13.6	S	73	NE	1962	74	4.4	14	7	9	6	0	5	2	7	0	0	0	925.8
OCT	65	72	44	46	13.6	S	63	SW	1949	74	4.3	15	8	8	5	*	2	2	2	*	3	0	925.9
NOV	73	78	54	58	13.7	S	59	NW	1948	67	5.1	12	7	11	4	1	1	1	0	2	18	0	925.1
DEC	72	76	56	60	13.5	N	57	NW	1957	65	5.4	12	7	12	4	1	*	3	0	6	28	1	925.1
	68	76	50	48	14.0	S	78	SW	JUL 1951	71	5.2	138	104	123	78	6	53	23	65	26	126	5	924.3

Means and extremes above are from existing and comparable exposures. Annual extremes have been exceeded at other sites in the locality as follows: Lowest temperature -26 in February 1899; maximum monthly precipitation 12.82 in May 1881; minimum monthly precipitation 0.00 in December 1889; maximum precipitation in 24 hours 6.03 in June 1899; maximum monthly snowfall 27.7 in February 1903; maximum snowfall in 24 hours 17.5 in March 1922.

(a) Length of record, years, through the current year unless otherwise noted, based on January data.
(b) 70° and above at Alaskan stations.
* Less than one half.
T Trace.

NORMALS - Based on record for the 1941-1970 period.
DATE OF AN EXTREME - The most recent in cases of multiple occurrence.
PREVAILING WIND DIRECTION - Record through 1963.
WIND DIRECTION - Numerals indicate tens of degrees clockwise from true north. 00 indicates calm.
FASTEST MILE WIND - Speed is fastest observed 1-minute value when the direction is in tens of degrees.

MADISON, WISCONSIN

Narrative Climatological Summary

Madison is set on a narrow isthmus of land between Lakes Mendota and Monona. Lake Mendota (15 sq. mi.) lies northwest of Lake Monona (5 sq. mi.) and the lakes are only two-thirds of a mile apart at one point. Drainage at Madison is southeast through two other lakes into the Rock River, which flows south into Illinois, and then west to the Mississippi. The westward flowing Wisconsin River is only 20 miles northwest of Madison. Madison lakes are normally frozen from December 17 to April 5.

Madison has the typical continental climate of interior North America with a large annual temperature range and with frequent short period temperature changes. The absolute temperature range is from 107° (City Office) to -37° (Truax Field). Winter temperatures (December-February) average 20° and the summer average (June-August) is 68°. Daily mean temperatures average below 32° for 118 days and above 42° for 206 days of the year.

Madison lies in the path of the frequent cyclones and anticyclones which move eastward over this area during fall, winter and spring. In summer, the cyclones have diminished intensity and tend to pass farther north. The most frequent air masses are of polar origin. Occasional outbreaks of arctic air affect this area during the winter months. Although northward moving tropical air masses contribute considerable cloudiness and precipitation, the true gulf air mass does not reach this area in winter, and only occasionally at other seasons. Summers are pleasant, with only occasional periods of extreme heat or high humidity.

There are no dry and wet seasons, but 59 percent of the annual precipitation falls in the 5 months of May through September. Cold season precipitation is lighter, but lasts longer. Soil moisture is usually adequate in the first part of the growing season. During July, August and September, the crops depend on current rainfall, which is mostly from thunderstorms and tends to be erratic and variable. Average occurrence of thunderstorms is just under 7 days per month during this period.

March and November are the windiest months. Tornadoes are infrequent. The average occurrence for Dane County is about one tornado in every three to five years.

The ground is covered with an inch or more of snow about 60 percent of the time from December 10 to February 25 in an average winter. The soil is usually frozen from the first of December through most of March with an average frost penetration of 25 to 30 inches. The growing season averages 175 days. The most probable period (50 percent of the years) for the last crop damaging freeze in spring is April 17 to May 2. The first frost/freeze in autumn is most probable from October 6 to 25. The latest recorded killing freeze was on June 10, 1972 and the earliest in fall was on September 12, 1955.

Farming is diversified with the main emphasis on dairying. Field crops are mainly corn, oats, clover and alfalfa, but barley, wheat, rye, and tobacco are also raised. Canning factories pack peas, sweet corn and lima beans. Fruits are mainly apples, strawberries and raspberries.

noaa NATIONAL OCEANIC AND / ENVIRONMENTAL DATA AND / NATIONAL CLIMATIC CENTER
ATMOSPHERIC ADMINISTRATION / INFORMATION SERVICE / ASHEVILLE, N.C.

Normals, Means, And Extremes

Madison, Wisconsin (Truax Field); Central Sandard Time.
Latitude: 43°08'N; Longitude: 89°20'W; elevation (ground) 858 feet.

Month	Temperatures °F — Normal — Daily maximum	Daily minimum	Monthly	Extremes — Record highest	Year	Record lowest	Year	Normal Degree days Base 65°F — Heating	Cooling	Precipitation in inches — Water equivalent — Normal	Maximum monthly	Year	Minimum monthly	Year	Maximum in 24 hrs.	Year	Snow, Ice pellets — Maximum monthly	Year	Maximum in 24 hrs.	Year
(a)				40		40					40		40		31		31		31	
J	25.4	8.2	16.8	55	1947	-37	1951	1494	0	1.25	2.45	1974	0.19	1961	1.27	1960	26.9	1979	11.6	1971
F	29.5	11.1	20.3	60	1976	-28	1959	1252	0	0.95	2.77	1953	0.08	1958	1.55	1953	20.9	1975	10.3	1950
M	39.2	21.2	30.2	81	1978	-29	1962	1079	0	1.93	5.04	1973	0.28	1978	2.52	1973	25.4	1959	13.6	1971
A	56.0	34.6	45.3	90	1952	9	1972	591	0	2.66	7.11	1973	0.96	1946	2.83	1975	17.4	1973	12.9	1973
M	67.3	44.6	56.0	93	1975	19	1978	297	18	3.41	6.26	1960	0.98	1971	3.64	1966	0.7	1966	0.7	1966
J	76.9	54.6	65.8	97	1953	31	1972	72	96	4.33	9.95	1978	0.81	1973	3.67	1963	0.0		0.0	
J	81.4	58.8	70.1	104	1976	36	1965	14	172	3.81	10.93	1950	1.38	1946	5.25	1950	0.0		0.0	
A	80.0	57.3	68.7	101	1947	35	1968	39	154	3.05	7.47	1948	0.70	1948	2.90	1965	0.0		0.0	
S	70.9	48.5	59.7	99	1953	25	1974	173	14	3.36	9.51	1941	0.11	1979	3.57	1961	T	1965	T	1965
O	60.9	38.9	49.9	90	1976	14	1952	474	6	2.16	5.55	1959	0.06	1952	2.01	1960	0.9	1967	0.9	1967
N	43.0	26.4	34.7	76	1964	-11	1947	909	0	1.87	3.94	1961	0.11	1976	2.32	1971	10.4	1977	6.8	1954
D	29.8	14.0	21.9	62	1970	-22	1962	1336	0	1.47	3.64	1971	0.25	1960	1.66	1971	24.6	1977	16.0	1970
YR	55.0	34.8	44.9	104	JUL 1976	-37	JAN 1951	7730	460	30.25	10.93	JUL 1950	0.06	OCT 1952	5.25	JUL 1950	26.9	JAN 1979	16.0	DEC 1970

Month	Relative humidity pct. — Hour 00	Hour 06	Hour 12	Hour 18 (Local time)	Wind — Mean speed m.p.h.	Prevailing direction	Fastest mile — Speed m.p.h.	Direction	Year	Pct. of possible sunshine	Mean sky cover, tenths, sunrise to sunset	Mean number of days — Sunrise to sunset — Clear	Partly cloudy	Cloudy	Precipitation .01 inch or more	Snow, Ice pellets 1.0 inch or more	Thunderstorms	Heavy fog, visibility ¼ mile or less	Temperatures °F — Max. — 90° and above (b)	Max. — 32° and below	Min. — 32° and below	Min. — 0° and below	Average station pressure mb. Elev. 866 feet m.s.l.
	20	20	20	20	33	14	33	33		33	31	33	33	33	31	31	31	33	20	20	20	20	7
JAN	77	78	69	73	10.5	WNW	68	E	1947	49	6.7	8	6	17	10	3	*	3	0	22	30	13	985.6
FEB	77	79	65	69	10.4	WNW	57	W	1948	52	6.5	7	6	15	8	2	*	2	0	16	28	7	985.9
MAR	79	82	64	67	11.2	NW	70	SW	1954	53	6.9	6	8	17	11	3	2	3	0	7	26	2	982.7
APR	77	82	56	57	11.4	NW	73	SW	1947	52	6.7	6	8	16	11	1	4	1	0	*	15	0	984.1
MAY	77	81	54	55	10.3	S	77	SW	1950	59	6.5	7	9	15	11	0	5	1	*	0	4	0	982.2
JUN	81	83	56	56	9.2	S	59	W	1947	65	6.1	7	10	13	11	0	7	1	3	0	*	0	982.7
JUL	84	86	56	58	8.1	S	72	NW	1951	69	5.7	9	12	10	9	0	8	1	5	0	0	0	984.5
AUG	87	91	59	62	8.0	S	47	W	1955	67	5.6	10	10	11	9	0	6	2	3	0	0	0	985.6
SEP	88	92	61	68	8.7	S	52	W	1948	62	5.6	10	9	11	9	0	4	2	1	0	1	0	985.9
OCT	82	87	59	68	9.5	S	73	SW	1951	56	5.8	10	8	13	9	0	2	2	*	0	10	0	985.7
NOV	83	86	68	75	10.7	S	56	SE	1947	40	7.1	6	6	18	9	1	*	3	0	4	21	0	985.0
DEC	82	83	73	77	10.2	W	65	SW	1949	40	7.2	6	6	19	10	3	*	3	0	18	29	6	984.5
	81	84	62	66	9.9	S	77	SW	MAY 1950	57	6.4	92	98	175	117	12	40	22	12	66	164	27	984.5

Means and extremes above are from existing and comparable exposures. Annual extremes have been exceeded at other sites in the locality as follows: Highest temperature 107 in July 1936; minimum monthly precipitation T in October 1889 and earlier; maximum precipitation in 24 hours 5.31 in September 1941; maximum monthly snowfall 31.8 in January 1929.

(a) Length of record, years, through the current year unless otherwise noted, based on January data.
(b) 70° and above at Alaskan stations.
* Less than one half.
T Trace.

NORMALS - Based on record for the 1941-1970 period.
DATE OF AN EXTREME - The most recent in cases of multiple occurrence.
PREVAILING WIND DIRECTION - Record through 1963.
WIND DIRECTION - Numerals indicate tens of degrees clockwise from true north. 00 indicates calm.
FASTEST MILE WIND - Speed is fastest observed 1-minute value when the direction is in tens of degrees.

INSIDEOUT

PHOENIX, ARIZONA

Narrative Climatological Summary

Phoenix is located in about the center of the Salt River Valley, a broad, oval-shaped, nearly flat plain. The Salt River runs from east to west through the valley but, owing to impounding dams upstream, it is usually dry. The climate is of a desert type with low annual rainfall and low relative humidity. Daytime temperatures are high throughout the summer months. The winters are mild. Nighttime temperatures frequently drop below freezing during the three coldest months, but afternoons are usually sunny and warm.

At an elevation of about 1100 feet, the station is in a level or gently sloping valley running east and west. The Salt River Mountains, or South Mountains, as they are commonly called, are located 6 miles to the south and rise to 2600 feet m.s.l. The Phoenix Mountains lie 8 miles to the north with Squaw Peak rising to 2400 feet. The famous landmark of Camelback Mountain lies 6 miles to the north northeast and rises to 2700 feet. Eighteen miles to the southwest lie the Sierra Estrella Mountains with a maximum elevation of 4500 feet, and 30 miles to the west northwest are found the White Tank Mountains with a maximum elevation of 4100 feet. The Superstition Mountains are approximately 35 miles to the east and rise to 5000 feet.

The central floor of the Salt River Valley is irrigated by water from dams built on the Salt River system. To the north and west of the gravity flow irrigated district there is considerable agricultural land irrigated by pump water. There is no evidence that the irrigation has in any way affected the relative humidity in the valley. The average daytime relative humidity is about 30 percent based on observations at 11:00 a.m. and 5.00 p.m.

There are two separate rainfall seasons. The first occurs during the winter months from November to March when the area is subjected to occasional storms from the Pacific Ocean. While this is classed as a rainfall season, there can be periods of a month or more in this or any other season when practically no precipitation occurs. Snowfall occurs very rarely in the Salt River Valley, while snows occasionally fall in the higher mountains surrounding the valley. The second rainfall period occurs during July and August when Arizona is subjected to widespread thunderstorm activity whose moisture supply originates both in the Gulf of Mexico and along Mexico's west coast. These thunderstorms are extremely variable in intensity and location.

The spring and fall months are generally dry, although precipitation in substantial amounts has fallen on occasion during every month of the year.

Since the Phoenix area is primarily agricultural, minimum temperatures and their variation over the valley have been studied closely. During the winter months the temperature is marginal for some types of crops. However, the valley is subject to occasional killing and hard freezes in which no area escapes damage.

The valley floor, in general, is rather free of wind. During the spring months southwest and west winds predominate and are associated with the passage of low pressure troughs. During the thunderstorm season there are often local gusty winds, usually flowing from an easterly direction. Throughout the year there are periods, often several days in length, in which winds remain under 10 miles an hour.

Sunshine in the Phoenix area averages near 86 percent of the possible amount, ranging from a minimum monthly average of about 77 percent in January and December to a maximum of about 94 percent in June. During the winter, skies are sometimes cloudy, but clear skies predominate and temperatures are mild. During the spring, skies are also predominately clear with warm temperatures during the day and mild pleasant evenings. Beginning with June, daytime weather is hot. During July and August, there is often considerable afternoon cloudiness associated with cumulus clouds building up over the nearby mountains. Summer thunderstorms seldom occur in the valley before evening.

The autumn season, beginning during the latter part of September, is characterized by sudden changes in temperature. The change from the heat of summer to mild winter temperatures usually occurs during October. The normal temperature change from the beginning to the end of this month is the greatest of any of the twelve months in central Arizona. By November, the mild winter season is definitely established in the Salt River Valley region.

NATIONAL OCEANIC AND / ENVIRONMENTAL DATA AND / NATIONAL CLIMATIC CENTER
ATMOSPHERIC ADMINISTRATION / INFORMATION SERVICE / ASHEVILLE, N.C.

Normals, Means, And Extremes

Phoenix, Arizona (Sky Harbor Intl. Airport); Mountain Standard Time.
Latitude: 33°26'N; Longitude: 112°01'W; Elevation (ground) 1110 feet.

Month	Temperatures °F Normal - Daily maximum	Daily minimum	Monthly	Extremes - Record highest	Year	Record lowest	Year	Normal Degree days Base 65°F - Heating	Cooling	Precipitation in inches Water equivalent - Normal	Maximum monthly	Year	Minimum monthly	Year	Maximum in 24 hrs.	Year	Snow, Ice pellets - Maximum monthly	Year	Maximum in 24 hrs.	Year
(a)				42		42					42		42		42		42		42	
J	64.8	37.6	51.2	88	1971	17	1950	428	0	0.71	2.41	1955	0.00	1972	1.31	1951	T	1962	T	1962
F	69.3	40.8	55.1	89	1963	22	1948	292	14	0.60	2.23	1944	0.00	1967	1.22	1978	0.6	1939	0.6	1939
M	74.5	44.8	59.7	95	1972	25	1966	185	21	0.76	4.16	1941	0.00	1959	1.32	1941	T	1976	T	1976
A	83.6	51.8	67.7	104	1949	32	1945	60	141	0.32	2.10	1941	0.00	1962	1.38	1941	T	1949	T	1949
M	92.9	59.6	76.3	113	1951	40	1967	0	355	0.14	1.06	1976	0.00	1974	0.96	1976	0.0		0.0	
J	101.5	67.7	84.6	117	1979	50	1944	0	588	0.12	1.70	1972	0.00	1974	1.64	1972	0.0		0.0	
J	104.8	77.5	91.2	118	1958	61	1944	0	812	0.75	4.19	1955	T	1947	1.97	1955	0.0		0.0	
A	102.2	76.0	89.1	116	1975	60	1942	0	747	1.22	5.56	1951	T	1975	3.07	1943	0.0		0.0	
S	98.4	69.1	83.8	118	1950	47	1965	0	564	0.69	4.23	1939	0.00	1973	2.43	1970	0.0		0.0	
O	87.6	56.8	72.2	104	1979	34	1971	17	240	0.46	4.40	1972	0.00	1973	2.27	1972	0.0		0.0	
N	74.7	44.8	59.8	93	1975	25	1938	182	26	0.46	3.04	1952	0.00	1956	1.14	1978	0.0		0.0	
D	66.4	38.5	52.5	88	1950	22	1948	388	0	0.82	3.98	1967	0.00	1973	1.89	1967	T	1974	T	1974
YR	85.1	55.4	70.3	118	JUL 1958	17	JAN 1950	1552	3508	7.05	5.56	AUG 1951	0.00	JUN 1974	3.07	AUG 1943	0.6	FEB 1939	0.6	FEB 1939

Month	Relative humidity pct. - Hour 05	Hour 11	Hour 17	Hour 23	Wind - Mean speed m.p.h.	Prevailing direction	Peak Gust - Speed m.p.h.	Direction	Year	# Pct. of possible sunshine	$ Mean sky cover, tenths, sunrise to sunset	Mean number of days Sunrise to sunset - Clear	Partly cloudy	Cloudy	Precipitation .01 inch or more	Snow, Ice pellets 1.0 inch or more	Thunderstorms	Heavy fog, visibility ¼ mile or less	Temperatures °F Max. - 90° and above	32° and below	Min. (b) - 32° and below	0° and below	Average station pressure mb. Elev. 1107 feet m.s.l.
	19	19	19	19	34	18	42	42		84	34	42	42	42	40	42	40	42	19	19	19	19	7
JAN	67	45	32	57	5.2	E	51	SW	1974	78	4.8	14	7	10	4	0	*	1	0	0	6	0	978.4
FEB	60	38	26	48	5.9	E	49	SSE	1959	80	4.4	13	6	9	4	0	*	*	0	0	2	0	979.0
MAR	58	33	23	43	6.6	E	51	W	1977	83	4.3	15	8	8	3	0	1	*	2	0	1	0	974.5
APR	44	23	15	29	7.0	E	49	S	1977	88	3.3	17	7	6	2	0	1	0	8	0	0	0	972.9
MAY	36	18	13	23	7.1	E	59	SSE	1954	93	2.7	21	6	4	1	0	1	0	22	0	0	0	970.8
JUN	33	17	12	21	7.0	E	73	NE	1978	94	1.9	23	5	2	1	0	1	0	29	0	0	0	970.0
JUL	46	28	20	33	7.3	W	86	SE	1976	85	3.7	16	11	4	4	0	6	0	31	0	0	0	971.1
AUG	52	33	23	38	6.7	E	78	E	1978	85	3.2	18	9	4	5	0	7	0	31	0	0	0	971.5
SEP	51	31	23	40	6.4	E	75	SW	1950	89	2.1	22	5	3	3	0	3	0	27	0	0	0	971.7
OCT	52	30	22	42	5.9	E	54	E	1978	88	2.7	20	6	5	3	0	1	*	15	0	0	0	974.2
NOV	60	37	28	51	5.4	E	55	S	1975	84	3.5	18	6	6	2	0	*	0	*	0	*	0	976.8
DEC	67	46	33	57	5.2	E	68	W	1953	77	4.0	15	7	9	4	0	1	*	0	0	3	0	978.6
	52	32	22	40	6.3	E	86	SE	JUL 1976	86	3.4	212	83	70	35	0	23	2	165	0	11	0	974.1

Means and extremes above are from existing and comparable exposures. Annual extremes have been exceeded at other sites in the locality as follows: Lowest temperature 16 in January 1913; maximum monthly precipitation 6.47 in July 1911; maximum precipitation in 24 hours 4.98 in July 1911; maximum monthly snowfall 1.0 in January 1937 and earlier; maximum snowfall in 24 hours 1.0 in January 1937 and earlier.

♬ As observed Jan. 1938 through Oct. 1953; from recorder thereafter.

From City Office Aug. 1895 through Oct. 1953; from Sky Harbor Airport thereafter.

$ Broken record: 1940, 1941 and 1948 to date.

(a) Length of record, years, through the current year unless otherwise noted, based on January data.
(b) 70° and above at Alaskan stations.
* Less than one half.
T Trace.

NORMALS - Based on record for the 1941-1970 period.
DATE OF AN EXTREME - The most recent in cases of multiple occurrence.
PREVAILING WIND DIRECTION - Record through 1963.
WIND DIRECTION - Numerals indicate tens of degrees clockwise from true north. 00 indicates calm.
FASTEST MILE WIND - Speed is fastest observed 1-minute value when the direction is in tens of degrees.

SALEM, OREGON

Narrative Climatological Summary

Salem is located in the middle Willamette Valley some 60 airline miles east of the Pacific Ocean. The valley here is approximately 50 miles wide with the City about equidistant from the valley walls formed by the Coast Range on the west and the Cascade Range on the east.

The usual movement of very moist maritime air masses from the Pacific Ocean inland over the Coast Range produces near its crest some of the heaviest yearly rainfall in the United States. An annual total of nearly 170 inches has been recorded, and one station established a period-of-record annual average of approximately 130 inches. From the ridge crest of the Coast Range, approximately 3,000 feet above sea level, there is a gradual decrease of rainfall downslope to the valley floor where annual totals average between 35 and 45 inches. As these marine conditioned air masses continue to move farther inland they are forced to ascend the west slopes of the Cascades to approximately 5,000 feet above sea level and again rainfall amounts substantially increase with elevation.

Most of this precipitation in both the valley and its bordering mountain ranges occurs during the winter. At Salem 70 percent of the annual total occurs during the five months of November through March while only 6 percent occurs during the three summer months, with practically all of it falling in the form of rain. In the immediate area, on the average, there are only three or four days a year with measurable amounts of snow. Its depth on the ground rarely exceeds 2 or 3 inches, and it usually melts in a day or two. The few thunderstorms that occur each year are not generally severe and seldom do they, or the hail that occasionally accompanies them, cause any serious damage. A tornado in the immediate metropolitan area has never been recorded.

The seasonal difference in temperatures is much less marked than that of precipitation. There is only a range of about 28° between the mean temperature for January, the coldest month, and July, the warmest. Maximums as high as 100° or more seldom occur, and in only four years during the station's 84 years of record have minimums of 0° or lower been observed. A 30-year average for the dates of the last occurrence in spring and first in fall of temperatures of 32° or lower are respectively April 14 and October 28, though these have been recorded as late as May 28 and as early as September 12. This provides an average growing season of six and a half months.

The mild temperatures, long growing season and plentiful supply of moisture are ideally suited for a wide variety of crops. In dollar value of agricultural returns, this is the most productive land in Oregon. Large orchards of sweet cherries are grown and processed here for maraschino cherries. Hops, filberts, walnuts, cane, and strawberries each contribute many millions of dollars to the annual farm income. A wide variety of vegetables is raised for both the fresh market and to support a large number of processing plants located in Salem. This climate is also suitable for the production of a number of specialty crops including mint, several seed crops, and nursery stock, particularly roses and ornamental shrubs.

 **NATIONAL OCEANIC AND ATMOSPHERIC ADMINISTRATION** / **ENVIRONMENTAL DATA AND INFORMATION SERVICE** / **NATIONAL CLIMATIC CENTER ASHEVILLE, N.C.**

Normals, Means, And Extremes

Salem, Oregon (McNary Field); Pacific Standard Time.
Latitude: 44°55'N; Longitude: 123°00'W; Elevation (ground) 196 feet.

Month	Temperatures °F Normal Daily maximum	Temperatures °F Normal Daily minimum	Temperatures °F Normal Monthly	Temperatures °F Extreme Record highest	Year	Temperatures °F Extreme Record lowest	Year	Normal Degree days Base 65°F Heating	Normal Degree days Base 65°F Cooling	Precipitation Water equivalent Normal	Maximum monthly	Year	Minimum monthly	Year	Maximum in 24 hrs.	Year	Snow, Ice pellets Maximum monthly	Year	Maximum in 24 hrs.	Year
(a)				42		42					42		42		42		42		42	
J	45.3	32.2	38.8	64	1971	-10	1950	812	0	6.90	15.40	1953	0.57	1949	3.07	1972	32.8	1950	10.8	1943
F	51.4	34.4	42.9	72	1968	-4	1950	619	0	4.79	12.31	1949	0.78	1964	3.16	1949	8.4	1962	5.4	1962
M	54.9	35.4	45.2	80	1947	12	1971	614	0	4.33	8.42	1938	0.87	1965	3.03	1943	10.9	1951	8.5	1960
A	61.0	38.5	49.8	88	1957	23	1968	456	0	2.29	5.18	1955	0.39	1939	2.22	1971	0.1	1972	0.1	1972
M	68.1	43.3	55.7	95	1956	25	1954	295	7	2.09	4.58	1942	0.18	1947	1.84	1963	T	1978	T	1978
J	74.0	48.4	61.2	102	1942	32	1976	133	19	1.39	3.60	1947	0.01	1951	1.60	1950	0.0		0.0	
J	82.4	50.7	66.6	108	1941	37	1962	43	92	0.35	1.80	1974	0.00	1967	0.87	1961	0.0		0.0	
A	81.3	50.9	66.1	106	1978	38	1969	53	87	0.57	4.17	1968	T	1970	1.25	1943	0.0		0.0	
S	76.5	47.3	61.9	103	1944	26	1972	120	27	1.46	3.98	1971	0.00	1975	1.86	1951	0.0		0.0	
O	64.1	43.2	53.2	93	1970	23	1971	366	0	3.98	11.17	1947	0.37	1978	2.84	1955	T	1974	T	1974
N	53.0	37.4	45.2	72	1970	9	1955	594	0	6.08	15.23	1973	0.84	1939	2.82	1950	6.1	1977	6.1	1977
D	47.1	34.7	40.9	64	1958	-12	1972	747	0	6.85	12.40	1964	1.26	1976	2.72	1964	14.6	1972	9.4	1972
YR	63.3	41.3	52.3	108 JUL 1941		-12 DEC 1972		4852	232	41.08	15.40	1953	0.00 JAN 1975		3.16 FEB 1949		32.8 JAN 1950		10.8 JAN 1943	

Month	Relative humidity pct. Hour 04	Relative humidity pct. Hour 10	Relative humidity pct. Hour 16	Relative humidity pct. Hour 22 (Local time)	Wind Mean speed m.p.h.	Wind Prevailing direction	Wind Fastest mile Speed m.p.h.	Wind Fastest mile Direction	Wind Fastest mile Year	Pct. of possible sunshine	Mean sky cover, tenths, sunrise to sunset	Mean number of days Sunrise to sunset Clear	Partly cloudy	Cloudy	Precipitation .01 inch or more	Snow, Ice pellets 1.0 inch or more	Thunderstorms	Heavy fog, visibility ¼ mile or less	Temperatures °F Max. (b) 90° and above	Min. 32° and below	Temperatures °F Max. 32° and below	Min. 0° and below	Station pressure mb. Elev. 201 feet m.s.l.
	17	17	17	17	31	15	30	30		35	42	42	42	42	42	42	42	42	17	17	17	17	7
JAN	85	83	76	85	8.5	S	40	18	1966	8.3		3	4	24	19	1	*	6	0	2	14	0	1011.8
FEB	87	81	69	84	7.8	S	46	18	1958	8.1		3	5	20	17	*	*	4	0	*	11	0	1009.7
MAR	86	75	60	80	8.1	S	40	19	1971	7.9		4	6	21	17	*	*	2	0	0	7	0	1009.1
APR	85	69	57	78	7.2	S	44	18	1962	7.4		4	7	19	14	0	*	1	0	0	1	0	1011.2
MAY	85	65	52	75	6.6	S	28	25	1975	6.8		6	8	17	11	0	1	*	*	0	*	0	1010.9
JUN	85	61	49	73	6.5	S	25	18	1974	6.4		7	9	15	8	0	1	*	2	0	*	0	1010.7
JUL	84	57	40	68	6.5	N	26	24	1979	4.0		15	9	7	3	0	1	*	7	0	0	0	1009.9
AUG	85	59	41	71	6.3	NW	24	29	1969	4.7		13	9	9	4	0	1	*	6	0	0	0	1009.0
SEP	87	65	47	78	6.1	S	31	18	1969	5.1		11	9	10	7	0	1	2	2	0	*	0	1008.8
OCT	90	77	61	85	6.2	S	58	18	1962	6.8		6	8	17	13	0	*	*	7	0	3	0	1011.3
NOV	90	85	77	88	7.3	S	38	18	1951	8.1		3	5	22	18	*	*	7	0	*	8	0	1011.3
DEC	88	86	81	87	8.2	S	45	23	1953	8.8		2	4	25	20	1	*	7	0	1	12	*	1012.5
YEAR	86	72	59	79	7.1	S	58 OCT 1962	18		6.9		77	82	206	149	2	5	38	17	3	68	*	1010.5

Means and extremes above are from existing and comparable exposures. Annual extremes have been exceeded at other sites in the locality as follows: Maximum monthly precipitation 17.54 in December 1933; maximum precipitation in 24 hours 4.30 in December 1933; maximum snowfall in 24 hours 25.0 in February 1937.

(a) Length of record, years, through the current year unless otherwise noted, based on January data.
(b) 70° and above at Alaskan stations.
* Less than one half.
T Trace.

NORMALS - Based on record for the 1941-1970 period.
DATE OF AN EXTREME - The most recent in cases of multiple occurrence.
PREVAILING WIND DIRECTION - Record through 1963.
WIND DIRECTION - Numerals indicate tens of degrees clockwise from true north. 00 indicates calm.
FASTEST MILE WIND - Speed is fastest observed 1-minute value when the direction is in tens of degrees.

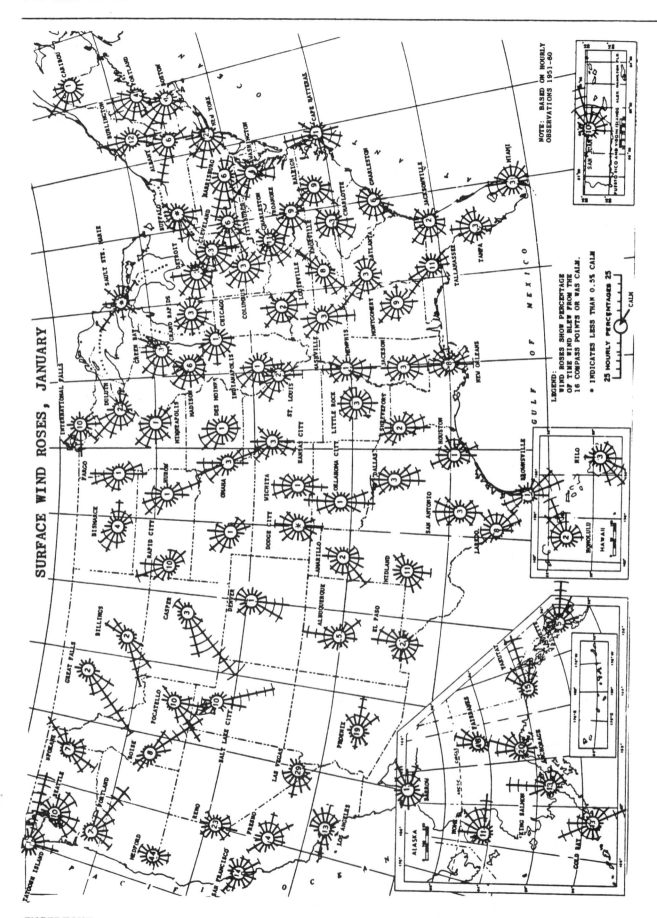

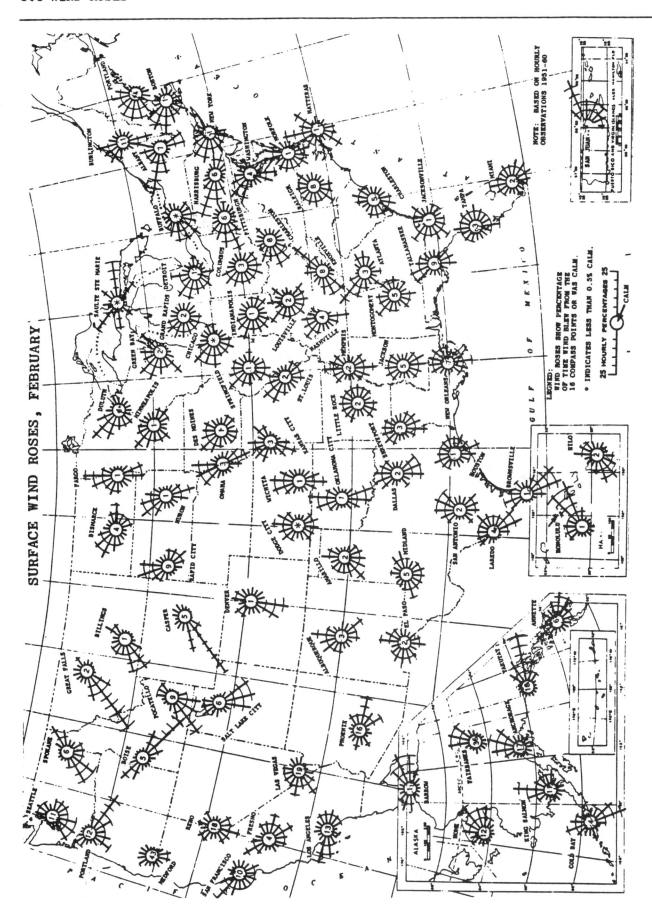

SURFACE WIND ROSES, FEBRUARY

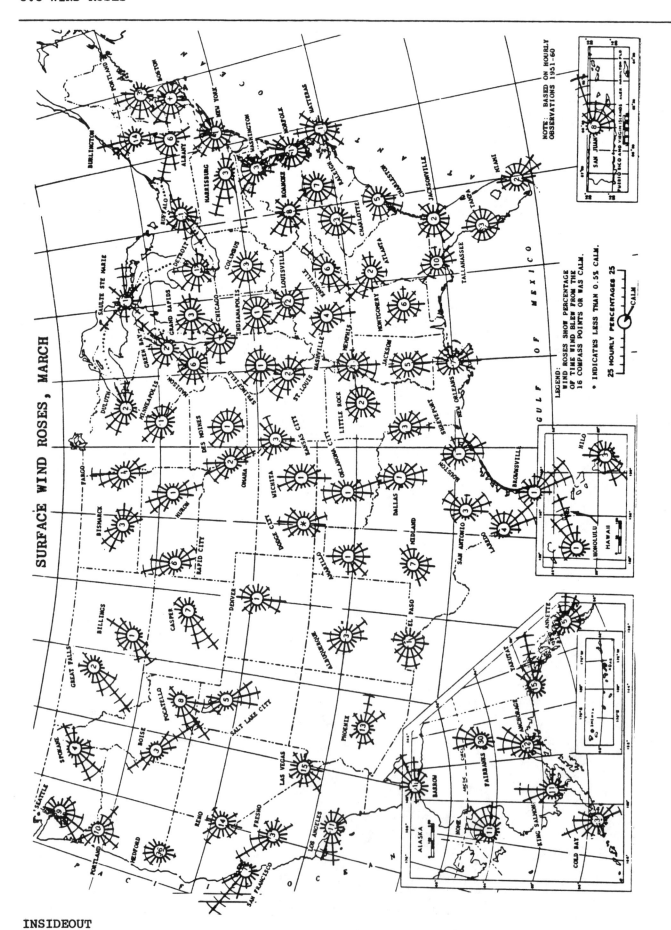

SURFACE WIND ROSES, MARCH

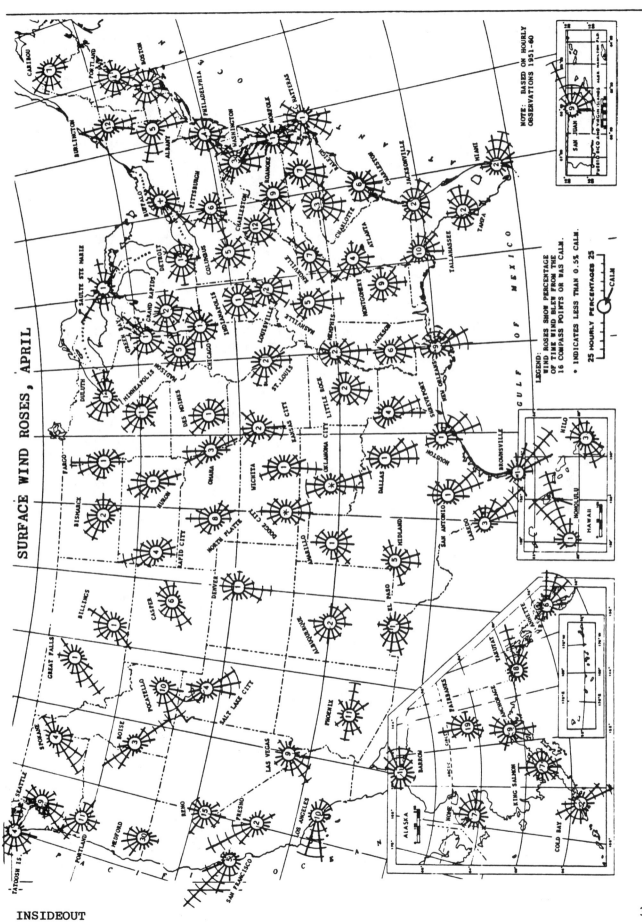

SURFACE WIND ROSES, APRIL

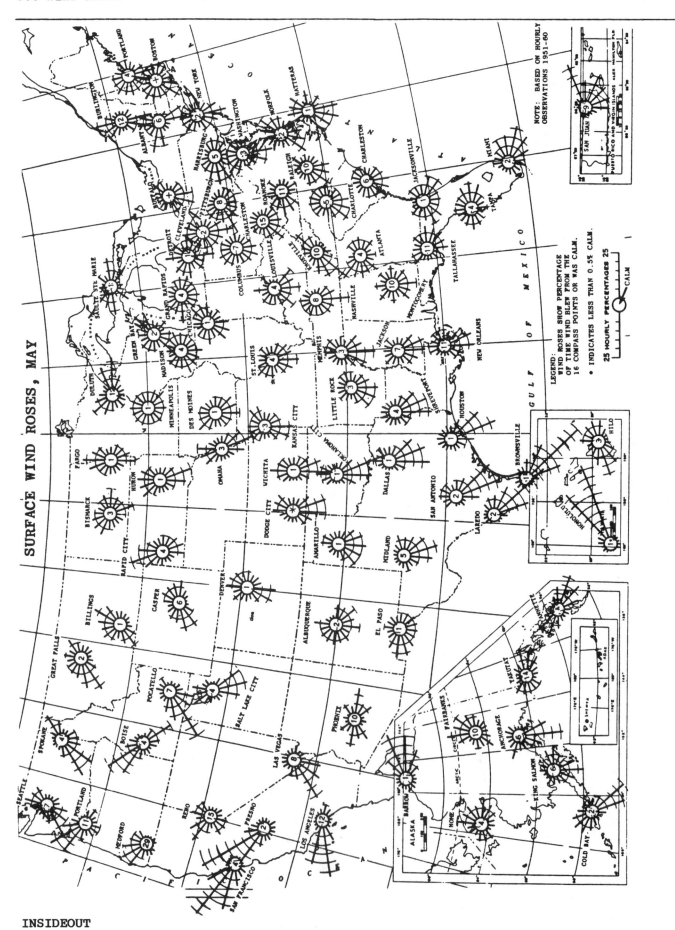

SURFACE WIND ROSES, MAY

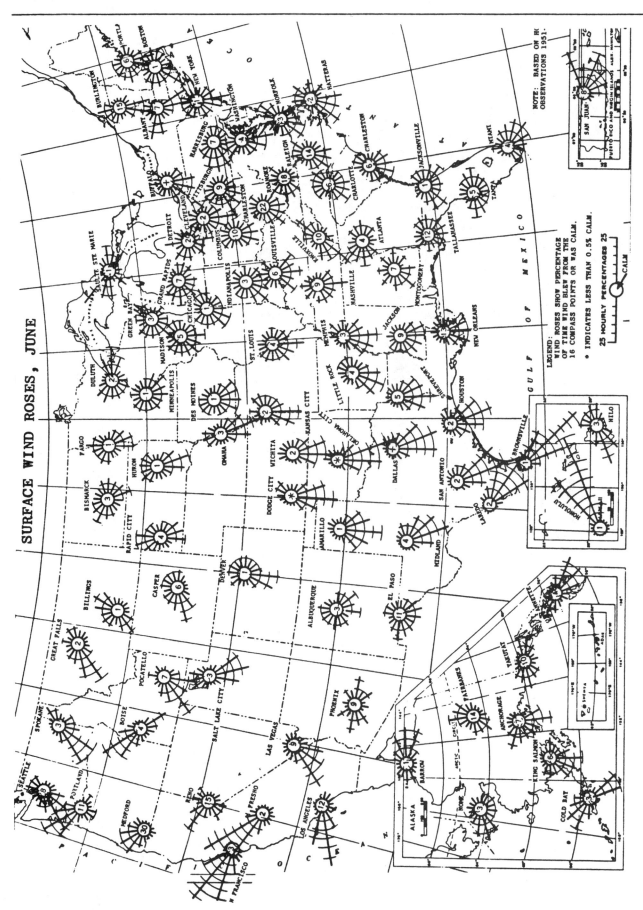

SURFACE WIND ROSES, JUNE

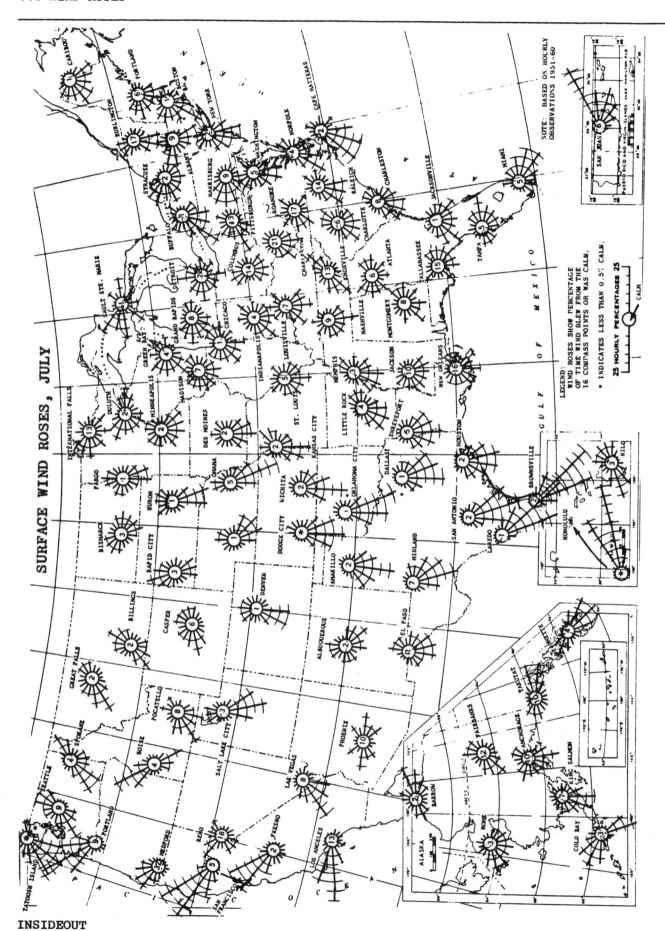

SURFACE WIND ROSES, JULY

NOTE: BASED ON HOURLY OBSERVATIONS 1951-60

LEGEND:
WIND ROSES SHOW PERCENTAGE OF TIME WIND BLEW FROM THE 16 COMPASS POINTS OR WAS CALM.
* INDICATES LESS THAN 0.5% CALM.

3.0 CLIMATE

3.8 WIND ROSES

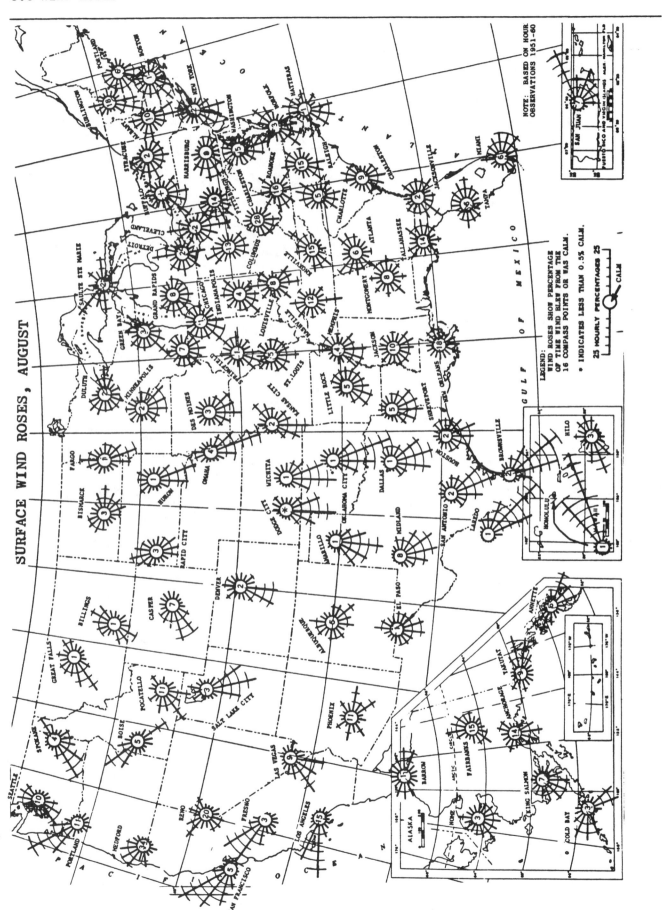

SURFACE WIND ROSES, AUGUST

NOTE: BASED ON HOUR OBSERVATIONS 1951-60

LEGEND:
WIND ROSES SHOW PERCENTAGE OF TIME WIND BLEW FROM THE 16 COMPASS POINTS OR WAS CALM.

* INDICATES LESS THAN 0.5% CALM.

25 HOURLY PERCENTAGES 25

CALM

INSIDEOUT

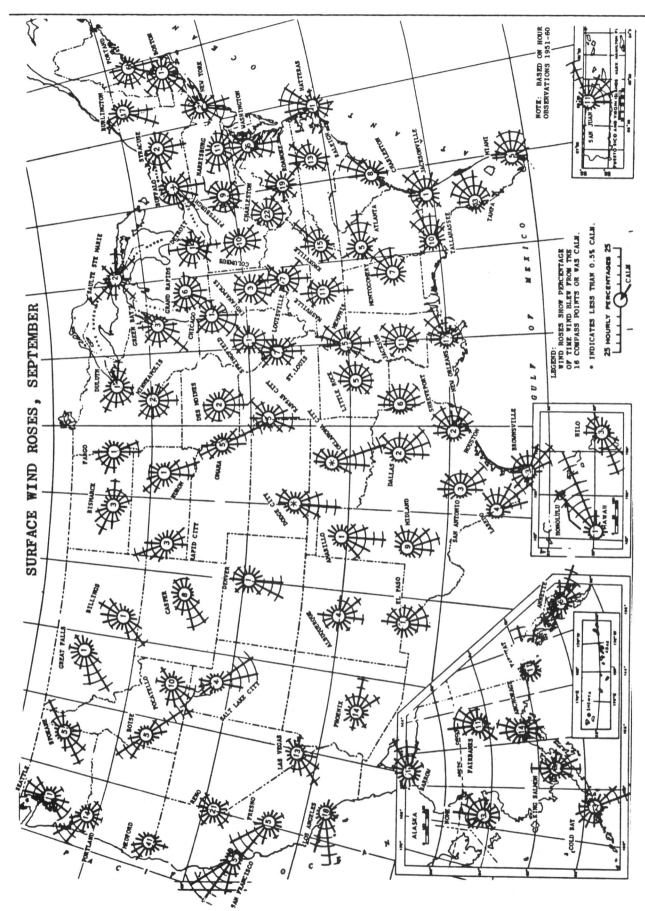

SURFACE WIND ROSES, SEPTEMBER

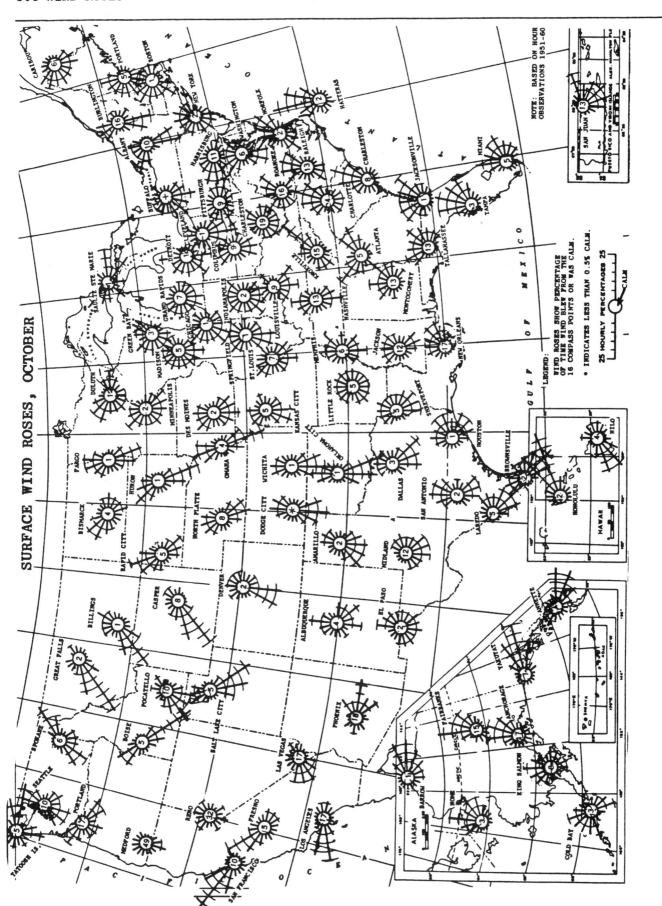

SURFACE WIND ROSES, OCTOBER

NOTE: BASED ON HOUR OBSERVATIONS 1951-60

LEGEND: WIND ROSES SHOW PERCENTAGE OF TIME WIND BLEW FROM THE 16 COMPASS POINTS OR WAS CALM.

* INDICATES LESS THAN 0.5% CALM.

25 HOURLY PERCENTAGES 25

CALM

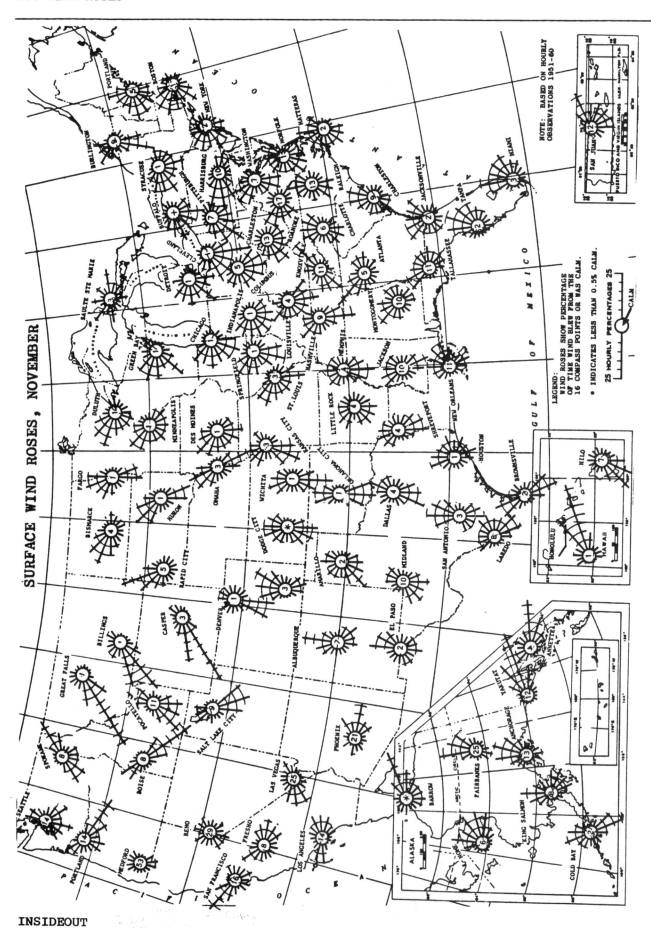

SURFACE WIND ROSES, NOVEMBER

LEGEND: WIND ROSES SHOW PERCENTAGE OF TIME WIND BLEW FROM THE 16 COMPASS POINTS OR WAS CALM.

* INDICATES LESS THAN 0.5% CALM.

25 HOURLY PERCENTAGES 25 — CALM

NOTE: BASED ON HOURLY OBSERVATIONS 1951-60

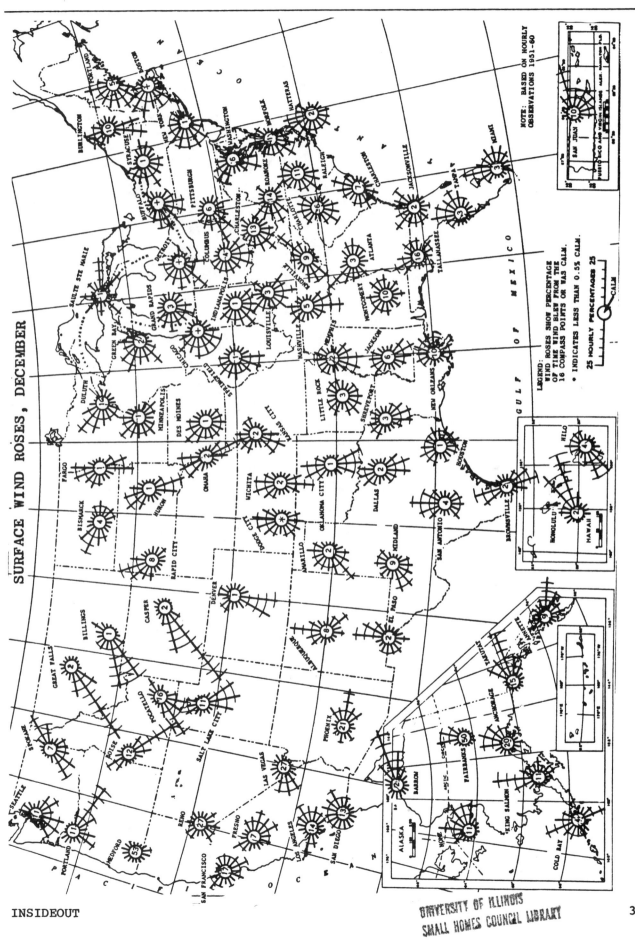

SURFACE WIND ROSES, DECEMBER

NOTE: BASED ON HOURLY OBSERVATIONS 1951-60

LEGEND:
WIND ROSES SHOW PERCENTAGE OF TIME WIND BLEW FROM THE 16 COMPASS POINTS OR WAS CALM.
* INDICATES LESS THAN 0.5% CALM.
25 HOURLY PERCENTAGES 25
CALM

INSIDEOUT

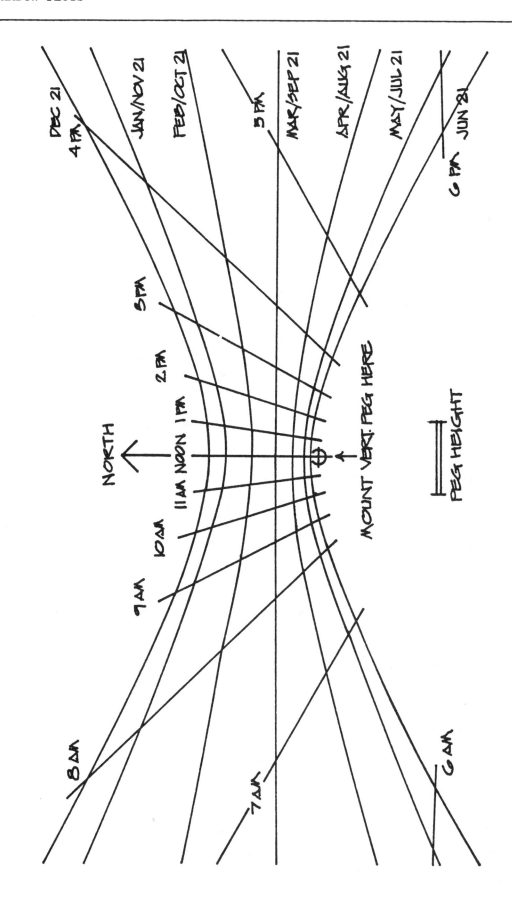

32° NL

36° NL

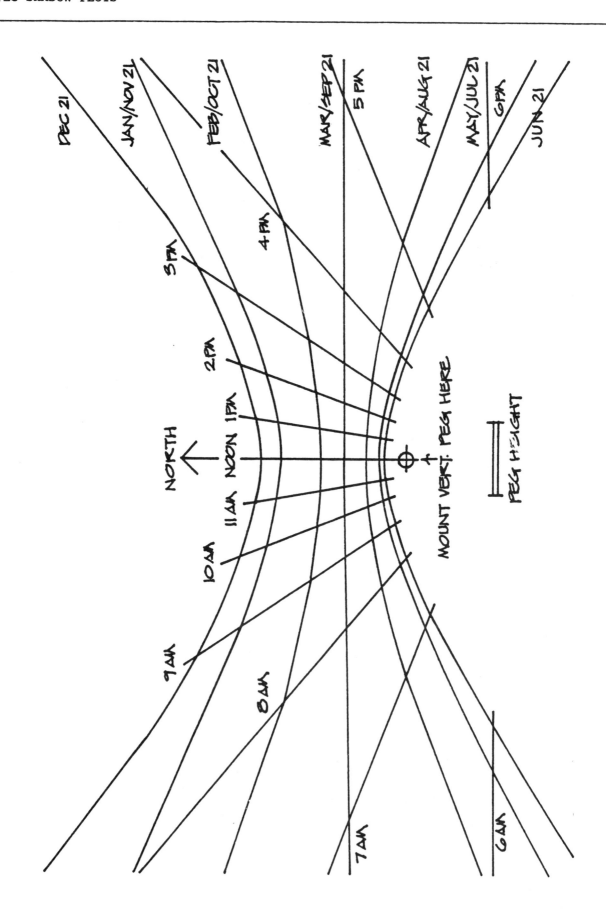

40° NL

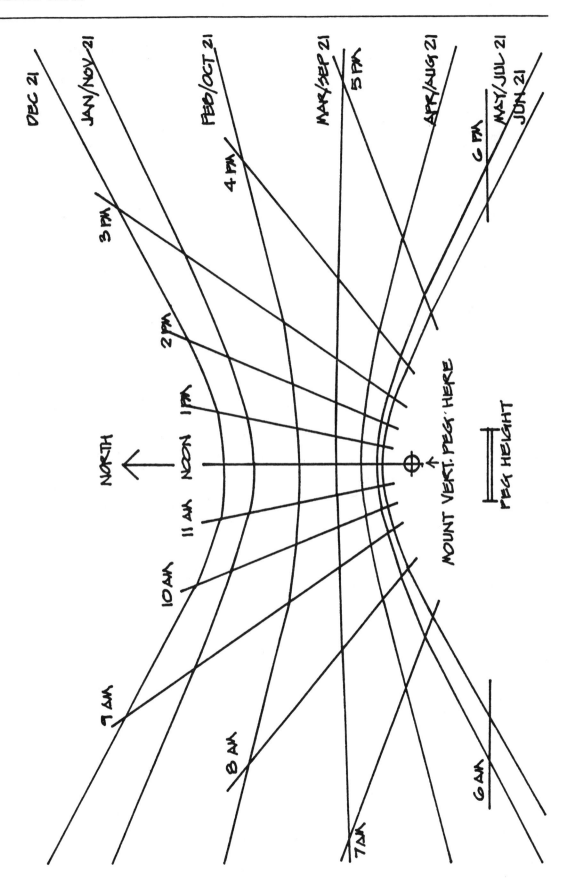

44° NL

The exercises in this section deal with heating and cooling. We have devoted such a large segment of the exercises to thermal considerations because of their potentially high impact on building design and because of the tremendous amounts of energy used in heating and cooling buildings.

It is useful to make an analogy between the response of organisms to their thermal environment and the behavior of buildings. There are three basic ways in which organisms respond to their thermal environment -- migration, form, and metabolism. In migration, they move from an environment that is too cold or too hot to one that is just right. This may happen seasonally (e.g. birds) or diurnally.[1]

Animals have large or small skin areas (form) in relation to their volume, to increase or decrease their rate of heat loss to the environment. For example, the form of an elephant's ears is primarily described by the ears' large surface area which helps to).

Metabolism refers to animals' internal chemical conversion work processes. These work processes require food energy and result in the production of heat. Consequently, those animals with a high heat loss eat large amounts of food, which is converted to heat within their bodies to balance their heat loss.[2]

These three forms of thermal response have their analogies in buildings. Migration -- moving from one area of a building to another, or having the building move to create thermally different areas, has numerous examples, especially in low technology indigenous architecture. An example in residential buildings is the use of the sleeping porch to cope with hot summer evenings.

Form, which includes the factors of size, shape, skin area, orientation, proportion, volume, openings, articulation, etc., is also dependent upon many considerations other than thermal response.

Metabolism, the fuel-sustained processes of the building, is largely concerned with maintaining thermal equilibrium. Current design practice usually separates form and metabolism in order to simplify preliminary design development and does not recombine them until later in the process, when building form must be used as the basis for determining the metabolic rate. Even though form and metabolism are dependent upon different sets of variables, they are intrinsically linked. Form is a function of aesthetic attitudes, land values, material availability, use patterns, circulation routes, etc. Metabolic rate is a function of how well the form uses available energies to modify climate.

In this section of the exercises we will be using all of these mechanisms to understand and design buildings for their thermal environment.

---

[1] Knowles, R. -- Energy and Form
See Chapter 1, "Adaptive Behavior in Nature".

[2] Thompson, D'Arcy -- Growth and Form
See especially Chapter 2, "On Magnitude".

Stevens, Peter -- Patterns in Nature
See especially Chapter 1 "Space and Size". This is an easy-to-understand introduction to the basic responses to forces which exist within the environment.

5.0 HEATING

5.1 INTRODUCTION

5.1 INTRODUCTION _____

This exercise deals with heating, which is a major end-use of energy for many buildings in most of the United States.

Your climate analysis in Exercise 3.0 helped to identify whether heating, cooling, or daylighting (or some combination) was the dominant concern in each building. This should give you a sense of to what degree you need to keep in mind cooling and daylighting objectives in the design for each building.

Note that the sun peg evaluation tool is used to check your design for a number of criteria in the scheming and design section. This tool requires both a model and sunshine, so plan your work accordingly.

Contents

5.0 HEATING

5.2 CONCEPTUALIZING

5.2 CONCEPTUALIZING

GOALS

A. Use the sun for heating.
B. Prevent losing the heat you have collected.

DESIGN STRATEGIES

A. Site Scale
Sunny spaces for winter places.
B. Cluster Scale
Make winter places warmer by arranging buildings to form
sun traps.

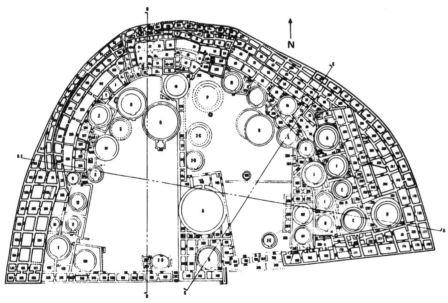

New Pueblo Bonito. Reprinted from Energy and Form by
Ralph Knowles by permission of the MIT Press, Cambridge,
Massachusetts.

C. Building Scale
Increase the south facing skin area.

St. George's School, Wallasey, England. From <u>Process</u>
<u>Architecture No. 6 "Solar Architecture"</u>.

D. Component Scale
Windows for solar gain are potentially different from
windows for light, view, or ventilation.
The greater the wall mass the less the indoor temperature
fluctuates.

DESIGN

Choose an existing building/site which has a clear conceptual approach, at any scale, to achieving these goals.
Document your choice as follows:
A. Identify the location, program, architect (if any),
and the source of your information.
B. Include photocopies or drawings (whatever is quick and
easy for you) to explain the design.

EVALUATION

Evaluation tool - "Conceptual Diagrams".
Diagram how this design is organized to achieve these
goals.

5.3 SCHEMING

GOALS

A. Expose yourself to the south <u>in proportion to</u> the heat you need.
B. Protect yourself from the wind to the extent you need heat.

CRITERIA

A. Buildings are located on the site so they don't cast shadows off the site.
B. The buildings are clustered and sited to create optimal microclimates for heating in the months you need heat.
C. The necessary quantity of south aperture for heating in your climate is provided.
D. The winter sun can reach all the south aperture you have provided for collection.
E. (If you have completed the cooling exercise) Your buildings still meet the schematic phase criteria for cooling, section 5.3.

DESIGN STRATEGIES

A. Site Scale
Use wind breaks to protect building clusters without shading them in winter.

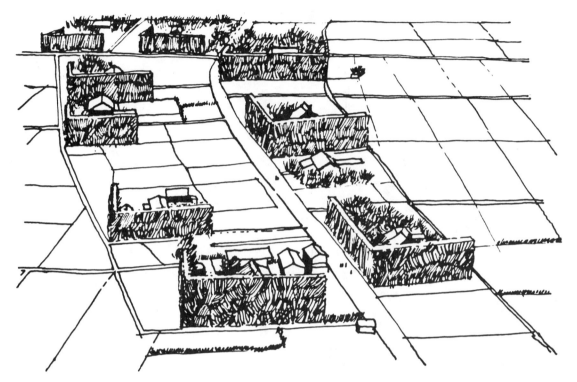

Windbreaks in Shi Mani prefecture, Japan. Reprinted by
permission of the publisher from Introduction to
Landscape Architecture by Michael Laurie, p. 176,
copyright 1975 by Elsevier North Holland, Inc.

B. Cluster Scale
Arrange buildings so that winter wind is blocked from
sunny exterior spaces.
C. Building Scale
Put the spaces with higher internal gains on the colder
side of the building.

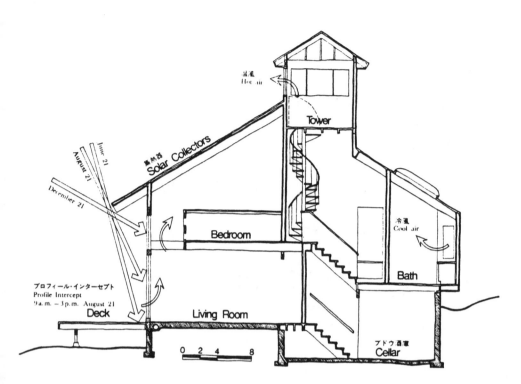

Interactive Resource Inc., Neihuser House (Process 6, 1978, p. 148)

Use spaces that can tolerate greater temperature variations as buffer areas.

Erskine, Villa Gadelius, 1961 (A.D., vol. 47, p. 784)

The orientation of the solar glazing should be between 20° east and 32° west of south (Balcomb, 1980, p. 28).

Recommended areas of south facing glazing ("solar aperture") for a small, SDL building (Balcomb, 1980, Table D-1) are:

Location	Aperture as % of floor area
Madison	20-40%
Dodge City	12-23
Salem	12-24
Phoenix	6-12
Charleston	7-14

Figure 5-1 Solar Aperture Sizing for Heating

Note: a similar rule of thumb does not exist for predominantly IDL buildings, although they may have a heating load under some conditions. The % solar aperture required would be proportionally smaller, as the % heat provided by internal sources was larger. One unsubstantiated way to adjust these numbers to reflect internal gains is to use a 24 hr. avg. Balance Point Temp. that is the assumed Balance Point for the Btuh daily average gain. (Use the total Btuh and balance Point Charts from Climate 3.6, Building Load.)

$$\frac{\text{Btuh daily}}{\text{avg. gain}} = \frac{(\text{Btuh gain while occupied}) \times (\text{\# hours occupied})}{24 \text{ hours}}$$

$$\% \text{ Adjusted aperture} = \% \text{ Aperture} \times \frac{\text{Balance Point Temp (24 hr. avg.)}}{65^\circ}$$

DESIGN

Design the office, factory and outside areas according to program requirements in 2.1 building/site description. Document your design as follows using the grids provided:

A. Site plan including parking and access drive(s)
B. Cluster plan including outdoor space
C. Floor plans of office and factory
D. Section(s) of office and factory
E. Roof plan and elevations, or paralines which show all sides and top of the buildings

Note: Your design may change in the course of the evaluation. The final schematic design which you show to meet all criteria should be fully documented here.

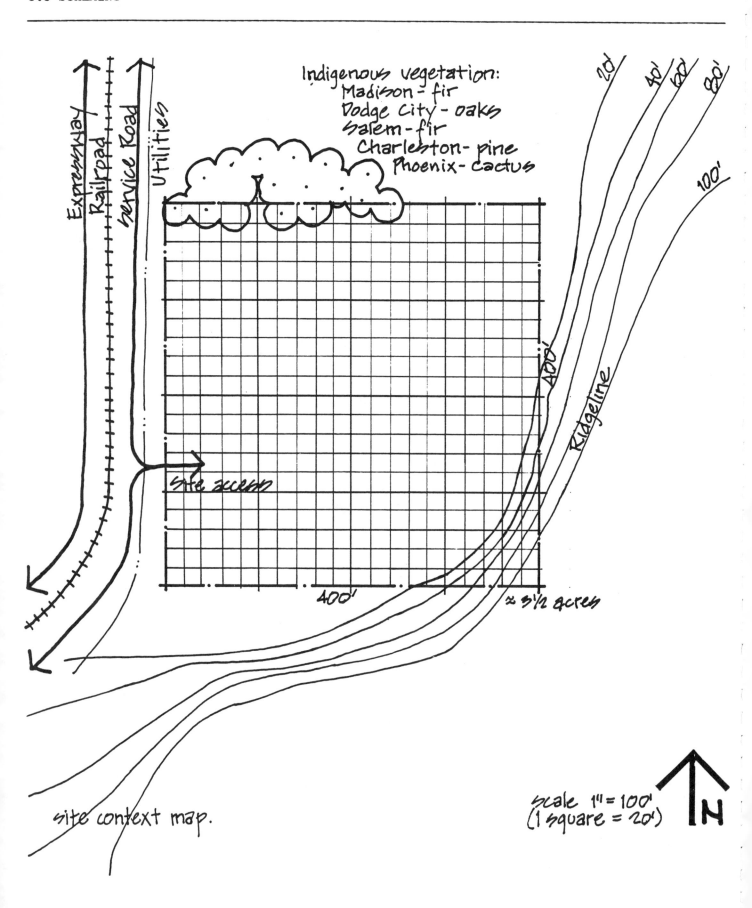

Indigenous vegetation:
Madison - fir
Dodge City - oaks
Salem - fir
Charleston - pine
Phoenix - cactus

Expressway
Railroad
Service Road
Utilities

20' 40' 60' 80'
100'
600'
Ridgeline
400'
≈ 3½ acres

site access

site context map.

scale 1" = 100'
(1 square = 20')
N

cluster site plan: 1"=40'-0".

N

for plans, sections, & elevations: if   1/16" = 1'-0",   1 square = 10'.

for plans, sections, & elevations: if   1/16"=1'-0",   1 square = 10'.

for plans, sections, & elevations: if 1/16"=1'-0", 1 square = 10'.

for plans, sections, & elevations: if   1/16"=1'-0",   1 square = 10'.

for plans, sections, & elevations: if 1/16"=1'-0", 1 square = 10'.

for plans, sections, & elevations: if   1/16"=1'-0",   1 square = 10'.

for plans, sections, & elevations: if   1/16"=1'-0",   1 square = 10'.

for plans, sections, & elevations: if 1/16"=1'-0", 1 square = 10'.

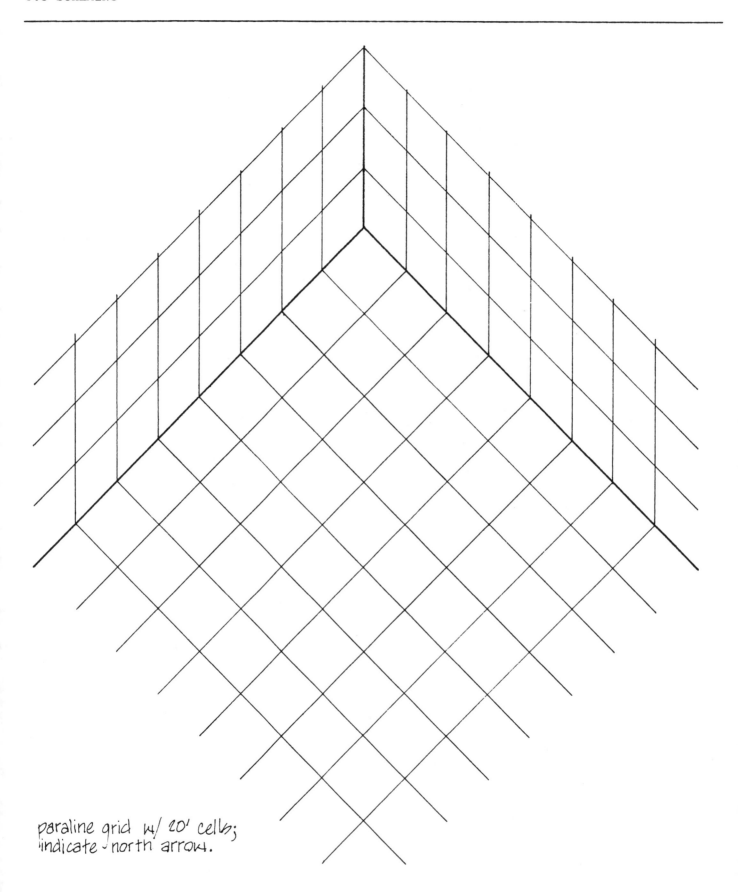

paraline grid w/ 20' cells;
indicate north arrow.

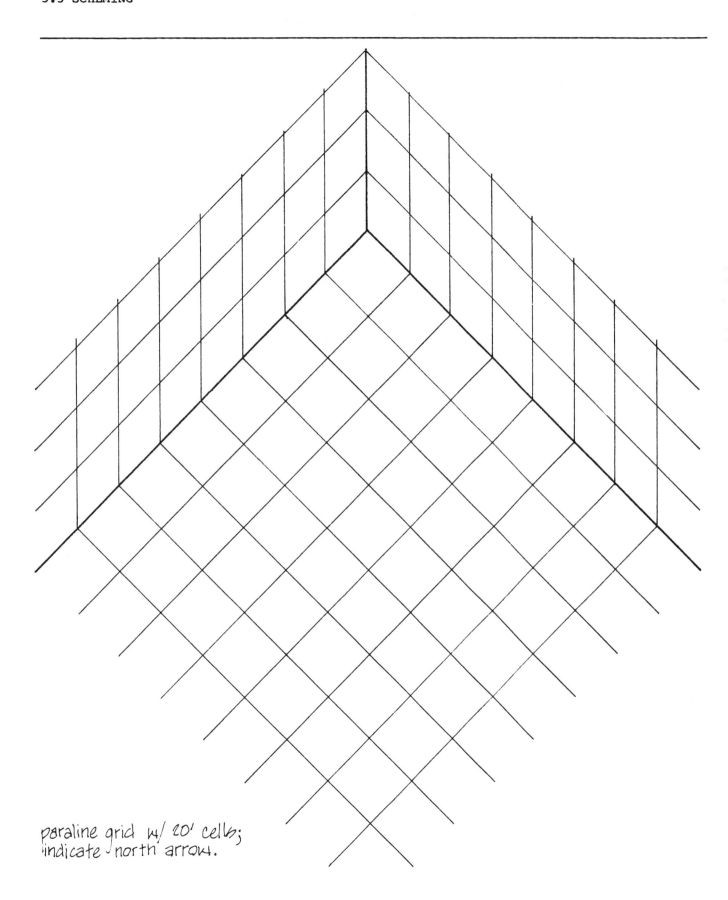

paraline grid w/ 20' cells;
indicate north arrow.

EVALUATION

Criterion A: Buildings are located on the site so they don't cast shadows off the site.

Evaluation Tool for Criterion A - "Sun Peg Check"
1. Make a model of your buildings at 1"=40' of clay or cardboard to go on your site model from 3.3 Climate Exercise. The models must be accurate in size and massing, as well as location on the site.
2. Using the shadow plot and sun peg (as described in 3.3 Climate Exercise) look at the shadows cast by your buildings throughout the year.
3. Draw the shadows cast on December 21st at 10 a.m. and 2 p.m. on the site plan.

Does your first trial design meet the criteria?

If not, what will you change in your second trial design?

Remember that the design documented for this phase should be the trial which you have shown to meet all criteria.

Criterion B: The buildings are clustered and sited to create optimal microclimates for heating in the months you need heat.
According to the climate analysis, what months will you need to block wind/admit sun to heat:
the outdoor area?
office building?
factory?

Evaluation Tool for Criterion B - "Draw Wind Flows"
Draw the wind flows on your site plan and cluster plan during the months you need heating.

Evaluation Tool for Criterion B - "Sun Peg Check"
1. Use the 1/40th model of the site and buildings to check your design for the months you need heating, 10 a.m. to 2 p.m.
2. Draw and clearly label the shadows cast during these times on your cluster plan.

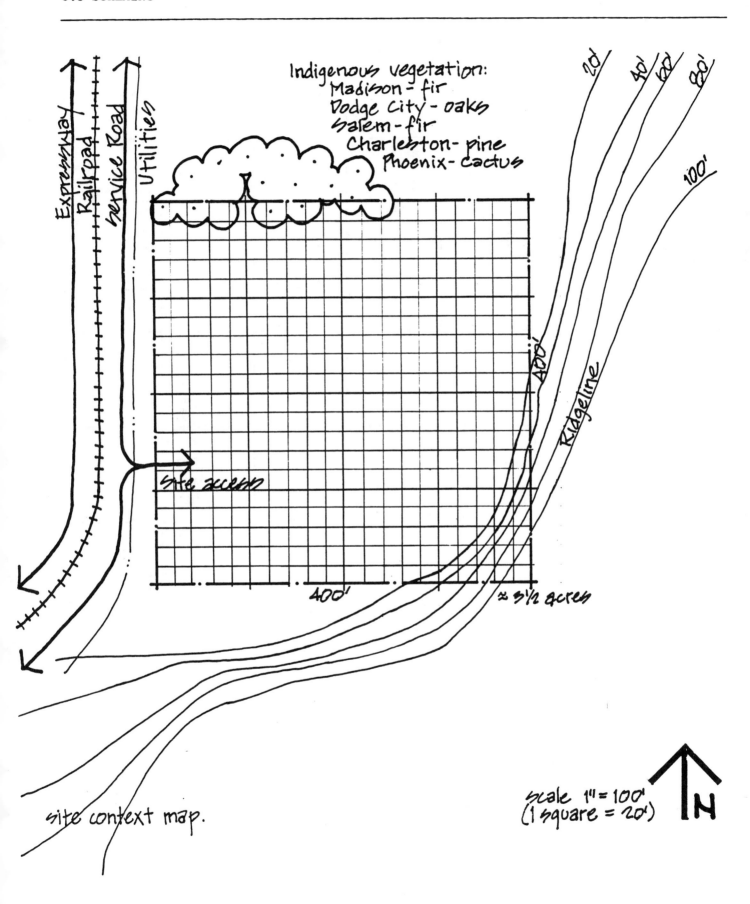

Indigenous vegetation:
Madison - fir
Dodge City - oaks
Salem - fir
Charleston - pine
Phoenix - cactus

Expressway
Railroad
service Road
Utilities

20' 40' 60' 80'
100'

400'

Ridgeline

site access

400'

≈ 3½ acres

site context map.

scale 1" = 100'
(1 square = 20')

N

cluster site plan: 1" = 40'-0".

N

Does your first trial design meet the criteria?

If not, what will you change in your second trial?

Criterion C: The necessary quantity of south (solar collecting) aperture is provided in your office design for heating in your climate.

Evaluation Tool for Criterion C - "Rule of Thumb - Solar Aperture Sizing"
Providing the recommended aperture to floor area ratios for your climate in Figure 5-1 will allow you to achieve a reasonable % solar heating without overheating on clear days.
We have included this as a schematic criterion to help organize your basic building form and orientation. Assumptions involved in the table and further references are in the developing section of this exercise, 5.4 Criterion B.

If you have adjusted the criterion to reflect internal gains, your new criterion is:
office: _____%
factory: _____%

	Trial 1		Trial 2	
	Office	Factory	Office	Factory
Total floor area	.		.		.
Total south aperture	.		.		.
sq.ft. south aperture sq.ft. floor area	.		.		.

Does your first trial design meet the criteria?

If not what will you change in the next design trial?

Criterion D: The winter sun can reach all the south aperture you have provided for solar collection.
From the climate exercise, what months will you need to use sun to heat
the office?
the factory?

Evaluation Tool for Criterion D - "Sun Peg Check"
1. Clearly identify the areas of solar aperture on your 1/40th models.
2. Use the sun peg and model to see if the sun reaches these areas between 10 a.m. and 2 p.m. during the months you have identified above.
3. Draw the shadows on your cluster plan for December 21, 10 a.m. and 2 p.m. or include photographs of the study.

Does your first trial design meet the criteria?

If not, what will you change for your second trial?

Criterion E: Cooling
If you have already done the cooling exercise your design must still meet the criteria in the Developing section 5.4 for cooling.

If you have not yet done the cooling exercise but you know that cooling will be a concern in either building in your climate, check these criteria now so that you don't make things impossible in the next exercise.

5.4 DEVELOPING

GOALS

A. Use the sun and mass together to heat buildings as much as necessary.

B. Use materials and construction techniques to control heat loss as necessary in your climate.

CRITERIA

A. The sun reaches the thermal mass during months your buildings need heating.

B. Mass and aperture in the office building are sized so solar heat will reduce fuel consumption by the following percentage over a non-solar conserving building:

Madison	50%
Dodge City	50%
Salem	40%
Phoenix	40%
Charleston	30%

These values are derived from a range of values given in Balcomb, 1980, Table D-1. Values for Madison, Dodge City, and Salem are low-end values assuming night insulation; for Phoenix and Charleston, low-end values without night insulation. We tabulate the range of values in Fig. 5-7 on p. 5-37.

C. The heat loss from the office building must be less than or equal to the following:

Heating Degree Days	Btu/DD SF
Less than 1,000	9
1,000 - 3,000	8
3,000 - 5,000	7
5,000 - 7,000	6
Greater than 7,000	5

DESIGN STRATEGIES

A. Building Scale

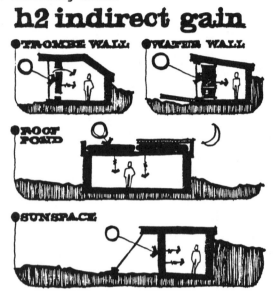

From "Natural Solar Cooling" by David Wright, 3rd
National Passive Conference Proceedings 1979 by
permission of the author.

| Location | Mass Compared to Aperture Area | |
	sq.ft. of 4" Masonry sq.ft. of Aperture	Lbs. of Water sq.ft. of Aperture
Madison	4	36
Dodge City	3	30
Salem	2.5	24
Phoenix	3	30
Charleston	2.5	24

Figure 5-2 Mass Sizing for Heating (Balcomb, 1980, p. 26)

Note: Because you have already adjusted your % aperture
to reflect internal gains you will not have to adjust
these numbers.

Lesser U values mean slower heat flow through walls, an
energy conservation benefit in cold climates.

	BEST ORIENTATION	VIEW IN FROM SOUTH	WARM PLACES	WARM WHEN	NATURAL VENTILATION	SOUTH LIGHT		% HEAT / % FLOOR
TROMBE		poor	near south wall	rad. at night / warm air-sun	good still air / poor cross	some	in between	35% / 30%
DIRECT GAIN		good	outdoor or in sun	radiant night + day / over-heats	good cross	possible glare	least	25% / 50% HEAT
GREENHOUSE		ok	green-house or in sun.	warm air / some overheat	good cross	filtered	most	50% / 30% HEAT

Figure 5-3 Passive Space Heating Systems

B. Component Scale
Consider the possibility of reducing the need for heating by hours of use; if the work schedule corresponds to the hours of greatest solar heat utilization less auxiliary heat would be needed.

Nearly horizontal reflectors in front of south glass can substantially increase winter solar collection, but may also become sources of glare.

Thermal mass should be kept dark in color, and inside wall or floor insulation whenever solar heating is a design objective.

Winter heat conservation and summer heat rejection are both aided by well-insulated roofs.

Note (warning!):
The next two design tools help you choose wall, window, and roof constructions that are energy conserving. They are based on ASHRAE 90-75, a set of recommendations drawn up in the mid-70's, and the first

serious U.S. building design response to energy conservation. They are for conventionally designed buildings. The criteria you will be asked to meet, however, are based on much tougher and newer ideas about energy conservation. Although a direct comparison is impossible, you can expect the tougher criteria to allow about 1/3 as much heat loss as do the ASHRAE 90-75 standards.

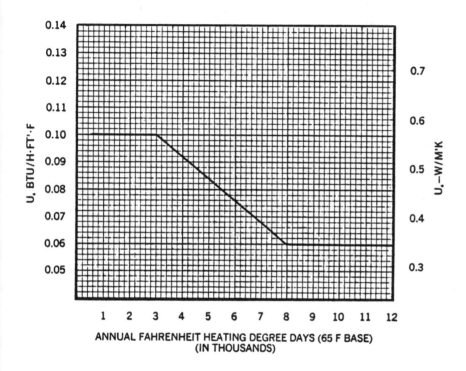

Figure 5-4   Uo - Roofs, Commercial Buildings.   From ASHRAE 90-75.

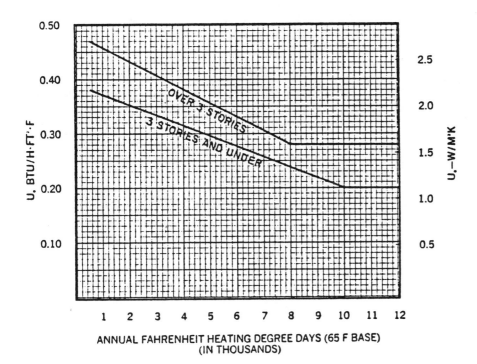

ANNUAL FAHRENHEIT HEATING DEGREE DAYS (65 F BASE)
(IN THOUSANDS)

Figure 5-5   Uo - Walls, Commercial Buildings.   From
ASHRAE 90-75.

Wall insulation is common to nearly all buildings.   The
maximum U-value recommended (which is averaged for walls
and windows and called U_o,) is listed below.   Since a
double glazed window has a relatively high U-value, about
.60, we have also listed the U-values the wall itself
must have to still reach Uo and have windows in it:

Heating DD	U_o	U Wall, if 25% of area is double glazed window
2,000	0.35	.26
4,000	0.31	.21
6,000	0.28	.17
8,000	0.24	.11

Figure 5-6   Recommended Wall U-values

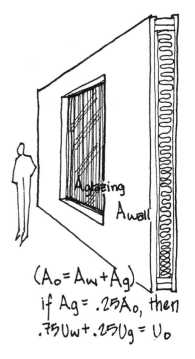

DESIGN

Develop your design for the office and factory to achieve
the goals.   Document your design as follows:

A. State your choice of passive heating system or combination of systems (if any) for each building and draw a sectional perspective to show the relation of south aperture, mass and room. Include at least one person for scale.
B. Draw typical sections at 3/4" = 1'-0" indicating materials of the following:
 - non south wall at opening
 - south wall at opening
 - roof
 - thermal mass
C. State what hours per day and months per year that thermal shutters, if you have any, will be open and closed.

Note: Your design may change in the course of the evaluation. The final development design which you show to meet all criteria should be fully documented here.

for daylighting contours, enlarged plans: if 1/8"= 1'-0", 1 square = 10'.

construction sections: if 3/4" = 1'-0", 1 square = 1'-0".

construction sections: if 3/4" = 1'-0", 1 square = 1'-0".

EVALUATION

Criterion A: The sun reaches the thermal mass during months you need heating.

Evaluation Tool for Criterion A - "Sun Peg Check"
1. Build a quick cardboard model at 1/8"=1'-0" of the building(s) which use the sun for heat. The location and size of the openings and thermal mass must be accurate.
2. Use the sun peg and shadow plot to check sun penetration for the months you need heating, at 10 a.m. and 2 p.m.
3. Show the results by drawing the penetration patterns on your plans and section or your perspective.

Does your first trial design meet the criteria?

If not, what will you change for your second trial?

(Be sure that the final design you document is the one that meets all developing criteria.)

Criterion B: Mass and aperture in your buildings are sized so that solar heat will reduce heating fuel consumption over a non-solar conserving building by:

Madison	50%
Dodge City	50%
Salem	40%
Phoenix	40%
Charleston	30%

Evaluation Tool for Criterion B - "Aperture and Mass Sizing for Solar Savings"

for plans, sections, & elevations: if   1/16"=1'-0",   1 square = 10'.

for plans, sections, & elevations: if 1/16"=1'-0", 1 square = 10'.

5.4 DEVELOPING

---

1. Complete the following for your building:

	Trial 1		Trial 2	
	Office	Factory	Office	Factory
Floor area (sq.ft.)	.	.	.	
South aperture (sq.ft.)	.	.	.	
Masonry mass (sq.ft. surface)	.	.	.	
Water mass (lbs)	.	.	.	
$\dfrac{\text{sq.ft. south aperture}}{\text{sq.ft. floor area}}$	.	.	.	
$\dfrac{\text{sq.ft. masonry mass}}{\text{sq.ft. aperture}}$	.	.	.	
$\dfrac{\text{lbs water mass}}{\text{sq.ft. aperture}}$	.	.	.	

2. Is insulation used for the south aperture?
3. According to the following rules of thumb (Balcomb, 1980, pp. 20-26), the office building will use the sun to reduce fuel costs by what % (over a conventional design)?

4. Does this meet Criterion B?

Location	$\dfrac{\text{sq.ft. so. aperture}[2]}{\text{sq.ft. floor area}}$	$\dfrac{\text{sq.ft. masonry}}{\text{sq.ft. aperture}}$	$\dfrac{\text{lbs water}}{\text{sq.ft. aperture}}$	% Reduction in Heating Load (SSF) w/o night insulation	w/night insulation
Madison	20-40%	4	36	15-17%	51-74%
Dodge City	12-23	3	30	27-42	46-73
Salem	12-24	2.5	24	21-32	37-59
Phoenix	6-12	3	30	37-60	48-75
Charleston	7-14	2.5	24	25-41	34-59

Figure 5-7 Sizing for Solar Savings[1]

Note 1: These rules of thumb were developed assuming that solar glazing was half direct gain and half water wall, well insulated walls and double glazing. The upper limit was chosen to prevent overheating on clear January days and internal gains were assumed to be low. See Balcomb for further details.

Note 2: If you have adjusted these percentages to reflect internal gains and have provided the correct aperture as adjusted, you can use the rest of this table as it is to answer to questions above.

Criterion C: The heat loss from your office building must be less than or equal to the following (Balcomb, 1980, p. 24):

Heating Degree Days	Heat Loss in Btu/DD SF
Less than 1,000	9
1,000 - 3,000	8
3,000 - 5,000	7
5,000 - 7,000	6
Greater than 7,000	5

If your buildings have greater internal gains than a small residential building for which these numbers were generated, you can adjust the criteria as follows using the average Balance Point Temperature found (for sizing solar aperture) on p. 5-7.

$$\text{Required Btu/DD sq.ft.} \times \frac{65^\circ}{\text{Balance Point Temp (24 hr. avg.)}} = \text{adjusted criterion}$$

Office criterion: _____
Factory criterion: _____

Note: When your design meets this criteria your building will achieve the Solar Savings Fraction specified in Criterion B above.

Evaluation Tool for Criteria C - "Overall Heat Loss - ASHRAE 90-75 Procedure" (See ASHRAE Standard 90-75 for a more complete derivation and explanation.)

Note: This heat loss calculation is done for the whole building rather than room by room and will be done for a 24 hour period in order to use the DD data and calculate internal temperatures in the Thermal Finalization Exercise, 7.2. If you used earth berms in your design to reduce heat loss, you will not be able to calculate their effect using the procedure as written or use them to meet Criterion C numerically. However, we have included more information at the end of this tool so that you can estimate the potential contribution.

1. Document the U Values
a. Find the U Values for your walls and roof:

WINDOW U-values are listed in MEEB, Table 4.16, p. 128-129. For windows utilizing <u>night insulation</u>, as is typical for passive solar buildings in colder climates, these U-values may be used (Balcomb, 1980, p. 142):

	No Night Insulation	Night Insulation: R 4	R 9
Direct Gain	0.55	0.30	0.24
Trombe Wall	0.22	0.15	0.12
Water Wall	0.33	0.20	0.17

Note: R4 insulating shades are rather common; R9 are more expensive and available in less variety.

WALL U-values are listed in MEEB, Tables 4.7-4.9, p. 119-121 (with added insulation in Table 4.13, p. 125)

Note: Berms are not calculated by their U-value but by their effect on the ΔT. The procedures used in this workbook use Degree Days rather than ΔT so you will not be able to calculate the effect of berming on heat loss in your buildings.

SKYLIGHT U-values are found in MEEB, Table 4.16, p. 128-129. (Also see U-values, above, for windows using night insulation.)

ROOF U-values are listed in MEEB, Tables 4.10-4.12, p. 122-124, with added insulation in tables 4.13-4.14.

b. Document these by labelling the 3/4" sections in the design documentation. If you use values not found in MEEB, please reference the source.

---

2. Calculate Heat Loss Through Building Components

a. Heat Loss Through Walls

	Trial 1		Trial 2	
1. Determine A_o	Office	Factory	Office	Factory
Percent wall area in windows (including clerestories)	%	%	%	%
Percent wall area in opaque (typical) construction (including doors)	%	%	%	%
(If more than one type, list separately)				
Vertical Surface Area Total (A_o wall)	s.f.	s.f.	s.f.	s.f.

Determine U_o

$$\text{Wall } U_o = \frac{(U \text{ wall})(\% A \text{ wall}) + (U \text{ window})(\% A \text{ window})}{100\%}$$

	Trial 1		Trial 2	
	Office	Factory	Office	Factory
Wall Uo				

Heat Loss

$$\text{Wall heat loss } (\text{Btuh}/^\circ F) = \text{Wall } U_o \times A_o \text{ Wall}$$

	Trial 1		Trial 2	
	Office	Factory	Office	Factory
Wall Heat Loss				

b. Heat Loss Through Roofs

Determine A_o	Trial 1		Trial 2	
	Office	Factory	Office	Factory
Percent roof area in sky-lights	%	%	%	%
Percent roof area in opaque (typical) construction(s)	%	%	%	%
(If more than one type, list separately.)				
Horizontal/Sloped Surface Area Total (A_o roof)	_____ s.f	_____ s.f	_____ s.f	_____ s.f

Determine U_o

$$\text{Roof } U_o = \frac{(\text{U roof})(\% \text{ A roof}) + (\text{U skylight})(\% \text{ A skylight})}{100\%}$$

	Trial 1		Trial 2	
	Office	Factory	Office	Factory
Roof Uo				

Heat Loss

$$\text{Roof heat loss (Btuh/}^{\circ}\text{F)} = \text{roof } U_o \times \text{roof } A_o$$

	Trial 1		Trial 2	
	Office	Factory	Office	Factory
Roof Heat Loss				

---

c. Heat Loss Through Floors

A simplifying technique is to assume that only the perimeter of the floor slab loses heat, and that the ground
temperature below the floor slab stabilizes to the extent
that all slab losses can be ignored except for the edge.
(ASHRAE Fundamentals, 1972, P. 354)

Assuming a 2" slab edge insulation at R = 4 per inch,
(MEEB, 6th ed., Table 4.6, p. 113), the approximate
U-value at the slab edge =

$$\frac{1}{2 \times 4} = .125$$

The resulting slab edge heat loss per lineal foot for a
6" slab would be
heat loss per foot = UxA = .125 x $\frac{6"}{12"}$ = .06 Btuh/$^{\circ}$F

Floor heat loss (Btuh/$^{\circ}$F) = heat loss per foot x lineal
feet floor slab

	Trial 1		Trial 2	
	Office	Factory	Office	Factory
Floor Heat Loss				

d. Heat Loss by Infiltration

At this development stage, we can assume a rule-of-thumb
of 3/4 air change per hour (ACH) for infiltration heat
loss calculation. (Balcomb, 1980, p. 37)

Infiltration heat loss (Btuh/$^{\circ}$F) = 3/4 (ACH) x volume x
.018*
(in Btu/hr $^{\circ}$F)

*A constant, relating specific heat and density of air
(MEEB, 1980, p. 135-136).

	Trial 1		Trial 2	
	Office	Factory	Office	Factory
Infiltration Heat Loss				

3. Approximate Total Heat Loss

Add the heat losses due to individual components of the design:

	Trial 1		Trial 2	
	Office	Factory	Office	Factory
Wall Heat Loss				
Roof Heat Loss				
Floor Heat Loss				
Infiltration Heat Loss				
Total loss in Btuh/^{o}F				

Total Loss in Btu/DDSF
$$= \frac{(Btuh)(^{o}F)(24)}{floor\ area}$$

Note: If you choose to use earth berms to meet the criterion you must show that they significantly reduce the heat loss of your designs. Earth rather than air on the outside wall surface effects the ΔT factor of the heat loss calculation. The room to air ΔT is already assumed in the Degree Day data, however you can also look at heat loss on an hourly basis:

heat loss (Btuh) = $UA\Delta T$

To calculate the ΔT, the earth temperature can be estimated as the ground water temperature (see Figure 12. Water and Waste Exercise), and you can use 65°F or whatever you have chosen as the inside temperature.

Does your first trial design meet the criteria?

_____

If not, what will you change in the second trial?

The design you document (for the DESIGN section in DEVELOPING) should be the final trial design which meets all criteria.

6.0 COOLING

6.1 INTRODUCTION

_____

6.1 INTRODUCTION _____

In this set of design phases you will explore ways of
naturally cooling your buildings. Even in the severe
winter climates, such as Madison, summer overheating of
buildings can occur. As you design for cooling, it is
obviously important to keep your building in good
condition for winter solar heating performance. In some
climates, for some functions, this may create a very
different-looking building in winter when compared to
summer. For example you may have minimal openings on all
walls but south in winter, but generous north and south
openings for summer ventilation.

Contents

6.2 CONCEPTUALIZING

GOALS

A. Use the wind for cooling.
B. Use heat sinks for cooling. (A heat sink is any cold area that is available for the dumping of excess heat. This includes the atmosphere, such as the night sky, water bodies, ground, or massive building materials.)
C. Minimize heat gains.

DESIGN STRATEGIES

A. Site Scale
Cool places for summer spaces.

MORNING   EVENING

From Landscape Planning for Energy Conservation, ed. by G. Robinette, 1977 by permission of Environmental Design Press.

B. Cluster Scale
Use building arrangement to provide shading of people and buildings.
Preserve each building's access to cooling breezes during overheated months.

PERFORATED CANOPY   SINGLE LAYER CANOPY   MULTIPLE LAYER CANOPY

From Robinette, 1977 by permission of Environmental
Design Press.

C. Building Scale
Orient buildings to the direction of cooling breezes
and/or to heat sinks.
D. Component Scale
Openings for ventilation can be separate from those for
view and light.

DESIGN

Choose an existing building/site which has a clear
conceptual approach, at any scale, to achieving these
goals.
Document your choice:
A. Identify the location, program, architect (if any),
and the source of your information.
B. Include photocopies or drawings to explain the design.

EVALUATION

Evaluation Tool - "Building Response Diagrams"
Diagram how this building achieves the goals. If the
climate of this documented example is different from
yours, speculate on how the design responses might change
for your climate.

6.3 SCHEMING

GOALS

A. Use ventilation and/or mass for cooling, as suitable for your climate.
B. Expose yourself to the cool source in proportion to the cooling you need.
C. When overheated protect yourself from the sun by shading.

CRITERIA

A. The cooling strategy chosen will work in your climate.
B. Enough aperture is provided to ventilate the building as needed.
C. Ventilation flows remove heat from the buildings without leaving hot, dead areas.
D. Glazing is shaded from the sun during overheated months.
E. The building's heat performance has not been damaged by cooling design decisions.

DESIGN STRATEGIES

A. Site Scale
Use trees for shading, particularly on the east and west building faces, and for outdoor spaces.
Schedule building use periods to avoid the hottest times of the day.

Design with the wind and heat, considering that wind tends to keep moving in the same direction, wind flows from high pressure to negative pressure areas, and hot air rises while cool air falls.
B. Cluster Scale
Provide courts which are open to heat sinks.

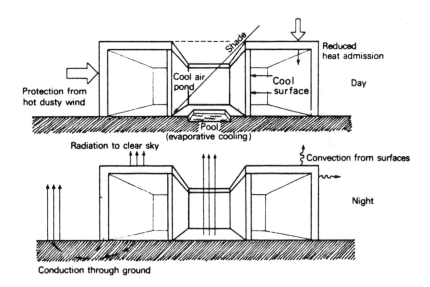

(Koenigsberger, 1974, p. 205)

C. Building Scale
Move activities to cooler places during the hottest
periods of the day (or season).
Adopt a "closed", "open" or mixed strategy regarding
ventilation:
"Open" buildings are those using natural ventilation.
"Closed" buildings are those using high mass without
simultaneous ventilation, those using evaporative
cooling, and those using mechanical refrigeration or
"air-conditioning".
Mixed buildings can be "closed" during very hot hours,
then "opened" for night ventilation to remove stored
heat.

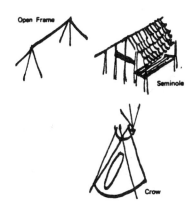

The open building depends upon internal air-flow for
interior cooling, and this works only if the inside
temperature is higher than the outside temperature. For
some climates, natural ventilation can be used even
during the hottest months, producing indoor temperatures
near the upper limit of comfort with moving air, about
87° in drier climates, less in humid climates (refer to
the Bioclimatic Chart, Climate exercise).   In other
climates, it is preferable to close up the building in
the   early   morning   to   prevent
hot air from entering. In this case, the indoors will
remain cooler than outside, although conditions indoors
might not be within the comfort zone unless you have
auxiliary refrigeration equipment.   This is a closed
building. The thermal mass you may have provided when
solar heating the building can be a useful heat sink for
the closed building.

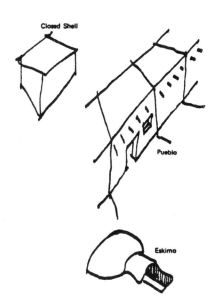

Two basic ways to ventilate a building are:

<u>Cross Ventilation</u> (w<u>in</u>d) - depending on the force of the wind to expell hot air from the leeward side of the building, to be replaced by cooler air forced in on the windward side.
<u>Stack (gravity)</u> - depending on the principle of hot air rising to expell hot air from openings high up in a building, to be replaced by cooler air drawn in at much lower openings.   This is an important strategy if you have calm wind conditions in overheated months.

Building codes generally allow commercial buildings to be built without mechanical ventilation systems if <u>at least 5%</u> of the floor area is provided in fully openable window area.   The more you depend upon natural ventilation for cooling, the larger these window openings should become; approximate sizes are graphed in the Evaluation section.

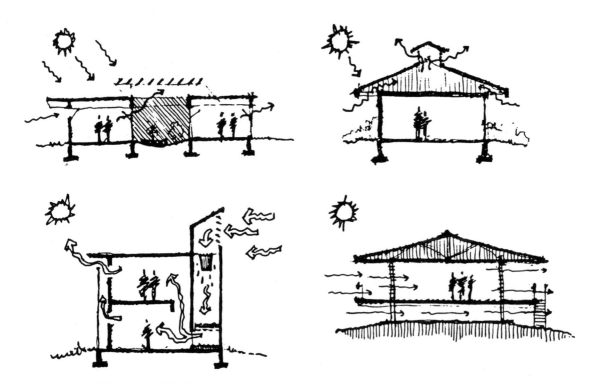

(Holmes, 1979, p. 452-3)

Keep clear paths through buildings for unobstructed ventilation.
While excluding direct sun, utilize daylighting instead of electric lighting.

D. Component Scale

Glass areas in the hottest summer exposures (east and west walls, and roofs) should be minimized, or fully shaded from direct sun.

Clerestories protected from direct summer sun are potential sources of daylight, as well as openings to allow heated air to escape. They are especially effective ventilators if they open on the leeward (downwind) side.

DESIGN

Design the office, factory and site to meet the program requirements in 2.0 Site and Building Description. Document your design using the grids provided.

A. State whether your buildings will be open, closed or mixed:

Office:

Factory:

B. Site plan including parking and access drives, and vegetation.

C. Cluster plan including outdoor spaces and vegetation.

D. Floor plans of office and factory.

E. Sections of office and factory.

F. Roof plan and elevations (or paralines) which clearly show all sides and tops of the buildings.

G. Key each glazed aperture to the plan and state at which scale (site, cluster, building or component) you will shade each:

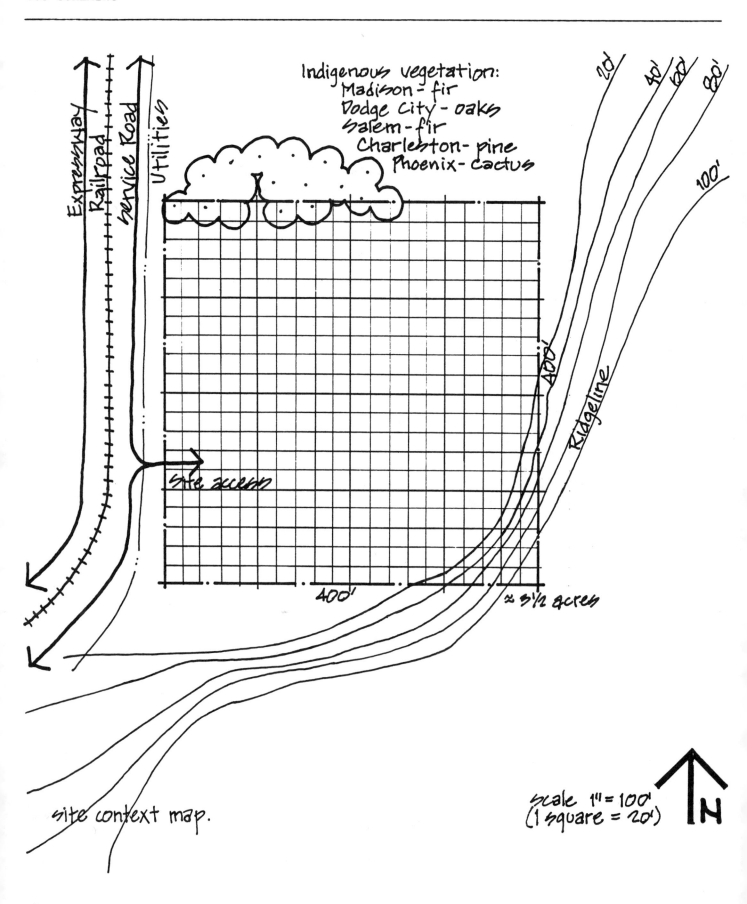

Indigenous vegetation:
 Madison - fir
 Dodge City - oaks
 Salem - fir
 Charleston - pine
 Phoenix - cactus

Expressway
Railroad
Service Road
Utilities

20' 40' 60' 80'
100'
400'
Ridgeline

site access

400'

≈ 3½ acres

site context map.

scale 1" = 100'
(1 square = 20')

N

cluster site plan: 1"=40'-0".

for plans, sections, & elevations: if 1/16"=1'-0", 1 square = 10'.

for plans, sections, & elevations: if 1/16"=1'-0",   1 square = 10'.

for plans, sections, & elevations: if 1/16"=1'-0", 1 square = 10'.

for plans, sections, & elevations: if 1/16"=1'-0", 1 square = 10'.

for plans, sections, & elevations: if 1/16"=1'-0",   1 square = 10'.

for plans, sections, & elevations: if 1/16"=1'-0",   1 square = 10'.

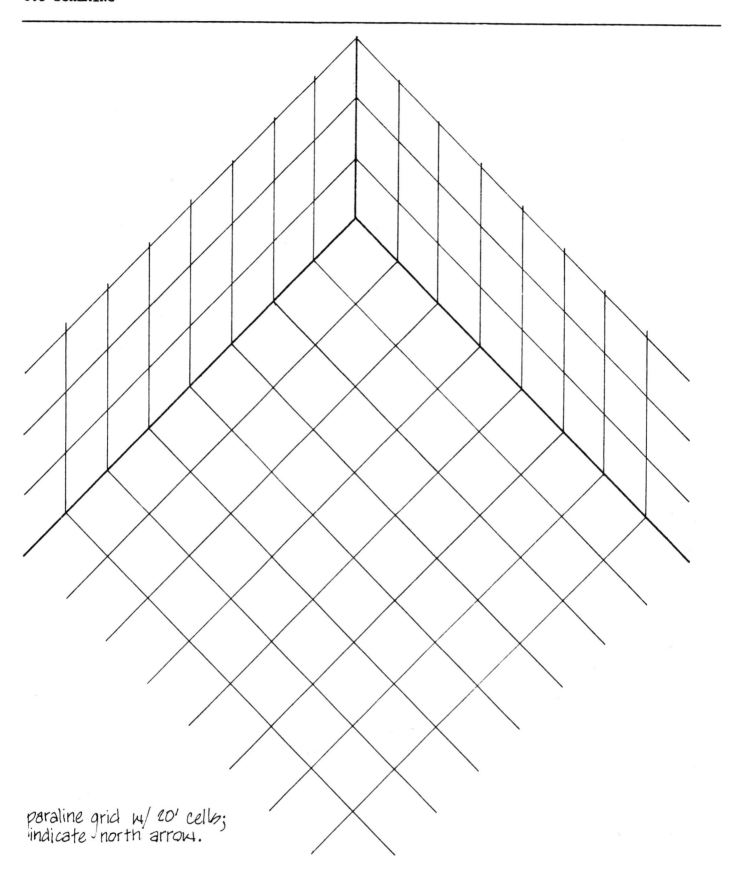

paraline grid w/ 20' cells;
indicate north arrow.

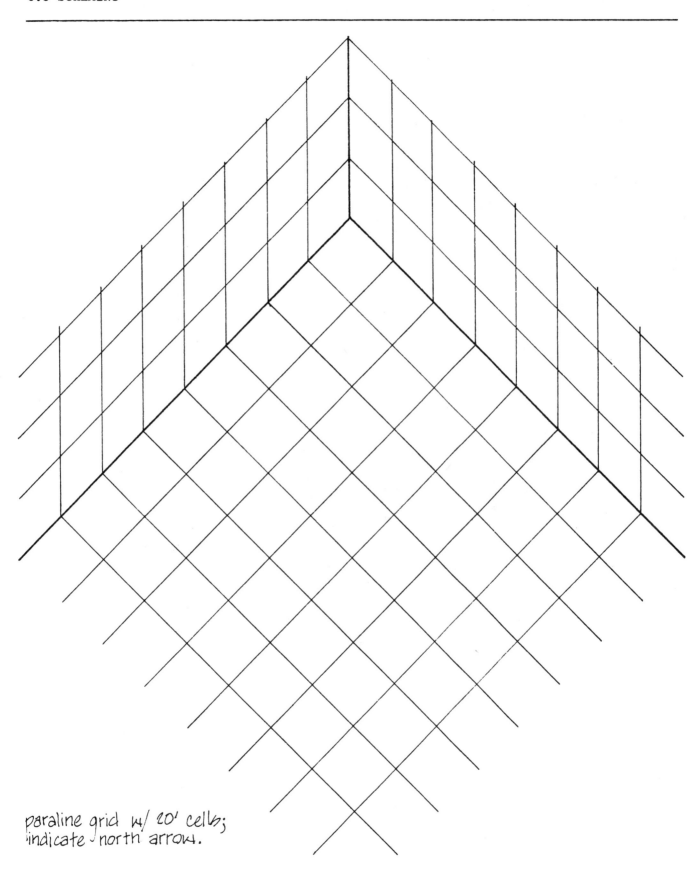

paraline grid w/ 20' cells;
indicate north arrow.

EVALUATION

A. Criterion A: The cooling strategy chosen works in your climate.

Evaluation Tool for Criterion A - "Bioclimatic Chart of Cooling Strategies"
In the Climate exercise section 3.2 you plotted average monthly temperatures on a chart similar to Fig. 6.1. This chart tells you whether buildings in your climate should be open or closed. In the comfort zone, either of these strategies is possible.

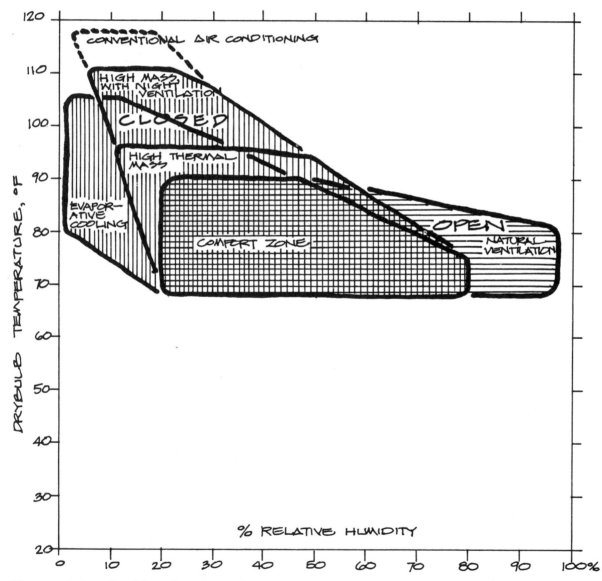

Figure 6-1 Bioclimatic Chart for Cooling Strategies

A. Use overlays to transfer your climate as plotted in the Climate exercise, 3.2 to Fig. 6-1.
B. Is the cooling strategy appropriate for your climate an open or closed building?

Is this the strategy you have chosen?

If not, what will you do to meet Criterion A?

For a detailed explanation of Fig. 6-1, see Milne and Givoni, "Architectural Design Based on Climate" in Watson, Energy Conservation Through Building Design, 1979.

B. Criterion B: Enough aperture is provided for ventilation.

Evaluation Tool for Criterion B - "Aperture Sizing for Ventilation"
1. Complete the table below. From the Climate exercise, section 3.6, find the Btuh internal gain for each building. For open buildings, use the Btuh gain while occupied. For closed buildings, you will need to ventilate stored heat in addition to heat being produced while venting, so use two times the Btuh while occupied.

Building	Open or Closed	Cross-ventilation or stack ventilation?	Internal gains Btuh	sq.ft. floor area	Revised Btuh/sq.ft.
Office					
Factory					

Btuh/sq.ft. gain	sq.ft. Aperture as % of floor area*			
	6 mph Salem	7 mph Phoenix	8 mph Madison Charleston	13 mph Dodge City
10	2%	2%	1%	1%
20	3	2	2	1
30	5	4	3	2
40	6	5	4	3
50	7	6	5	3
60	9	8	7	4
70	10	9	8	5
80	12	10	9	5
90	13%	11%	10%	6%

*needed on each of two sides of a building

Figure 6-2 Aperture Sizing For Cross-Ventilation

Btuh/sq.ft. gain	sq.ft. Aperture as% of floor area*
10	11%
20	22
30	34
40	45
50	56
60	67
70	78
80	90
90	100%

*needed in three places: lower openings, stack cross-section, and openings at top of stack.

Figure 6-3 Sizing for Stack Ventilation

Technical note: These rules of thumb were generated with these assumptions: $\Delta T=3^{o}$, a conservative choice. Wind effectiveness factor = 0.4 (see discussion in Thermal Finalization 7.2, p.). Stack height = 30'.
Heat gain used here is attributable only to estimated internal gains. It seems likely that the additional solar gains will be controlled by shading devices such that they will essentially equal the decrease in internal gains allowed by daylighting.

2. Use Figs. 6-2 and 6-3 to complete this table.

	Trial 1		Trial 2	
	Office	Factory	Office	Factory
Sq.ft. aperture necessary as either inlet or outlet				
Sq.ft. aperture provided for inlets				
Sq.ft. aperture provided for outlets				

Does your first trial design provide the necessary aperture?

If not, what will you change for trial 2?

C. **Criterion C:** Ventilation flows remove heat from the buildings without leaving hot, dead areas.

Evaluation Tool for Criterion C - "Heat Flow Diagrams"

STREAM OF WIND (open buildings)
On your site plan and your building plans and sections, draw the flow pattern of air around and through the buildings for the months you need to cool your buildings. In rendering those flows, remember: Wind tends to keep going in the same direction. Wind flows from positive to negative pressure areas. Hot air rises, cool air falls.

MASS COOLING (closed buildings, including night ventilation)
In your building plan and section show the locations of thermal mass. If the mass will be cooled primarily by radiation, draw the path of radiation transfer, from mass to sink, and label the sink. If the mass is cooled by conduction/convection (as in night ventilation), draw the path of air currents over the face of the mass, both in plan and section.

Does your first trial design meet Criterion C?

If not, what will you change for the second trial?

D. Criterion D: Glazing is shaded from the sun in the overheated months. (From the balance point analysis in Climate section 3.6, the months your buildings are overheated are when the balance point is lower than the outdoor temperature.)

Factory overheated months are:
Office overheated months are:

Evaluation Tool for Criterion D - "Sun Peg and Site Model"
If you have opted to shade at the site or cluster scale, use the sun chart and peg with your 1/40th scale model to evaluate this decision. Building and component scale shading will be evaluated at a more detailed level in 6.4 DEVELOPING.
1. Check the shading of all glazed apertures for 9 a.m., noon, and 3 p.m. of the overheated months above.
2. Document the shading with photos of your model or elevations or paralines with shadows drawn.

Does your first trial design meet Criterion D?

If not what will you change for your second trial?

E. Criterion E: Heating Check
If you have already done the heating exercise your design must still meet the criteria for 5.3 SCHEMING on p. 5-4. If you have not yet done the heating exercise, but you know that heating will be a concern in either building in your climate, check the criteria for 5.3 SCHEMING now so that you don't make things impossible in the next exercise.

6.4 DEVELOPING

GOALS

A. Minimize peak daily heat gains during overheated months.

CRITERIA

A. Glazed apertures are fully shaded during overheated months.
B. Daily heat peaks are minimized during overheated months by avoiding simultaneous peaks from more than one source.
C. Heat gain peaks do not occur during the hours of the day when it is hottest outside.

DESIGN STRATEGIES

A. Site/Cluster Scales
Where evaporative cooling is a strategy, incoming air can be brought across the surface of a water body, lowering its temperature while raising its relative humidity.

B. Building Scale

From "Natural Solar Cooling" by David Wright, <u>3rd National Passive Conference Proceedings</u>, 1979. Used by permission of the author.

Organize your most severely overheated spaces so that they can be cooled without overheating other spaces.

In <u>closed buildings</u>, single-story, high-ceiling spaces
with thermal mass allow heat build-up higher in the
space, while cooler air remains at occupancy levels
nearer the ground.

In <u>open buildings</u>, shape ceilings so that heat from the
highest part(s) of the space can readily escape to the
outdoors. Consider prevailing wind direction so that
heat from one space is not carried to another.

C. Component Scale

The openings for incoming ventilating air should be
adjustable so that when such air is colder than comfortable in contact with people, it can be admitted without
directly chilling the occupants. Yet when moving air
across the body is a requirement for comfort, it can be
brought in directly across the occupants.

Choose the shading devices that allow daylighting while
preventing direct sun from overheating your space.

The months with highest temperatures (such as July and
August) have the same sun paths as much cooler months
(such as May and April). To respond to the problem of
identical sun paths/different temperatures, use <u>variable</u>
shading devices. Deciduous vegetation is one seasonal
response (see MEEB p. 54-55). Note that vines may be
trimmed back to allow 100% solar gain; deciduous trees
are likely to shade 50% of solar gain even in winter.
Moveable shading devices (perhaps incorporating moveable
thermal insulating devices) can allow daily responses to
sunny or overcast weather.

(The chart below is diagrammatic. You should refer to
the LOF sun charts for your specific latitude in section
6.5)

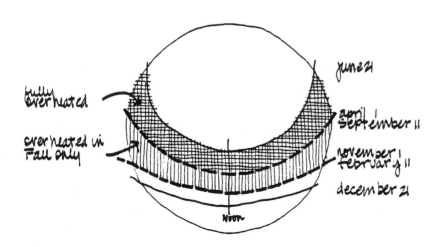

DESIGN

Develop your design for the office and factory to achieve the goals. Document the design using grids provided:

A. Describe the natural cooling system you have chosen for each building: cross or stack ventilation, high mass, or evaporative cooling.
B. Draw sectional perspectives of the rooms which are passively cooled. These should illustrate:
1. size and location of the cooling system components (aperture, mass, etc.) and their relation to the space
2. how the system changes diurnally (if it does)
3. how the heat is removed from the room
C. Draw all the types of shading devices, including vegetation, that you are using in both section and elevation at 1/8th scale. Key them to paralines or elevations and label the devices as fixed or movable.
D. Show your office and factory occupancy schedule on the graph below.

Factory

Office

        Midnight   Noon   Midnight

E. If you have completed the HEATING section 5.4, include your wall sections design documentation. (Revise anything you have changed in the design).
If you have not completed HEATING section 5.4 you must make preliminary material and construction decisions. Specify what basic materials (wood, brick, concrete block, insulation, etc.) your walls and roofs will be. (See MEEB Table 4.22c p. 149 for the level of detail required in this exercise.)

for daylighting contours, enlarged plans: if 1/8"= 1'-0", 1 square = 10'.

_____

for daylighting contours, enlarged plans: if 1/8"= 1'-0", 1 square = 10'.

for daylighting contours, enlarged plans: if 1/8"= 1'-0", 1 square = 10'.

for daylighting contours, enlarged plans: if 1/8" = 1'-0", 1 square = 10'.

construction sections: if 3/4" = 1'-0", 1 square = 1'-0".

construction sections: if 3/4" = 1'-0", 1 square = 1'-0".

6.0 COOLING

6.4 DEVELOPING

EVALUATION

A. Criterion A
All glazed apertures are fully shaded during overheated
months. From 6.3 Criterion D:
Factory overheated months are:

Office overheated months are:

Evaluation Tool for Criterion A - "Shading Masks"
1. Plot the overheated months on photocopies of the
sunchart for your latitude (found in section 6.5).
a. For each building the Balance Point (from section 3.6
climate exercise) is:
  office:
  factory:
b. On the graph of your climate on the Bioclimatic Chart
the max temperature occurs approximately 4 p.m., the min
temperature approximately 4 a.m. Temperatures can be
matched to the remaining hours of the day along the line
drawn between the two points with 10 a.m. and 10 p.m.
occurring at mid-point. Note that you will graph the
temperature pattern for August in the next criterion.
c. Plot a separate sun chart for each building. For each
overheated month between January and June, use vertical
lines to shade hours of the day in which the exterior
temperature is GREATER than that of the balance point.
d. For each overheated month between July and December
use horizontal lines to shade in hours of the day in
which the exterior temperature is GREATER THAN the
Balance Point.

When both months in a pair are overheating, this shows up
as a cross-hatched area, signalling the appropriateness
of fixed shading devices. Where only one of the paired
months is overheated, movable shading devices are needed;
for example, to admit sun in cold March, exclude it in
warm September.

2. Shading Masks. External shading devices produce
distinct zones of full shade. These patterns of sun
protection are called "masks" and can be compared to the
plot's overheated periods. Some masks are illustrated in
Olgyay, Design with Climate, p. 82-83 and in Olgyay and
Olgyay, Solar Control and Shading Devices, p. 88.
a. For all apertures not shaded at the building and site
scale draw the mask of the shading device on tracing
paper at the same scale as the LOF suncharts (label by
aperture and key to elevations or paralines).

---

b. Evaluate the performance of each shading device by superimposing the mask over the overheated plot for the right building. Whenever the building is overheated the mask should cover the plot. Conversely when the building is not overheating the mask should not cover the plot.

c. Each of your shading devices has a "shading coefficient" (S.C.) that indicates what percent of the window's <u>unshaded</u> solar gain is admitted through (or around) the shading device. The S.C. is listed both in Olgyay <u>Design with Climate</u> p. 68-71, and in MEEB (Table 4.28 a, b).

List the shading coefficient for each of your shading devices.

Does your first trial design meet Criterion A?

If not, what will you change for your second trial?

B. Criterion B: Daily heat peaks are minimized during overheated months by avoiding simultaneous peaks from more than one source.

Evaluation Tool for Criterion B - "Heat Gain Calculation"
The following heat gain calculation is an abbreviated version of the procedure you will follow in 7.0 THERMAL FINALIZATION. This calculation looks at the daily patterns of heat gain in your buildings, and will be done for August 21st.

1. Roof and Wall Gains
Opaque skin elements receive and store heat from the sun as well as heat from the air. To get an approximate 24-hour average of the heat passed to the interior through these elements, heat transfer factors (HTF's) are used. Heat transfer factors combine the U-value of a construction with various temperature differences, or differentials, between inside and outside.

A more precise nonresidential calculation would use Total
Equivalent Temperature Differentials (TETD's) rather than
HTF's. However, most of the heat gain in this exercise
is solar gain or internal gain, so HTF's provide a
reasonable approximation.
Use MEEB Table 4.22(c) p. 149 to find the HTF's for your
design. For all buildings, assume the lowest (20°)
temperature differential.

	Trial 1			Trial 2		
	Area sq.ft.	X HTF =	Daily average gain Btuh	Area sq.ft.	X HTF =	Daily average gain Btuh
a. OFFICE						
opaque wall						
opaque roof						
		TOTAL:			TOTAL:	
b. FACTORY						
opaque wall						
opaque roof						
		TOTAL:			TOTAL:	

2. Solar Gains

To look at the pattern over a day you first find the gain through east-facing glazed
apertures for morning, then through south glazed apertures at noon, then west-facing for
afternoon. For horizontal skylights, calculate gains at all three times. For sloped
skylights, calculate for the hour corresponding to the direction it faces, and interpolate
between vertical and horizontal solar gains. Shading coefficients (column 3) are from the
previous evaluation tool. Solar Heat Gains for August 21 (column 2) are as follows
(Mazria, 1979, Appendix I):

		32°	36°	40°	44°
	Northeast	101	91	80	70
9am	East	195	194	192	189
	Southeast	176	184	191	197
noon	South	109	127	145	151
	Southwest	165	171	175	179
4pm	West	211	210	208	206
	Northwest	137	130	122	115
Horizontal		216	211	205	197

6.0 COOLING

6.4 DEVELOPING

---

a. OFFICE

	Col. 1	Col. 2		Col. 3	Col. 4		Col. 5
Trial 1	aperture	sq.ft. area	X	S.C. X	Btuh/sq.ft. Heat Gain Factor	=	Btuh Total

East 9 am

South noon

West 4 pm

Trial 2

East 9 am

South noon

West 4 pm

b. FACTORY

	Col. 1	Col. 2		Col. 3	Col. 4		Col. 5
Trial 1	aperture	sq.ft. area	X	S.C. X	Btuh/sq.ft. Heat Gain Factor	=	Btuh Total

East 9 am

South noon

West 4 pm

Trial 2

East 9 am

South noon

West 4 pm

---

3. Internal Gains

Use the internal gain rate from p. 6.19 to calculate total while occupied. (This should be the same value found in 3.6 Climate.)

Office: _____ Btuh/sf x _____ sf = _____ Btuh

Factory: _____ Btuh/sf x _____ sf = _____ Btuh

Evaluation Tool for Criterion B - "24 Hour Gain Patterns"

Note: This tool will also be used to assess performance for Criterion C. Look ahead and consider this in any redesign.

1. Plot the gains from the preceeding tool below. Be sure to use different colors or hatch patterns to distinguish the gain sources. (See Salem Example as a model)
a. First, plot roof and wall gains as constant over the 24 hour period.
b. Next, add the internal gains for the hours of building use you specified in the design presentation.
c. Finally, plot each of the solar gains at the hour for which it was calculated. Plot zero gain at sunrise and sunset (approx. 5:30 am, 6:30 pm on August 21). Connect the points.

2. Below the gains plot the 24 hour temperature pattern for August estimated for the shading masks in the previous criterion. The high temperature generally occurs at 4 pm, the low at 4 am.

Trial 1

Trial 2

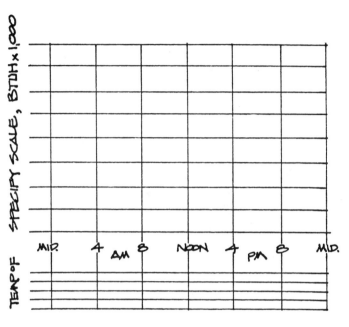

Office Heat Gain Pattern
Pattern

Office Heat Gain

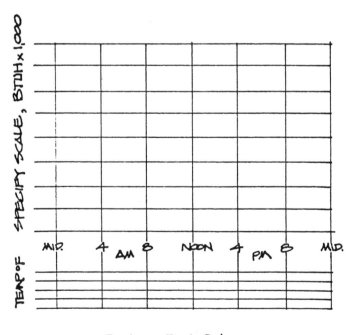

Factory Heat Gain Pattern
Pattern

Factory Heat Gain

Does your first trial design meet Criterion B by avoiding
simultaneous heat gain peaks from more than one source?

If not, what will you change for your second trial?

C. Criterion C: Heat gain peaks do not occur during the
hours of the day it is hottest outside. Use the previous
Evaluation Tool "Daily Heat Gain Patterns" to see if your
design meets this criterion.

Does your first trial design meet Criterion C?

If not, what will you change for your second trial?

32° SUN CHART
(32° N. Lat.)

© LIBBEY-OWENS-FORD COMPANY

36° SUN CHART
(36° N. Lat.)

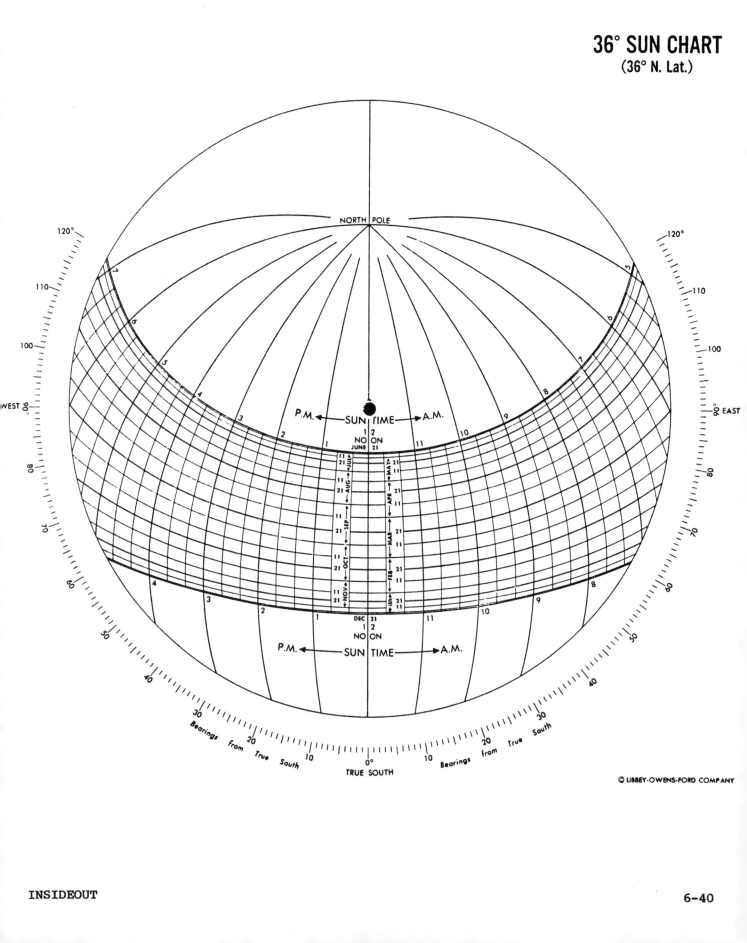

© LIBBEY-OWENS-FORD COMPANY

40° SUN CHART
(40° N. Lat.)

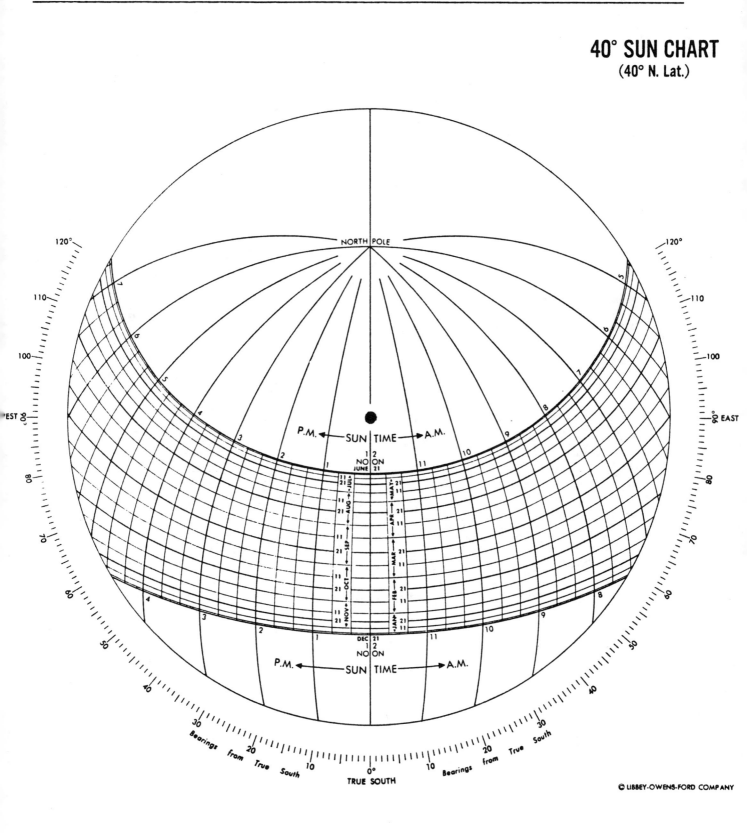

44° SUN CHART
(44° N. Lat.)

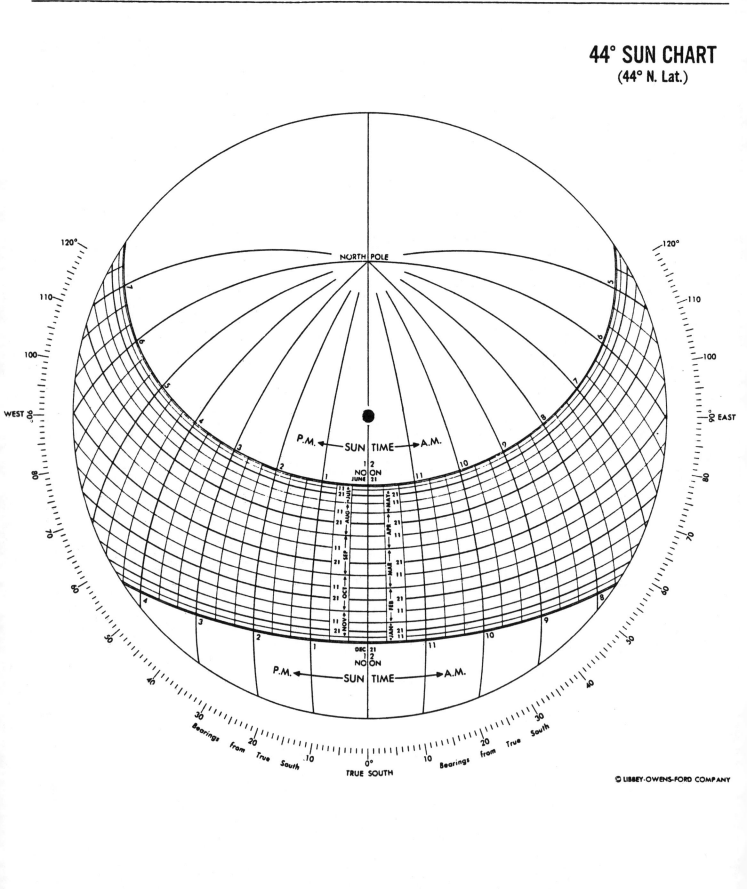

© LIBBEY-OWENS-FORD COMPANY

7.0 THERMAL FINALIZATION

7.1 INTRODUCTION

7.1 INTRODUCTION

In this finalization exercise you will see how your building design will work for both cooling and heating. The first step for each section is to figure the quantity of heat that needs to be supplied in the heating season (heat loss or heating load) and needs to be removed in the cooling season (heat gain, or cooling load). The calculations may seem relatively long, but they are the only way you can evaluate your design at this level. The Btu totals you calculate are not a goal in themselves, but are used to evaluate thermal performance in ways that tell you something about the building.

Contents

7.2 HEATING PERFORMANCE

GOAL

Keep occupants thermally comfortable throughout the year
by using the natural energies available in your climate,
thereby minimizing the use of electrical or fossil fueled
auxiliary equipment.

CRITERIA

A. Your buildings will use less fuel for heating in the
winter than comparable "energy conserving" buildings in
your climate.
B. The temperatures inside are between $60^{\circ}F$ and $80^{\circ}F$ when
the buildings are occupied in the winter months.
C. The capacity of the heating backup system is
sufficient to keep you warm when there is no sun or
internal gain.

DESIGN

Fully document the design which you are evaluating in
this exercise.

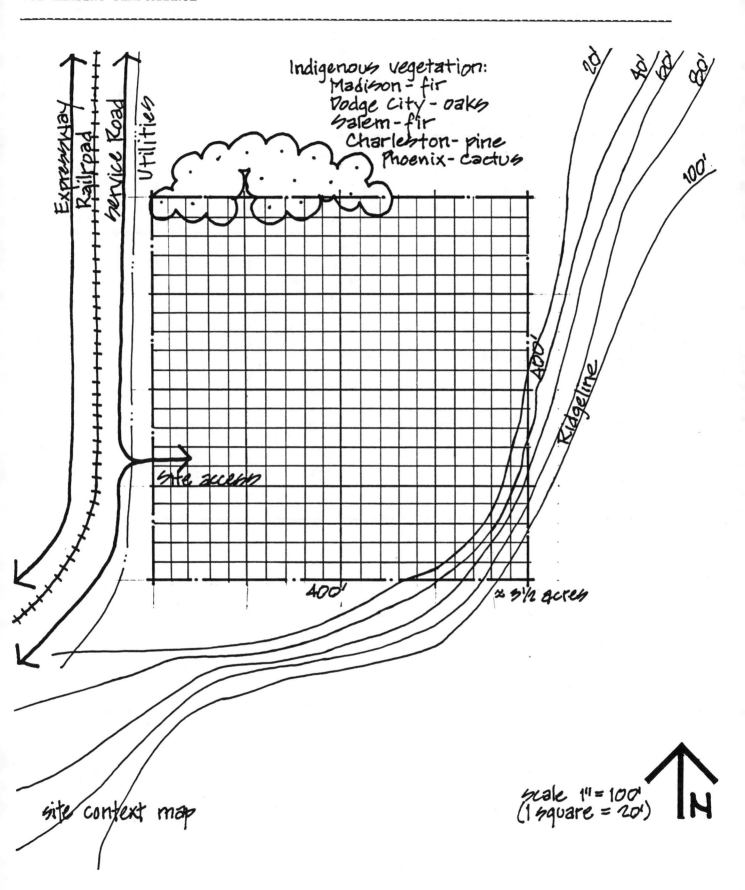

Indigenous vegetation:
Madison - fir
Dodge City - oaks
Salem - fir
Charleston - pine
Phoenix - cactus

Expressway
Railroad
Service Road
Utilities

site access

Ridgeline

20' 40' 60' 80'

100'

400'

400'

≈ 3½ acres

site context map

scale 1" = 100'
(1 square = 20')

N

cluster site plan: 1" = 40'-0".

N

---

for plans, sections, & elevations: if   1/16"=1'-0",   1 square = 10'.

7.2 HEATING PERFORMANCE

for plans, sections, & elevations: if 1/16"=1'-0", 1 square = 10'.

----

for plans, sections, & elevations: if   1/16"=1'-0",   1 square = 10'.

for plans, sections, & elevations: if   1/16"=1'-0",   1 square = 10'.

----

for plans, sections, & elevations: if   1/16"=1'-0",   1 square = 10'.

for plans, sections, & elevations: if 1/16"=1'-0",   1 square = 10'.

----

for plans, sections, & elevations: if   1/16"=1'-0",   1 square = 10'.

construction sections: if 3/4" = 1'-0", 1 square = 1'-0".

construction sections: if 3/4" = 1'-0", 1 square = 1'-0".

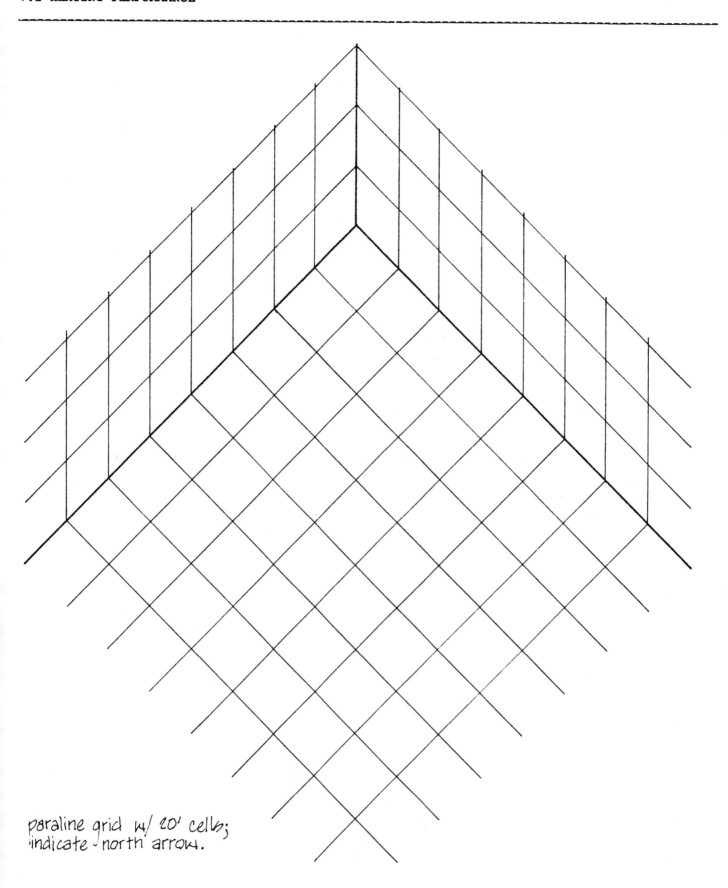

paraline grid w/ 20' cells;
indicate north arrow.

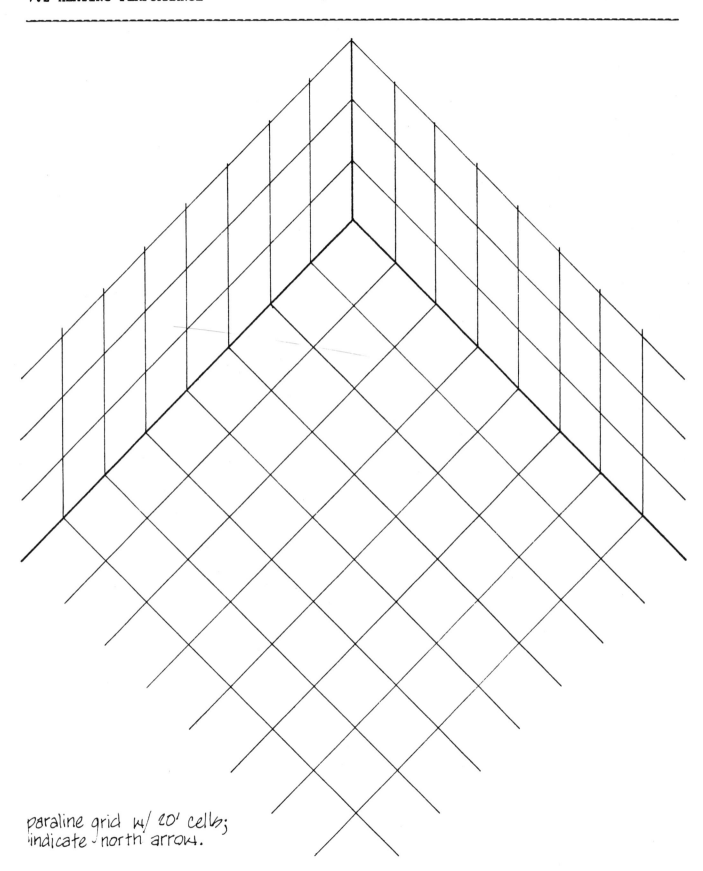

paraline grid w/ 20' cells;
indicate north arrow.

EVALUATION

A. Criterion A: Your buildings use less fuel for heating in the winter than comparable "energy conserving" buildings in your climate. (criteria from Balcomb, p. 24)

Heating Degree-days	"Energy Conserving", Conventionally fuelled	Solar Heated, not including the solar walls heat losses
less than 1,000	9 Btu/DD sq.ft.	7.6 Btu/DD sq.ft.
1,000-3,000	8	6.6
3,000-5,000	7	5.6
5,000-7,000	6	4.6
greater than 7,000	5	3.6

Because these values assume no internal gain, the criteria may be adjusted for Internally Dominated Load (IDL) buildings similarly to the adjustment for solar aperture in 5.3 Scheming:

$$\text{Adjusted Btu/DDSF} = \text{Btu/DDSF} \times \frac{65}{\text{Balance Point Temp (24 hr avg)}}$$

HEAT BALANCE ON A CAT

Office Building:

Factory:

From "Thermal and Economic Performance..." by J. Augustyn et al, 4th National Passive Conference Proceedings, 1979. Used by permission of the author.

Evaluation Tool - "Heat Loss Calculation"

Use the format below to establish UxA for all skin elements except the south facing solar aperture. (South aperture is excluded because it is assumed that there will be no net loss through this heat-admitting surface area.)

---

1. Office Building Trial 1
Include all spaces that are heated.

	Materials	Area (sq.ft.)	$U(Btuh/sq.ft./^{\circ}F)$	$UA(Btuh/^{\circ}F)$
Exterior opaque walls				
Roof: horizontal 45° slope				
Slab edge				
Non-south glass: vertical				
45° slope				
horizontal				
Exterior doors				
Infiltration (N.A.)	volume in cu.ft. x .018 x ACH =			

TOTAL UA =
(excluding solar aperture)

For the "Building Load Coefficient" (BLC), multiply this
UA total by 24 hours:

BLC Office = _____ Btu/DD

To relate to the initial criteria, divide by floor area:

$$\frac{\text{Btu/DD}}{\text{floor area in sq.ft.}} = \text{_____ Btu/DD sq.ft.}$$

Office Building Trial 2
Include all spaces that are heated.

	Materials	Area (sq.ft.)	U(Btuh/sq.ft./$^{\circ}$F)	UA(Btuh/$^{\circ}$F)
Exterior opaque walls				
Roof: horizontal				
45° slope				
Slab edge				
Non-south glass: vertical				
45° slope				
horizontal				
Exterior doors				
Infiltration (N.A.)	volume in cu.ft. x .018 x ACH =			

TOTAL UA =
(excluding solar aperture)

For the "Building Load Coefficient" (BLC), multiply this
UA total by 24 hours:

BLC Office = _____ Btu/DD

To relate to the initial criteria, divide by floor area:

$$\frac{\_\_\_\_\_ \text{Btu/DD}}{\text{floor area in sq.ft.}} = \_\_\_\_\_ \text{Btu/DD sq.ft.}$$

This should be no more than the criteria listed for your
type of building in your climate. How does it compare?

7.0 THERMAL FINALIZATION

7.2 HEATING PERFORMANCE

Factory Building Trial 1	Materials	Area (sq.ft.)	U(Btuh/sq.ft.°F)	UA(Btuh/°F)
Exterior opaque walls				
Roof - horiz: 45° slope:				
Slab edge				
Non-south glass: vertical 45° slope horizontal				
Exterior doors				
Infiltration	N.A.	volume cu.ft. x .018 x ACH		

TOTAL UA =
(excluding south aperture)

For the Building Load Coefficient (BLC), multiply this UA total by 24 hours:

BLC Factory = Total UA _____ x 24 = _____ Btu/DD

and to relate to the initial criteria, divide by the floor area:

$$\frac{\text{Btu/DD}}{\text{floor area in sq.ft.}} = \underline{\hspace{2cm}} \text{ Btu/DD s.f.}$$

Factory Building Trial 2	Materials	Area (sq.ft.)	U(Btuh/sq.ft.$^\circ$F)	UA(Btuh/$^\circ$F)
Exterior opaque walls				
Roof - horiz:				
45° slope:				
Slab edge:				
Non-south glass: vertical				
45° slope				
horizontal				
Exterior doors				
Infiltration	N.A.	volume cu.ft. x .018 ACH =		

TOTAL UA = _____
(excluding south aperture)

For the Building Load Coefficient (BLC), multiply this UA
total by 24 hours:

BLC Factory = _____ Btu/DD

and to relate to the initial criteria, divided by the
floor area:

$$\frac{\text{Btu/DD}}{\text{floor area in sq.ft.}} = \text{_____ Btu/DD sq.ft.}$$

This should be no more than the criteria listed for your
type of building in your climate. How does it compare?

**Evaluation Tool for Criterion A - "Solar Savings
Fraction"**

Now that you know your heat loss and building load
coefficient, you can see whether using passive solar
heating makes much difference. The Solar Savings

Fraction, or SSF, is a means of comparing your passively solar heated building to a non-solar, but energy-conserving building. (See Balcolm, 1980, p. 10).

Comparing the non-solar energy needed by above buildings, the solar building needs 25 units, while the conserving building needs 70. The difference is 70-25=45, which = 64% of the conserving buildings' total. Therefore, the solar building has a SSF (Solar Savings Fraction) of 64%. (It is, however, 75% solar heated. It needs more total heat, because it has substantially greater glass areas than does the energy conserving building.)

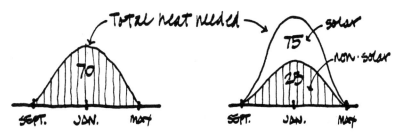

A "conserving" building.   A nearly identical building
                            except that it is solar heated.

1. Determine the LCR (Load-Collector Ratio)
A comparison between the buildings' need for heat, and its heat-admitting south aperture is called the load-collector ratio (LCR). Small LRC's mean large window areas and higher SSF's.

LCR = Building load coefficient
      South aperture area

	1st Trial	2nd Trial
LCR Office	_____	_____
LCR Factory	_____	_____

2.a. What passive systems are you using, in what portions of the total, for each building?

Office:

Factory:

----

b. Find your SSF in the charts below, using the LCR by interpolating where necessary.

WW = water wall system
TW = trombe wall system
DG = direct gain system
NI = night insulation

7.2 HEATING PERFORMANCE

MADISON WISCONSIN

43.1°NL
7730 DD LCR
T(Jan)=17

	SSF =	.1	.2	.3	.4	.5	.6	.7	.8	.9
WW		62	25	13	7	-	-	-	-	-
WWNI		137	62	39	27	20	15	12	9	6
TW		67	27	15	8	3	-	-	-	-
TWNI		130	59	36	25	19	14	11	8	5
DG		-	-	-	-	-	-	-	-	-
DGNI		133	59	36	24	17	12	8	5	2

DODGE CITY KANSAS

37.8°NL
5046 DD LCR
T(Jan) = 31

	SSF =	.1	.2	.3	.4	.5	.6	.7	.8	.9
WW		140	64	39	27	19	15	10	6	-
WWNI		228	107	67	48	36	29	23	18	13
TW		138	62	37	25	17	11	8	5	2
TWNI		215	100	63	45	34	26	20	15	10
DG		132	54	27	10	-	-	-	-	-
DGNI		236	109	68	47	35	26	20	14	8

SALEM OREGON

44.9°NL
4852 DD LCR
T(Jan)=39

	SSF =	.1	.2	.3	.4	.5	.6	.7	.8	.9
WW		137	57	34	22	15	9	5	-	-
WWNI		226	101	62	43	32	24	18	14	9
TW		133	56	32	20	13	8	4	-	-
TWNI		212	95	58	40	29	22	16	11	7
DG		121	41	-	-	-	-	-	-	-
DGNI		231	102	62	42	30	21	15	10	5

PHOENIX ARIZONA

33.4°NL
1552 DD LCR
T(Jan)=51

	SSF =	.1	.2	.3	.4	.5	.6	.7	.8	.9
WW		467	219	139	100	75	58	45	34	22
WWNI		620	293	188	136	104	82	65	51	36
TW		436	202	126	87	63	46	34	25	16
TWNI		583	275	176	126	95	73	56	42	28
DG		555	256	157	107	75	53	37	24	12
DGNI		673	316	201	143	107	82	62	45	29

CHARLESTON SO. CAROLINA

32.9°NL
2146 DD LCR
T(Jan)=49

	SSF =	.1	.2	.3	.4	.5	.6	.7	.8	.9
WW		252	118	72	51	38	29	22	16	9
WWNI		358	173	108	76	58	46	37	29	21
TW		238	110	67	46	32	23	17	11	7
TWNI		339	161	101	71	54	42	32	24	16
DG		276	124	73	47	30	19	10	-	-
DGNI		384	180	113	79	59	45	34	24	16

Fig. 7-1 Solar Savings Fraction Charts (Balcomb, 1980)

                        1st Trial   2nd Trial

SSF Office building: _____

SSF Factory: _____

Does your first trial design use less non-solar fuel for
heating than a conserving building to meet Criterion A?

If not, what will you change in your second trial?

B. Criterion B: Your buildings stay between 60°F and 80°F
inside while occupied in the winter months.

Note: At 60°F air temperature, for a person to be comfortable, a higher radiant temperature than 60° should be
available from walls or floors which have stored solar
heat.

**Evaluation Tool for Criterion B - "Internal January
Temperature Estimate"**

The average interior temperature is the sum of three
temperatures:

      1. Average outside January temperature, T out
    + 2. Internal heat-generated temperature, T int
    + 3. Temperature increase due to solar energy,
         called Δ T solar
    ───
    = 4. Average interior January temperature

1. T out = avg. outside January temperature (from Fig.
7-1) = _____°F

2. T int = $\dfrac{\text{total daily internal gains}}{\text{BLC}}$

	Office	Factory
Total Internal gains (in 24 hrs)	____ Btuh	____ Btuh
BLC	____ Btu/DD	____ Btu/DD
T int	____ °F	____ °F

3. (ΔT - solar) depends on your latitude and your Load
Collector Ratio (LCR). (Balcomb, 1980, p. 45)

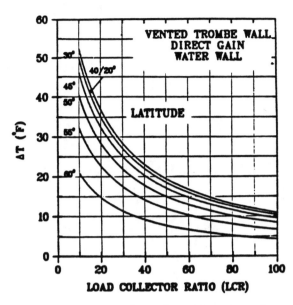

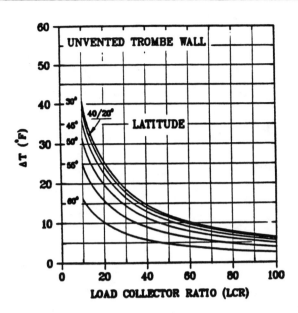

Fig. 7-2 △T Solar for Direct Gain (and Vented Trombe Wall.)

Fig. 7-3 △T Solar for Unvented Trombe Wall.

Note: These curves assume no night insulation, so your resulting △T solar may be higher than these conservative values. These curves are also for a <u>clear</u> day. If your climate rarely has clear January days, you may wish to <u>adjust</u> the △T solar value you obtain below, to account for a prevalence of cloudy weather. This approximate adjustment involves comparing your <u>average</u> January insolation on a vertical surface with the <u>clear day</u> insolation for your latitude.

```
CHARLESTON, SOUTH CAROLINA                          LAT = 32.9      ELEV =   39
       JAN   FEB   MAR   APR   MAY   JUN   JUL   AUG   SEP   OCT   NOV   DEC  YEAR
HS     744   995  1339  1732  1860  1844  1799  1585  1394  1193   934   721  1345
VS    1058  1135  1128  1006   832   753   759   808  1010  1226  1251  1083 12049
TA    48.6  50.5  56.5  64.6  72.1  77.9  80.2  79.6  75.2  66.1  56.3  49.3  64.7
D50    120    81    23     2     0     0     0     0     0     2    24   108   360
D55    222   161    75    10     1     0     0     0     0     7    75   205   756
D60    360   275   157    36     5     0     0     0     0    26   156   339  1355
D65    521   419   300    69     5     0     0     0     0    74   271   487  2146
D70    664   547   422   192    66    16     9    10    31   165   414   642  3178


DODGE CITY, KANSAS                                  LAT = 37.8      ELEV = 2582
       JAN   FEB   MAR   APR   MAY   JUN   JUL   AUG   SEP   OCT   NOV   DEC  YEAR
HS     827  1122  1476  1886  2090  2358  2295  2055  1687  1301   894   732  1560
VS    1345  1466  1436  1273  1063  1065  1084  1212  1417  1535  1369  1258 15523
TA    30.8  35.2  41.2  54.0  64.0  73.7  79.2  78.1  68.9  57.9  42.8  33.4  54.9
D50    596   417   289    48     4     0     0     0     1    20   239   517  2132
D55    750   555   432   116    15     1     0     0     4    63   373   670  2980
D60    905   695   584   210    50     4     0     0    15   139   518   825  3945
D65   1060   834   738   344   115    21     0     0    41   247   666   980  5046
D70   1215   974   893   482   218    51    14    19   118   382   816  1135  6318


MADISON, WISCONSIN                                  LAT = 43.1      ELEV =  860
       JAN   FEB   MAR   APR   MAY   JUN   JUL   AUG   SEP   OCT   NOV   DEC  YEAR
HS     515   804  1136  1398  1743  1948  1934  1708  1299   911   504   389  1191
VS     971  1214  1285  1109  1037  1014  1061  1185  1275  1244   893   776 13064
TA    16.8  20.3  30.2  45.3  56.0  65.8  70.1  68.7  59.7  49.9  34.7  21.9  44.9
D50   1029   832   614   167    19     1     0     0     6    86   460   871  4086
D55   1184   972   769   297    70     4     1     2    26   182   609  1026  5143
D60   1339  1112   924   442   157    19     5     8    87   318   759  1181  6352
D65   1494  1252  1079   591   297    72    14    39   173   474   909  1336  7730
D70   1649  1392  1234   741   436   156    83   106   314   623  1059  1491  9284


PHOENIX, ARIZONA                                    LAT = 33.4      ELEV = 1112
       JAN   FEB   MAR   APR   MAY   JUN   JUL   AUG   SEP   OCT   NOV   DEC  YEAR
HS    1021  1374  1814  2355  2676  2739  2486  2293  2015  1576  1150   932  1869
VS    1472  1589  1552  1388  1211  1128  1059  1186  1561  1643  1561  1419 16692
TA    51.2  55.1  59.7  67.7  76.3  84.6  91.2  89.1  83.8  72.2  59.8  52.5  70.3
D50     78    29     9     1     0     0     0     0     0     0     9    60   187
D55    162    85    36     4     0     0     0     0     0     1    34   137   459
D60    285   168   100    16     0     0     0     0     0     5    96   250   919
D65    428   292   185    60     0     0     0     0     0    17   182   388  1552
D70    584   419   327   130    24     2     0     1     3    64   314   544  2411


SALEM, OREGON                                       LAT = 44.9      ELEV =  200
       JAN   FEB   MAR   APR   MAY   JUN   JUL   AUG   SEP   OCT   NOV   DEC  YEAR
HS     332   588   947  1370  1738  1849  2142  1775  1328   769   410   277  1127
VS     659   933  1126  1148  1093  1014  1240  1301  1103  1103   764   582 12336
TA    38.8  42.9  45.2  49.8  55.7  61.2  66.6  66.1  61.9  53.2  45.2  40.9  52.3
D50    348   204   163    68    10     1     0     0     1    26   158   285  1265
D55    502   340   306   168    57     8     1     1     6    97   296   437  2220
D60    657   479   459   308   151    48     7     9    39   217   444   592  3411
D65    812   619   614   456   295   133    43    53   120   366   594   747  4852
D70    967   759   769   606   444   267   130   141   247   521   744   902  6496
```

HS = Normal daily value of total hemispheric solar radiation on a horizontal surface (Btu/ft^2 day)

VS = Normal daily value of total solar radiation on a vertical south-facing surface (Btu/ft^2 day)

TA = (T min. + T max.)/2 where T min. and T max. are monthly (or annual) normals of daily minimum and maximum ambient temperature ($^\circ$F)

Dxx = monthly (or annual) normal of heating degree-days below the base temperature xx ($^\circ$F days)
Base 65° is usually used in calculating heating loads.

Figure 7-4 Monthly Average Solar Radiation, Temperature, and Degree-Days

Source: Balcomb, 1980, Appendix A

Average insolation: see "VS" data, Fig. 7-4.
Clear Day insolation: approximate values may be obtained
by using MEEB table 4.27 for 40°N.Lat., pages 170-171.
(ASHRAE Fundamentals also lists these for 24°, 32°, 48°
and 56°N.Lat.)

Adjusted ΔT Solar = clear day ΔT solar x $\dfrac{\text{average insolation}}{\text{clear day insolation}}$

Trial 1

	LCR	ΔT solar	Adjusted ΔT solar, if applicable
Office			
Factory			

Trial 2

Office			
Factory			

4. Average interior January temperature = T out + T int +
ΔT solar Trial 1:   Trial 2:

Office: _____°F   _____

Factory:_____°F   _____

Evaluation Tool for Criterion B - "January Temperature Swing Estimate"
The temperature swing estimates how far your building's temperature will fluctuate above and below that average temperature on a clear January day. The total range from high to low is roughly proportional to the ΔT solar, as follows for various systems (remember, cloudy climates might use an adjusted ΔT solar):

Direct gain	T swing = 0.74 x ΔT solar
Vented trombe wall	T swing = 0.65 x ΔT solar
Water wall	T swing = 0.39 x ΔT solar
Unvented trombe wall	T swing = 0.13 x ΔT solar

7.0 THERMAL FINALIZATION

7.2 HEATING PERFORMANCE

High temperature = Av. Jan. interior temp. + $\dfrac{\text{T swing}}{2}$

Low temperature = Av. Jan. interior temp. − $\dfrac{\text{T swing}}{2}$

	Trial 1		Trial 2	
	Office	Factory	Office	Factory
high temperature:	___ $^{\circ}$F	___ $^{\circ}$F	___ $^{\circ}$F	___ $^{\circ}$F
low temperature:	___ $^{\circ}$F	___ $^{\circ}$F	___ $^{\circ}$F	___ $^{\circ}$F

If you assume high temperature occurs at 4 p.m. and low temperature at 4 a.m., what are the approximate high and low temperatures during operating hours?

	Trial 1		Trial 2	
	Office	Factory	Office	Factory
high temperature:	___ $^{\circ}$F	___ $^{\circ}$F	___ $^{\circ}$F	___ $^{\circ}$F
low temperature:	___ $^{\circ}$F	___ $^{\circ}$F	___ $^{\circ}$F	___ $^{\circ}$F

Does your first trial design meet the 60°F-80°F criteria range?

If not, is more or less solar aperture, or another passive system, appropriate?

What will you change on your second trial design?

C. Criterion C: The capacity of the heating backup system is sufficient to heat the building both when there is no sun, and when there is no sun nor gains from people, lights or machines.

Evaluation Tool for Criterion C - "Heating Backup System Sizing"

1. Figure your building heat loss in Btuh, <u>including</u> all south facing solar aperture. If you have no solar gains, the south facing glazing becomes an additional area which loses heat.

		Office	Factory
a.	Area of south aperture	_____ sq.ft.	_____ sq.ft.
b.	U value of south aperture	_____	_____
c.	Is this using insulating shades?	_____	_____
d.	UA for south aperture	_____ Btuh/$^\circ$F	_____ Btuh/$^\circ$F
e.	UA total excluding south aperture (from Criterion A)	_____ Btuh/$^\circ$F	_____ Btuh/$^\circ$F
f.	Total building UA (d + e = f)	_____ Btuh/$^\circ$f	_____ Btuh/$^\circ$f

2. The "Conservative" approach assumes no contributions from either solar or internal gains. This may be highly unrealistic, but quite "safe" in supplying lots of auxiliary heat in extremely cold and cloudy conditions. Use the maximum heat loss directly in sizing heater(s). Such heating units will be vastly oversized, and may never be used to full capacity.
Determine maximum hourly heat loss, under the coldest hour usually encountered in your climate. This "design temperature" can be determined from MEEB Appendix C, Table C-2a, Winter DB temperature column.

Maximum Heat Loss = U x AΔT
Where ΔT is the difference between interior temeperature and exterior winter DB, and UA have been determined in the previous step.

<u>Office</u> maximum hourly heat loss:

UAΔT = _____ Btuh/$^\circ$F x _____ $^\circ$F = _____ Btuh

<u>Factory</u> maximum hourly heat loss:

UAΔT = _____ Btuh/$^\circ$F x _____ $^\circ$F = _____ Btuh

3. The "Reasonable Risk" approach reduces the maximum heat loss by the internal gains (averaged over 24 hours). We can call this the "Assumed Heat Loss."

Assumed Heat Loss = Maximum Heat Loss - Average Internal Heat Gain

Office: Factory:
Average Internal Heat Gain = _____ Btuh   _____ Btuh
(as found in Heating p. 5-7.)

Office assumed heat loss = (_____ Btuh)-(_____ Btuh)= _____ Btuh

Factory assumed heat loss = Maximum hourly loss - average internal heat gain = (_____ Btuh)-(_____ Btuh)= _____ Btuh

4. Auxiliary heat sources are specified by Btuh delivered to the space. You have just calculated two different needs above. What capacity will your backup system be? Does this meet Criterion C?
Office: Factory:

With the Btuh heat needs determined by either of these approaches, you now can find catalogs of manufacturers, and choose the auxiliary equipment. This isn't required here, but it's a useful experience. Additional information about choosing these systems:

Several decisions are necessary when choosing a means for supplying the non-solar portion of your building's space heat need. One central heat source or several zones? The more separate spaces in your building, the more likely that each space will need heat in different quantities, and at different times. Each space could be called a "zone" for heating. Small heaters in each room, each controlled by its own thermostat, are an appropriate way to respond to several zones.

The disadvantage of this approach is that electric resistance heaters are almost always the economical choice for this task, and electrical resistance heat mismatches a very high grade energy source with a low grade energy task. (See MEEB, chapter 1.)

One central heat source is particularly appropriate for
simple, one-space buildings, and can utilize a lower
grade energy source more appropriate for space heating.
Since it is controlled by only one thermostat, the location of this control device is particularly important:
when the thermostat calls for heat, the majority of the
building should also need heat.

If you choose one central auxiliary heat source:
Natural gas, oil, wood or other combustion fuel heaters
can be chosen by the Btuh delivered to the space. (The
total Btuh burned includes some heat that is wasted, up
the flue.) Btuh delivered should be approximately equal
to the Maximum (or assumed, if you take risks) heat loss
in Btuh.

Electric resistance heaters can be chosen by converting
the Btuh needed into Kw:

$$\frac{Btuh}{3,412} = Kw$$

Heat pumps, while an attractive option for non-solar
buildings, are more rarely an economic choice for
passively solar heated buildings. The most likely time
for auxiliary heaters to be utilized is during the
coldest weather; in most climates where auxiliary heaters
are important, this cold weather is not conducive to
efficient heat pump operation, because there is less heat
available in the outdoor air for the heat pump to deliver
to the building.

After determining the Btuh (or Kw) to be delivered by
your heater, consult manufacturers' literature (such as
found in Sweets Files) to determine the range of choices
available.

If you chose individual zone heaters:
An approximate way to size the heaters in each zone is
simply use a ratio of its floor area to total floor area:

Maximum Heat Loss
(or assumed,
if you take risks)
$\times \dfrac{\text{Zone floor area}}{\text{Total floor area}}$ = Zone Heat Loss

A more exact method would be to break down the UxA
calculation by zone instead of for the whole building.
Choice of heaters then proceeds as outlined for one
central source, above.

A TYPICAL ACTIVE SOLAR THERMAL CAT SYSTEM

From J. R. Augustyn et al, 1979.

7.3 COOLING PERFORMANCE

GOAL

Keep occupants thermally comfortable throughout the year by using the natural energies available in your climate, thereby minimizing the use of electrical or fossil fueled auxiliary equipment.

CRITERIA

A. The internal temperature in your buildings is no greater than 3° hotter than outside during the overheated months.

B. The capacity of your cooling backup system is sufficient to keep you cool during overheated months.

DESIGN

Use the design documentation from 7.2 for your evaluation in this section.

A. **Criterion A:** The internal temperature in your buildings is no greater than 3° hotter than outside during the overheated months. You must of course make certain that this internal temperature is within the comfort zone. Many overheated months have an air temperature low enough to maintain human comfort.

Evaluation Tool for Criterion A - "Heat Gain Calculation"
Heat gain is calculated for the worst condition - people, lights and machines producing heat inside and sun striking glazed openings. Calculation procedures below are for both open buildings (such as naturally ventilated buildings) and for closed buildings (buildings with minimal untreated outdoor air, relying upon thermal mass or evaporative cooling while occupied).

The open building procedure will include internal heat sources, direct solar gains through glass, and some "stored solar" gains through roof and wall surfaces.

The closed building procedure will include all of the above, plus gains due to heat transfer from hot outside air to colder inside air, such as through glass, and by infiltration.

You should use the procedure for open or closed depending on which condition describes each of your buildings while occupied. It must be emphasized that these are still approximate calculation procedures. Much more exact and lengthy calculations can be done, which incorporate the hourly change in solar load on buildings. The ASHRAE Handbook of Fundamentals describes these in detail.

The calculations have three parts:

Heat gain through skin, regardless of its orientation, where differences between open and closed building procedures are accounted for;

Instantaneous solar gain through glass (same procedure for both open and closed buildings), which is highly dependent upon the choice of date and time for which calculation is performed, and is also greatly influenced by the shading devices which accompany windows;

Internal gain (same procedure for both open and closed buildings), as previously calculated.

Information necessary to do the calculations will be found as follows:

Heat Transfer Factors: if your construction hasn't changed since the developing phase of Cooling, you can use those. If it has, go back to MEEB Table 4.22c, p. 149 and use the lowest temperature differential of 20°, as previously.

Solar gains through glass area are listed in Table 4.27, MEEB, for 40° N. Latitude. (ASHRAE Handbook of Fundamentals also lists these for 24°, 32°, 48° and 56° N. Latitude.) Your choice of date and time can greatly change the resulting hourly Btu to be removed. After considering your climate and your function, aim for a date/time which approaches the worst hour (greatest hourly gain), condition. A fairly common date/time choice is late summer afternoon (4 pm August 21) when relatively low altitude sun can penetrate unshaded west glass at the time when peak daily air temperatures are occuring. On the other hand, a building with mostly un-shaded south facing glass would experience highest solar gains near noon. Obviously, you may choose different days/times for the office building as compared to the factory.

Shading coefficients are from Cooling Exercise 6.4 in the "Shading Masks" Evaluation Tool.

ΔT for air changes is the same as that specified for the HTF choice.

1. OFFICE BUILDING   Specify the date and hour you are calculating for:

Why is this your "worst" hour?

a. Gains Through Skin Trial 1

All spaces combined:	Closed Building			Open Building		
	Area (sq.ft.)	HTF (UΔT)	Hourly Gain (Btuh)	Area (sq.ft.)	HTF (UΔT)	Hourly gain (Btuh)
Exterior opaque walls						
Roof: horizontal 45° sloped						
All glass				(note #1 below)		
Exterior doors						
Infiltration				(note #2 below)		
cu.ft. .018xΔTxACH						
	Subtotal _____ Btuh			Subtotal _____ Btuh		

Notes to above calculations:
#1 open buildings are hotter inside than outside, so the heat will flow, if at all, from inside to outside through the glass, therefore contributing no gain. Opaque skin contributes a small gain because it can store and later release heat.
#2 the air will be cooler outside than inside so it contributes no heat.

Gains Through Skin, Trial 2

All spaces combined:	Closed Building			Open Building		
	Area (sq.ft.)	HTF (UΔT)	Hourly Gain (Btuh)	Area (sq.ft.)	HTF (UΔT)	Hourly Gain (Btuh)
Exterior opaque walls						
Roof: horizontal 45° sloped						
All glass				(note #1 above)		
Exterior doors						
Infiltration	Volume c.f.	.018 x ΔT x ACH		(note #2 above)		
			Subtotal _____ Btuh			Subtotal _____ Btuh

---

b. Instantaneous Solar Gains Through Glass, Trial 1

All Spaces Combined	Area (sq.ft.)	Solar Heat Gain Factor	Shading Coefficient	Hourly Gain (Btuh)
Horiz. glass				
Vertical glass: N				
NE				
E				
SE				
S				
SW				
W				
NW				

45° sloped
glass:
interpolate (between horizontal, vertical values)

Subtotal _____Btuh

Instantaneous Solar Gains Through Glass, Trial 2

All Spaces Combined	Area (sq.ft.)	Solar Heat Gain Factor	Shading Coefficient	Hourly Gain (Btuh)
Horiz. glass				
Vertical glass: N				
NE				
E				
SE				
S				
SW				
W				
NW				

45° sloped
glass:
(interpolate between horizontal, vertical values)

Subtotal _____Btuh

c. Internal Gains

You may use the hourly gains you found in the Climate Exercise, section 3.6. However, the electric lighting assumptions may be reduced, either based on your available daylight in this "worst month" compared to needed fc illumination for the tasks, or if you have completed Lighting Finalization 10.0. If you adjust your lighting gains, show your procedure and assumptions. Ordinarily you will be calculating gains for a time when the building is occupied, so you should use the higher internal gains while occupied, rather than the lower averaged internal gains (over a 24 hour period).

Subtotal: _____ Btuh internal gains while occupied.

d. Approximate Maximum Hourly Heat Gain, Trial 1

	Closed	Open
PART I. Skin	_____ Btuh	_____ Btuh
II. Solar	_____	_____
III. Internal	_____	_____
Office TOTAL Gains	_____ Btuh	_____ Btuh

Approximate Maximum Hourly Heat Gain, Trial 2

	Closed	Open
PART I. Skin	_____ Btuh	_____ Btuh
II. Solar	_____	_____
III. Internal	_____	_____
Office TOTAL Gains	_____ Btuh	_____ Btuh

How does this compare to the estimated peak gain from the Cooling Exercise 6.4 "Heat Gain" calculation?

2. FACTORY BUILDING Specify the date and hour you are calculating for:
Why is this your "worst" hour?

---

a. Gains Through Skin, Trial 1

	Closed Building			Open Building		
All spaces combined:	Area (sq.ft.)	HTF (UΔT)	Hourly Gain (Btuh)	Area (sq.ft.)	HTF (UΔT)	Hourly Gain (Btuh)
Exterior Opaque walls						
Roof: horizontal 45° sloped						
All glass				(note #1 below)		
Exterior Doors						
Infiltration	Volume c.f.	.018 x ΔT x ACH		(note #2 below)		
		Subtotal	_____ Btu/h		Subtotal	_____ Btu/h

Notes to above calculations:
#1 open buildings are hotter inside than outside so that the heat will flow, if at all, from inside to outside through the glass, therefore contributing no gain. Opaque skin contributes a small gain because it can store and later release heat.
#2 the air will be cooler outside than inside so it contributes no heat.

Gains Through Skin, Trial 2

All spaces combined:	Closed Building			Open Building		
	Area (sq.ft.)	HTF (UΔT)	Hourly Gain (Btuh)	Area (sq.ft.)	HTF (UΔT)	Hourly Gain (Btuh)
Exterior Opaque walls						
Roof: horizontal 45° sloped						
All glass				(note #1 above)		
Exterior Doors						
Infiltration	Volume c.f.	.018 x ΔT x ACH		(note #2 above)		
		Subtotal _____ Btuh			Subtotal _____ Btuh	

b. Instantaneous Solar Gains Through Glass, Trial 1

All spaces combined	Area (sq.ft.)	Solar Heat Gain Factor	Shading Coefficient	Hourly Gain (Btuh)
Horiz. glass				
Vertical glass: N				
NE				
E				
SE				
S				
SW				
W				
NW				
45° sloped glass: (interpolate between horizontal, vertical values)				
			Subtotal	_____ Btuh

7.0 THERMAL FINALIZATION

7.3 COOLING PERFORMANCE

Instantaneous Solar Gains Through Glass, Trial 2

All spaces combined	Area (sq.ft.)	Solar Heat Gain Factor	Shading Coefficient	Hourly Gain (Btuh)
Horizontal glass				
Vertical glass: N				
NE				
E				
SE				
S				
SW				
W				
NW				

45° sloped
glass: (interpolate between horizontal, vertical values)

Subtotal _____ Btuh

c. Internal Gains
See the discussion in the office calculation above about
Internal Gains. If you adjust your lighting gains show
your procedure and assumptions.

Subtotal: _____ Btuh internal gains while occupied.

d. Approximate Maximum Hourly Heat Gain, Trial 1

		Closed	Open
Part	I. Skin	_____ Btuh	_____ Btuh
	II. Solar	_____	_____
	III. Internal	_____	_____
Factory TOTAL GAINS		_____ Btuh	_____ Btuh

Approximate Maximum Hourly Heat Gain, Trial 2

Part	I. Skin	_____ Btuh	_____ Btuh
	II. Solar	_____	_____
	III. Internal	_____	_____
Factory TOTAL GAINS		_____ Btuh	_____ Btuh

How does this compare to the <u>estimated</u> peak gain from the calculation in the Cooling Exercise 6.4?

Evaluation Tool for Criterion A - "Mass Sizing for Closed Buildings"
Use this tool for buildings closed by day and cooled with mass storage, but open at night for ventilation. For buildings cooled by natural ventilation only ("open"), go ahead to the following Evaluation Tool "Aperture Sizing for Cooling", p. 7-47 (Procedure adapted from Crowther, "Night Ventilation Cooling of Mass", 1980)

1. Size Mass for Heat Storage
a. Combine your final heat gain pattern from the Cooling Exercise, p. 6-37 with a graph of your building schedule showing the hours the building is closed and storing heat and the hours the building will be open to ventilate the stored heat:
Ideally your building will close at the lowest temperature of the day and open again after the outside temperature has peaked and dropped to a bearable level, such as 82°F when it crosses into the comfort zone.

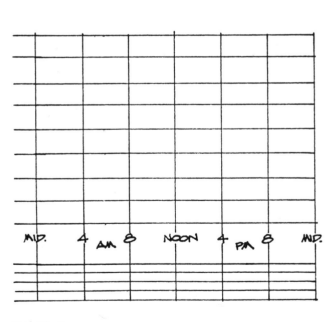

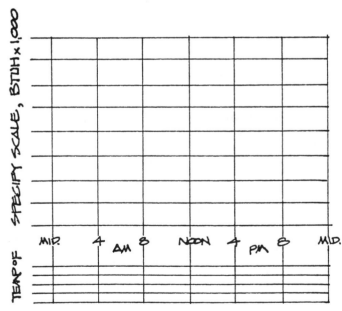

Trial 1
Office 24-hour Pattern

Trial 2
Office 24-hour Pattern

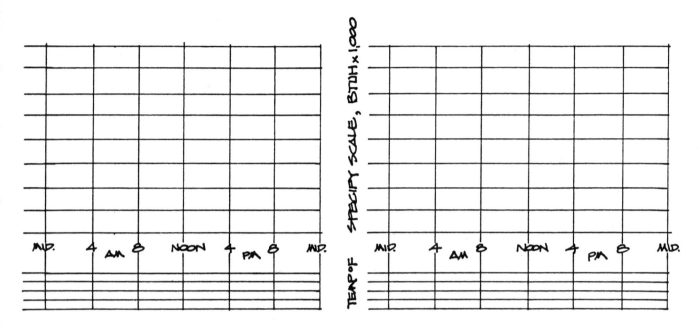

Trial 1
Factory 24-hour Pattern

Trial 2
Factory 24-hour Pattern

b. Find the total number of BTU's to be stored during the
hours the building will be closed. This total is the sum
of BTU's stored during each closed hour.

Office _____

Factory _____

c. Determine the amount of mass you have provided in your
design from the Heating Exercise, p. 5-34ff.

	Office	Factory
sf of mass exposed surface	_____	_____
material	_____	_____
thickness	_____	_____
cubic feet	_____	_____

---

d. Mass heat
storage capacity = (mass volume) x (density) x (specific heat)

Material	Density #/cuft	Specific Heat Btu/#/^{o}F
Ordinary Concrete	144	0.156
Common Brick	123	0.2
Gypsum	78	0.259
Limestone	103	0.217
Sand	95	0.191
Wood:		
Softwoods	27	0.45 approx.
Hardwoods	45	0.55 approx.
Water	62	1.0

Fig. 7-4 Density and Specific Heat of Materials. From The Passive Solar Energy Book by Ed Mazria, 1979, Appendix A.
(Other tables of density and specific heat can be found in Olgyay "Design With Climate".)

Office mass capacity_____Btu/^{o}F

Factory mass capacity_____Btu/^{o}F

Although you have only calculated the storage in mass specifically provided for that purpose, some heat will also be stored in the rest of the building materials. You may consider this additional cooling potential as a "safety factor" for extreme conditions, particularly if there is a lot of wall and floor area you have not included as mass above.

e. In column 2 below list the August hourly outside temperatures, from the Cooling Exercise 6.4, p. 6.37. If average figures are used, then average performance will be calculated. If worst case figures are used, then worst case performance will be calculated.

f. Begin calculating column 3 at the time when the outside temperature goes below 82°:

$$\text{Cooling BTU} = \left(\begin{array}{c} \text{mass temp. col. 4} \\ \text{preceeding hour} \end{array} - \begin{array}{c} \text{outside temp.} \\ \text{col. 2} \end{array} \right) \times \begin{array}{c} \text{mass} \\ \text{surface area} \end{array}$$

Assume the 1st mass temperature preceeding hour to be 80°

g. For column 4,

Mass temp. = mass temp. from $-$ $\frac{\text{(cooling from col. 3)}}{\text{(mass heat capacity)}}$
preceeding hour

h. Continue calculating columns 3 and 4. STOP when the mass temperature, col. 4, is lower than the outside temperature, col. 2

i. Add up the hourly Cooling Btu, in Column 3.

j. Very rough ventilation check: there must be apertures large enough to admit the outdoor air that is being used to cool down the storage mass. A rough approximation:

Inlet areas and outlet areas each = $\frac{1}{30}$ mass surface area

(Technical note: refer to the window sizing formula on p. 7-48. Assume that only 1 mph (88 fpm) of wind is available, and that it strikes the window diagonally (E=0.3) rather than head-on. With these conservative assumptions,

Inlet Area (sq.ft.) = $\frac{\text{Cooling Btu per hour}}{1.08 \times 0.3 \times \Delta T \times 88 \text{ fpm}}$

and since Cooling Btu = ΔT x Mass Surface Area,
Inlet Area (sq.ft.) =

$\frac{\text{Mass Surface Area}}{1.08 \times 0.3 \times 88}$, or about $\frac{\text{Mass Surface Area}}{30}$)

This procedure, steps c through i, is taken from Karen Crowther "Night Ventilation Cooling of Mass", in Passive Cooling Handbook, prepared for the Fifth National Cooling Conference, Amherst, MA, 1980.

7.3 COOLING PERFORMANCE

Trial 1

(1) Hour	(2) Outside Air Temp. ($^\circ$F)	OFFICE (3) Cooling Btu	OFFICE (4) Mass Temp. ($^\circ$F)	FACTORY (3) Cooling Btu	FACTORY (4) Mass Temp. ($^\circ$F)
8 pm	_____	_____	_____	_____	_____
9	_____	_____	_____	_____	_____
10	_____	_____	_____	_____	_____
11	_____	_____	_____	_____	_____
12	_____	_____	_____	_____	_____
1 am	_____	_____	_____	_____	_____
2	_____	_____	_____	_____	_____
3	_____	_____	_____	_____	_____
4	_____	_____	_____	_____	_____
5	_____	_____	_____	_____	_____
6	_____	_____	_____	_____	_____
7	_____	_____	_____	_____	_____
8	_____	_____	_____	_____	_____
9	_____	_____	_____	_____	_____

	Office	Factory
Total Btu's that can be cooled by mass: (sum of col. 3)	_____ Btu	_____ Btu
Final mass temperature (last temp. in col. 4)	_____ $^\circ$F	_____ $^\circ$F

Trial 2

		OFFICE		FACTORY	
(1)	(2)	(3)	(4)	(3)	(4)
Hour	Outside Air Temp. (^{o}F)	Cooling Btu	Mass Temp. (^{o}F)	Cooling Btu	Mass Temp. (^{o}F)
8 pm	_____	_____	_____	_____	_____
9	_____	_____	_____	_____	_____
10	_____	_____	_____	_____	_____
11	_____	_____	_____	_____	_____
12	_____	_____	_____	_____	_____
1 am	_____	_____	_____	_____	_____
2	_____	_____	_____	_____	_____
3	_____	_____	_____	_____	_____
4	_____	_____	_____	_____	_____
5	_____	_____	_____	_____	_____
6	_____	_____	_____	_____	_____
7	_____	_____	_____	_____	_____
8	_____	_____	_____	_____	_____
9	_____	_____	_____	_____	_____

	Office	Factory
Total Btu's that can be cooled by mass: (sum of col. 3)	_____ Btu	_____ Btu
Final mass temperature (last temp. in col. 4)	_____ ^{o}F	_____ ^{o}F

Check now to see that your mass is sized to meet Criterion A before continuing to size apertures:

Can you cool all the Btu's you need to with your mass?

Is your mass cooled down so the building can be closed at the coolest hour of the day?

If not, what will you change for your second trial?

Evaluation Tool for Criterion A - "Aperture Sizing for Cooling"
To check the size of openings you have provided for ventilation, complete the following table and then use either procedure 1. cross-ventilation or procedure 2. stack ventilation for each building.

"Heat gain to be removed" for open buildings will be the total Btuh hourly heat gain from the "Heat Gain" Evaluation Tool above.

"Heat gain to be removed" for closed buildings will be the hour with the largest total heat gain to be removed, including both stored BTU's (from column 3, from the preceeding evaluation tool), and internal, skin, and solar gains if any of these occur while the building is being ventilated in this worst hour:

Specify what is	Office	Factory
Worst Hour:	_____	_____
Stored Btu's:	_____	_____
Skin gains:	_____	_____
Solar gains:	_____	_____
Internal gains:	_____	_____
Total Btuh to be removed:	_____	_____

	Source of Information	Office	Factory
Heat gain (Btuh to be removed	See above		
Inlet area (sq.ft.) provided	your design		
Δ T	3° maximum (Criteria A)		
Average wind speed (mph)	Climate exercise		
Q (cu ft of air per minute required to remove heat gain)	$Q = \dfrac{Btuh}{1.08 \times \Delta T}$		

1. Procedure for cross-ventilation. Use the following equation:

$$A = \frac{Q}{EV}$$

Where:
A = inlet area required to remove heat gain, in sq.ft.
Q = cu ft air per minute from preceeding table.
E = effectiveness of opening due to angle of wind; for winds perpendicular to openings E = 0.5, for winds diagonal to openings E = 0.3
(ASHRAE Fundamentals, 1972, p. 344)
V = wind speed, in f.p.m. = mph x 88

A for office = _____ sq.ft.

A for factory = _____ sq.ft.

2. Procedure for stack ventilation. Use the following equation:

$$A = \frac{Q}{9.4 \ h(\Delta T)}$$

Where:
A = inlet area required to remove heat gain, in sq.ft.
h = height between centers of inlet and outlet openings
Q and ΔT are from the table above.

A for office = _____ sq.ft.

A for factory = _____ sq.ft.

Have you provided enough inlet area to keep your open buildings within the temperature range for Criterion A? If not, what will you change for your second trial?

If it is 3° hotter inside your buildings than outside, what is the inside temperature?

B. Criterion B: The capacity of your cooling backup system is sufficient to keep you cool during your worst overheated conditions.

Evaluation Tool for Criterion B - "Cooling Backup System Sizing"
If a building is using a mechanical system for cooling the air itself is being cooled to a lower temperature, rather than the heat moving to a cooler sink such as mass or outside air. This cooling process also involves removing moisture in the air to prevent an uncomfortable increase in humidity at the lower temperature. Doing so means more Btu's must be removed. (See MEEB 5.12 Psychrometry p. 226-234) These additional Btu's due to water vapor are termed "latent gains" and can be approximated as 0.3 x sensible heat gains, (MEEB 1980, p. 160). Sensible heat gains are those you calculated for Criterion A above.

Total cooling load = sensible heat gain + latent heat gain = 1.3 x sensible heat gain for mechanical system.

Total cooling load:

office = _____ Btuh

factory = _____ Btuh

Mechanical cooling systems can be sized using the above total (such as heat pumps) or in tons of refrigeration (Flynn & Segil, 1970, p. 224).

1 ton refrigeration capacity = 12,000 Btuh.

What capacity will your refrigeration system be for the office? _____ tons.

The factory?_____ tons

Does this meet Criterion B?

8.0 INTRODUCTION TO LIGHTING

Until only about 100 years ago, daylighting was clearly the most dominant consideration in the lighting of buildings. "Artificial" sources were installed, but provided far less light than daylight, and tended to be used only at night.

Following the development of mechanical cooling, it became possible to exclude daylight entirely from buildings; the resulting heat from electric lighting was rejected along with the heat of the occupants. The outdoors got hotter as the indoors stayed cool.

Why would anyone exclude daylight?

Daylight is highly variable; there is much less of it at 8 a.m. than at noon, and less in winter than in summer. Also, an overcast sky distributes daylight very differently than a clear sky - and sky conditions can shift several times a day.

If absolutely uniform, steady light is desirable, electric lighting is attractive.

Why would anyone utilize daylight?

Electric lighting is energy intensive; electricity production in fossil and nuclear thermal processes harnesses, at best, about 40% of the fuel's value; the remainder is dumped as waste heat to the environment. Then, as electric light is created, it brings about twice as much heat per unit of light to the space as does daylighting. If energy conservation is important, and the variability of outdoor climate is to be linked to indoor conditions, daylighting is attractive.

The interconnections between daylight, electric light, heat loss and heat gain come into focus here. More windows for more daylight mean more heat loss in winter. (Properly shaded, they can be relatively harmless to heat gain in summer.) More electric lights mean more internal heat gain, year round. They become the winter friend of the skin-dominated load building; the year-round enemy of the internal-dominated one.

9.0 DAYLIGHTING

9.1 INTRODUCTION

9.1 INTRODUCTION

This exercise begins with exploring organizational attitudes and then achieving the necessary quantity and distribution of daylight (enough light in the right places) through manipulation of building, room and window sizes and locations. This design is then developed through consideration of reflectances, obstructions and transmission properties. In Exercise 10 you will evaluate the annual performance of the design.

It is very important in starting this exercise to know if your climate is generally sunny or cloudy, or if it changes seasonally. In lower latitudes the light is brighter and the direct sun both more intense and less desirable from a thermal standpoint than farther north. These factors should be recognized in your design from the beginning. Vernacular buildings in your climate will provide good clues about where to start.

At this point, review the initial daylight explorations you made in the Climate Exercise, sections 3.4 and 3.5.

Contents

_____

9.2 CONCEPTUALIZING _____

GOALS

A. Use available daylight for lighting.
B. Prevent discomfort from excessive brightness or contrast.

DESIGN STRATEGIES

A. Site Scale
Locate buildings and spaces on site to either utilize or avoid obstructions of the skydome.
B. Cluster Scale
Recognize that buildings and ground surfaces can reflect light.
C. Building Scale
Consider zoning work by the visual difficulty of tasks. Put spaces which need light near openings in the skin.
D. Component Scale
Orient openings to recognize the brightest part of the skydome: zenith for overcast sky, horizon for clear sky.

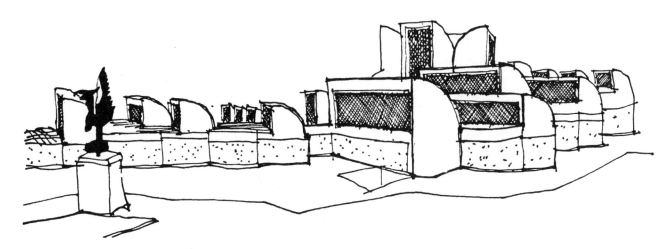

DAE Architects, Tehran Museum of Contemporary Art.

Consider openings for illumination as potentially different from openings for view, thermal gain, or ventilation.

DESIGN

Choose an existing building/site which has a clear conceptual approach, at any scale, to achieving these goals. Document your choice as follows:
A. Identify the location, program, architect (if any), and the source of your information.
B. Include photo copies or drawings (whatever is quick and easy for you) to explain the design.

EVALUATION

Evaluation Tool - "Building Response Diagrams"
Diagram how this design is organized to achieve these goals. If the typical sky condition for this building is different from yours, speculate on how the design responses might change your climate.

9.3 SCHEMING

GOAL

A. Open yourself to the sky in proportion to the brightness of the sky and the amount of light you need.

CRITERIA

A. The daylighting is distributed so there is light where people need it.
B. There are gradual changes in lighting levels between inside and out, and within rooms, to prevent glare problems.
C. Recommended Daylight Factors (DF) for tasks are achieved.
D. Solar collecting and ventilating apertures are maintained as required for heating and cooling.

DESIGN STRATEGIES

A. Cluster Scale
Place light-colored walls to the north of north-facing spaces to increase available reflected light.
Recognize the potential of thermal buffer zones (arbors, vestibules, courtyards, atria, greenhouses, etc.) to control glare and heat gain.
Snow is a highly effective reflector of light, but can also cause glare.

B. Building Scale
Consider which one of these strategies is most
appropriate to your climate and function:
1. Design for adequate daylight under average sky
conditions; below this average, supplementary lighting
will be necessary.
2. Design for minimum acceptable daylighting under
minimum daylight available conditions (such as 9 am in
December).
3. Design for adequate daylight in the great majority of
typical working hours.
4. Use supplementary artificial lighting for dark areas
rather than oversupplying daylight in lighter areas to
achieve the minimum everywhere.

Proportion rooms and/or locate tasks to use outer 15'
band of floor area where maximum light from side windows
is available.

Avoid over-daylighting your spaces in hot climates, where
darkest places are associated with cooler temperatures.

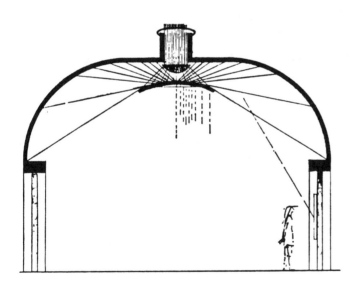

Kahn, Light Study for Kimball Art Museum (Ronner, 1977,
p. 345)

The resulting smaller daylight openings can be placed
near the ceiling, where ground reflected light can
illuminate a white ceiling without causing glare at eye
level.

C. Component Scale

Top lighting is brightest under the skylight and darkens in all directions.

Side lighting is brightest near the window and darkens toward the back wall.

Specular surfaces reflect light at an angle equal to the angle they receive it.

Tasks can be moved to use available light rather than moving light to tasks.

For standard lighting (office, lobby, circulation) the area of glazing = approximately 5%-10% floor area.

For intensive lighting (display, drafting, typing, factory work) area of glazing = approximately 25% floor area.

Proportion window height to equal 1/2 the depth of the room for good daylight penetration and distribution. (Hopkinson, 1966, p. 435)

Arrange shading devices so that the sun reflecting off their surfaces will not cause glare at eye level in the shaded spaces beyond.

DESIGN

Design the office, factory and outside areas according to program requirements in the 2.1 Building/site Description.
Document your design as follows using the grids provided:

A. State which sky condition (clear or overcast) you are designing for. Is this a typical or a worst condition?
B. Site plan, including parking and access drive(s). C. Cluster plan, including outdoor spaces and vegetation. Identify surfaces and their approximate light reflectances (see Olgyay, 1963, p. 33).
D. Floor plans of office and factory. Label the signifi-cant task areas on these plans.
E. Section(s) of office and factory.
F. Roof plan and elevations (or paralines) clearly showing all sides and the top of the buildings.

Important Note: Your design may change in the course of the following evaluation. The final schematic design which you show to meet all criteria should be fully documented here.

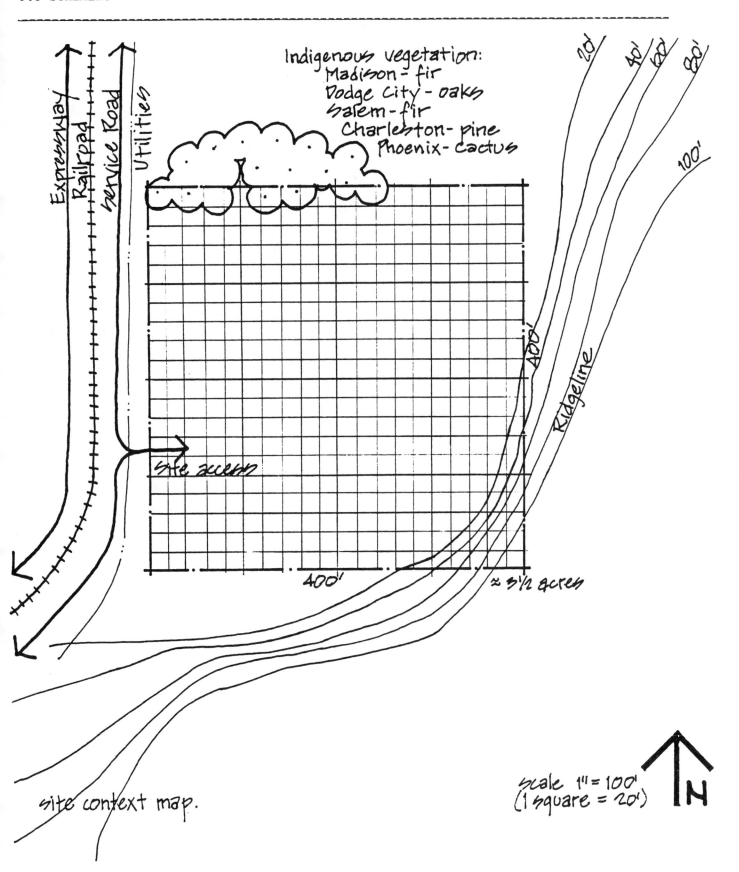

Indigenous vegetation:
Madison - fir
Dodge City - oaks
Salem - fir
Charleston - pine
Phoenix - cactus

Expressway
Railroad
Service Road
Utilities

20' 40' 60' 80'

100'

400'

Ridgeline

Site access

400'

≈ 3½ acres

site context map.

scale 1" = 100'
(1 square = 20')

N

cluster site plan: 1" = 40'-0".

N

for plans, sections, & elevations: if   1/16"=1'-0",   1 square = 10'.

for plans, sections, & elevations: if 1/16" = 1'-0", 1 square = 10'.

----

for plans, sections, & elevations: if   1/16"=1'-0",   1 square = 10'.

for plans, sections, & elevations: if 1/16"=1'-0", 1 square = 10'.

for plans, sections, & elevations: if 1/16"=1'-0", 1 square = 10'.

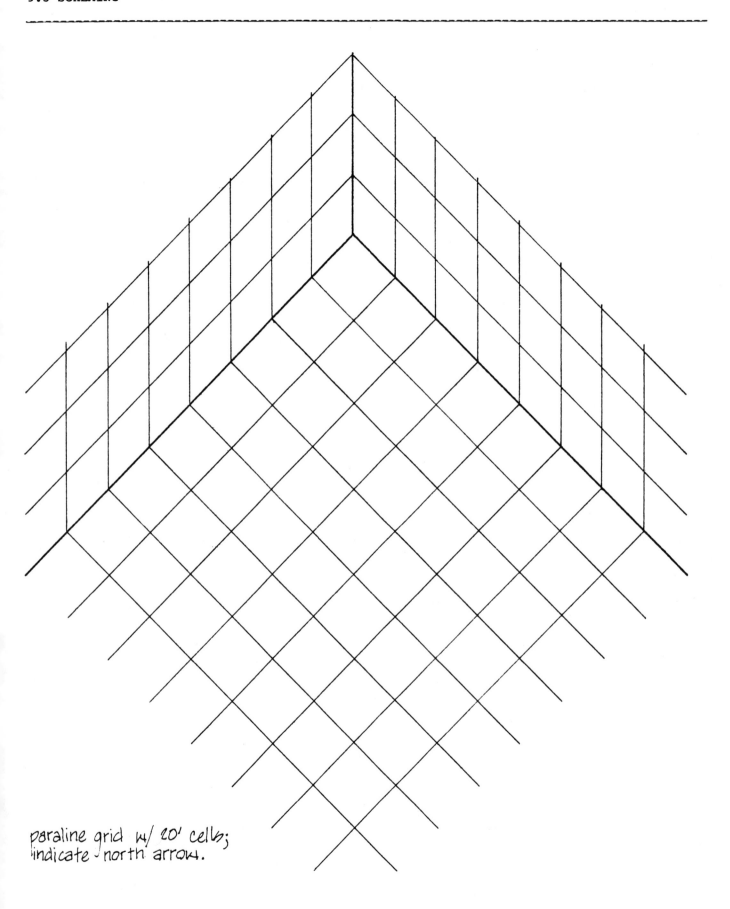

paraline grid w/ 20' cells;
indicate north arrow.

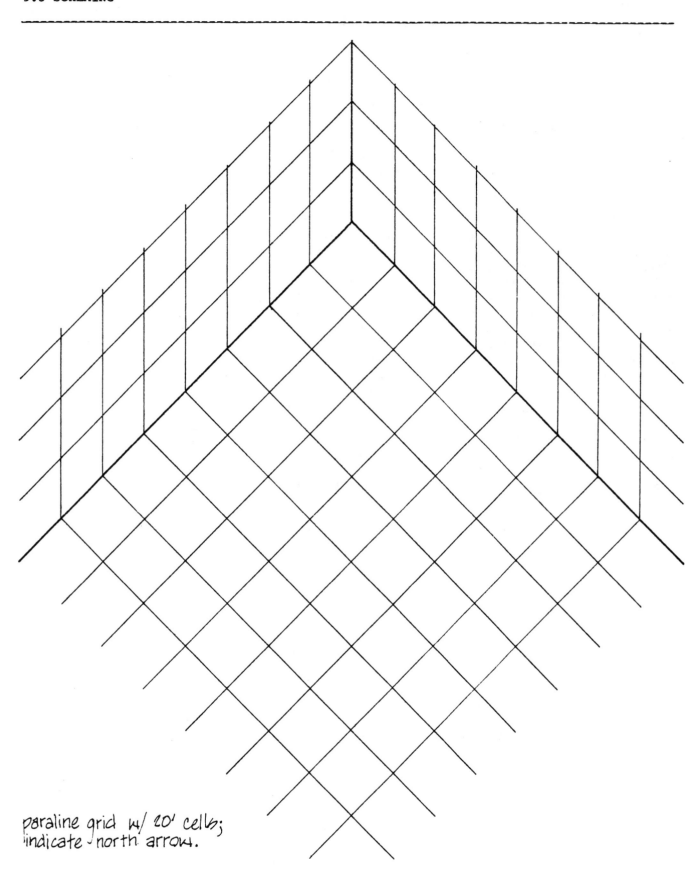

paraline grid w/ 20' cells;
indicate north arrow.

EVALUATION

A. Criterion A: Daylighting is distributed so there is light where people need it.

Evaluation Tool for Criterion A - "Light Distribution Rendering"

1. For the predominant sky condition in your climate, sketch shaded plans and sections of your office and factory buildings. Use media that allows you to quickly approximate the relative patterns of light and dark that result from your climate's daylight conditions and your methods of admitting that daylight to your buildings. (Pencil or charcoal on tracing paper is a common approach.) Each plan should have a corresponding section drawn on the same page for easy reference.
2. Clearly identify task areas, so that the light/task relationship is obvious.
3. Show how you have designed for gradual changes in brightness, especially at openings from inside to outside. Feel free to include a second trial evaluation.

Does your first trial design meet the criterion?

If not, what will you change for your second trial?

for plans, sections, & elevations: if   1/16"=1'-0",   1 square = 10'.

B. Criterion B: There are <u>gradual</u> changes in lighting levels between inside and out, and within rooms, to prevent glare problems.

Evaluation Tool for Criterion B - "Light Distribution Rendering"
Use the studies done for criterion A above to show you have also satisfied criterion B. If you have designed to achieve contrast rather than gradual changes in illumination levels, explain your design goals and show that glare will not be a problem.

Does your first trial design meet the criteria?

If not, what will you change in your second trial design?

C. Criterion C: Recommended Daylight Factors for tasks are achieved.

For <u>overcast</u> skies:

Task	Min. DF	Comments
office work	1%	With side lighting, min. 12' from window;
	2%	With top lighting, over whole area
typing, computing	4%	Over whole area
factory work	5%	General recommendation
public areas	1%	Depending on function, may be greater
drafting	6%	On drawing boards
	2%	Over rest of working area

Figure 9-1 Recommended Daylight Factors (Millet and Bedrick, 1978)

For <u>clear</u> skies, use the % from Climate section 3.5C., Summary questions A and B:

office   _____ %
factory   _____ %
circulation   _____ %

Evaluation Tool for Criterion C - "Aperture Sizing for Daylighting"
1. The following table gives you recommended aperture-to-floor-area ratios to achieve average Daylight Factors with various kinds of daylight openings.

This is included as a schematic step to help you further organize your basic building form. While still rough, it is a more precise step than you took back in the Climate Exercise when you first explored daylighting.

Daylight Factor (%) = $\dfrac{Ag}{Af}$ x 100   where Ag = area glazing, sf
                                              Af = area floor, sf

<u>Sidelighting</u>

average DF = $0.2 \dfrac{Ag}{Af}$

minimum DF = $0.1 \dfrac{Ag}{Af}$

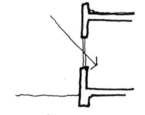

<u>Toplighting</u>

Vertical monitors: average DF = $0.2 \dfrac{Ag}{Af}$

North-facing
sawtooth openings: average DF = $0.33 \dfrac{Ag}{Af}$

Horizontal
skylights: average DF = $0.5 \dfrac{Ag}{Af}$

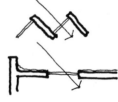

Fig. 9-2 Aperture Sizing for Daylighting (adapted from Hopkinson, 1966, p. 432)

2. Evaluate your office building and factory by completing the table below.

If more than one type of opening is used for a given space, <u>first</u> find the average DF for each, <u>then add</u> to obtain total average DF.

Room		Col. 1 Af	Col. 2 Type of opening	Col. 3 Ag	Col. 4 Applicable rule of thumb Average DF =		Col. 5 Average DF	Col. 6 Compare to recommended minimum DF
Office #1	Trial 1							
	Trial 2							
Office #2	Trial 1							
	Trial 2							
Display/Conf	Trial 1							
	Trial 2							
Factory	Trial 1							
	Trial 2							

Does your first trial provide DF (in col. 5) equal or greater than the recommended in Col. 6?

If not, what will you change for your second trial?

D. Criterion D: Apertures needed for solar collecting and ventilating are maintained.

If you have already done the heating and cooling exercises, your design must still meet the criteria for heating and cooling. If you have not yet done these exercises, but you know that either heating or cooling (or both) will be concern in either building in your climate, check the following criteria now so that you don't make things impossible in those exercises!

For solar heating: Aperture sizing, section 5.3.

For cooling: Aperture sizing, section 6.3.

Does your first trial maintain adequate heating and cooling performance?

If not, what will you change for your second trial?

9.4 DEVELOPING

GOALS

A. Admit daylight to achieve approximately uniform lighting within task areas, without sacrificing desirable views.
B. Use appropriate shading devices and interior surface finishes, together with the openings, to achieve recommended daylight factors.

CRITERIA

A. Provide recommended DF for each task area.
B. The range of DF within a task area should not exceed 3 to 1.

DESIGN STRATEGIES

A. Site Scale
Utilize a variety of landscape surfaces for their reflective characteristics. These will change seasonally in climates with snow. Remember that snow can cause glare, especially on sunny days.
B. Cluster Scale
Ground reflectance is potentially the major source of daylighting in hot, sunny conditions, when shading devices may be blocking direct sun and greatly reducing light from the skydome (including summertime in cold climates).
Exterior obstructions to daylight in overcast climates (such as other buildings or hills, etc.) will not have a major effect on distribution of daylight within spaces, unless they obscure a major portion of the glazing.
C. Building Scale
Proportion rooms to recognize the effect of overhangs:

the room effectively begins at the edge of the overhang,
not where the glazing is located.

Use side lighting for task lighting to minimize veiling
reflections and to supply useful modeling.

Use top lighting for more uniform, ambient lighting and
general illumination.

D. Component Scale

Be careful of glare problems from narrow vertical
windows, which create strong patterns of light and dark
contrasts.

Use light colors and higher reflectances near openings to
reduce glare by reducing the contrast between inside and
outside.

Choose internal surface reflectances to recognize the
major source of incoming light:

In temperate climates - direct and reflected skylight
hits the floor first.

In tropical climates - reflected sunlight hits the
ceiling first.

Use light colors/high reflectances on surfaces farthest
from openings to increase light in dimmer areas.

DESIGN

Develop your design to achieve these goals for either a
typical bay of your factory or the display conference
room in the office building.

Document the design:

A. Plan and section of the space, at 1/8"=1'0". (State
which room you are designing.)

B. Sectional perspective with finishes, materials,
colors, and areas (sf) of each called out. Include 1
person for scale.

C. State for which sky condition (clear or overcast) you
are designing.

D. Draw all daylight apertures in elevation and section
to show mullions; label the glazing material, and show
the shading devices in position for your chosen sky
condition.

----

for daylighting contours, enlarged plans: if 1/8"= 1'-0", 1 square = 10'.

for daylighting contours, enlarged plans: if 1/8"= 1'-0", 1 square = 10'.

----

for daylighting contours, enlarged plans: if 1/8"= 1'-0", 1 square = 10'.

for daylighting contours, enlarged plans: if 1/8"= 1'-0", 1 square = 10'.

EVALUATION

A. **Criterion A:** Provide recommended D.F. for each task area.
B. **Criterion B:** The range of D.F. will be less than 3 to 1 within a given task area.

Evaluation Tools for Criteria A & B - For overcast sky conditions: "Footprints"; for either clear sky or overcast: "Daylight Model" (model instructions start on p. 9-34.

"Footprints"

First determine whether the footprint prediction technique will work for your design. This technique is applicable to most spaces, however there are complications in using it with non-orthogonal openings, or in rooms of complex shape or abnormal proportions. It also does not evaluate the localized effect of interior or exterior room reflectances, the effect of exterior obstructions, or fixed-shading devices. (Fixed overhangs can be accommodated by considering the room as extending to the edge of the overhang; moveable shading devices, such as ventilation blinds, can be assumed to be clear of the opening on the darkest days you are evaluating).

If your design has one of these complications that makes the footprint method awkward, we urge you to use the model and light meter to evaluate your design. Believe it or not, you will <u>save</u> time by not trying to use the wrong technique to evaluate your design.

If you would like to use footprints, and fixed shading devices are your <u>only</u> complication, simply <u>estimate</u> the % daylight that gets through them. <u>Do not</u> use "shading coefficients" that are used in heat gain estimations, since these refer to <u>direct sun</u> more than to daylight from all sources. Thus, the % daylight will be somewhat higher than these "shading coefficients" would indicate.

The footprints included at the end of this exercise can be used <u>as is</u> for windows 5' x 10', 10' x 10', 10' x 5' <u>above</u> the workplane. For other sizes and shapes they must be re-scaled and or combined so the solid rectangle is the same size as your window. Any portion of the window <u>BELOW</u> the workplane will not contribute illumination

to the workplane so "your" window for footprint purposes
will be: _____.

This technique involves three basic steps:
1. Matching your openings with available footprints;
2. Drawing isolux contours that describe light
distribution and basic DF range in your space.
3. Correcting DF's applied to contours:
<u>increasing DF</u> due to factors which increase light in the
space (principally interior reflectances)
<u>reducing DF</u> due to factors which reduce light in the
space (glass transmission and dirt, and shading devices.)

The results give you an estimation of the distribution of
light and the quantity of light (in Daylight Factors) in
your space.

Note: The information and tables in this procedure are
from the two sources listed below. For more information,
distribution patterns, refinement steps, and procedure
for dealing with obstructions, see:
Millet and Bedrick, "Architectural Daylighting Design
Procedure", June 1978
Millet and Bedrick, "Manual: Graphic Daylighting Design
Method," 1981.

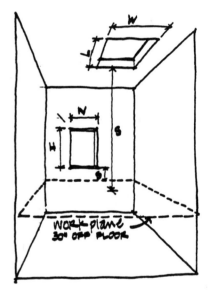

1. Match opening and footprints.
a. List your windows and skylights and find the
appropriate footprint in Fig. 9-4 on the following page.

H = height of window itself
W = width of window or skylight
L = length of skylight itself
S = distance from <u>workplane</u> to <u>sill</u>

Trial 1:

Opening	H/W window; L/W skylight	S/H window; S/W skylight	Footprint #

Trial 2:

Openings of shapes other than those below can be used by combining shapes that are catalogued. For example, a sloping skylight:

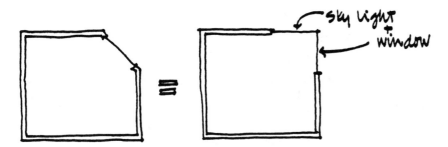

Figure 9-3 Sloped Skylight Simulation for Footprint Procedure. If a 45° skylight is square and has an area of 141 sf, the resulting window and horizontal skylight will each be 10 x 10. When estimating the IRC, use the actual area (141 sf) and Fig. 9-5, TOP-LIT Rooms.

Specify any assumptions you made to deal with your design:

Windows				Skylights			
H/W	S/H	# Footprint		L/W	S/W	#Footprint	
.5	0	A-7		1	1	A-30	
.5	.5	A-8		1	2	A-31	
.5	1	A-9		1	3	A-32	
.5	2	A-10		1	4	A-33	
.5	3	A-11					
.5	4	A-12		2	1	A-34	
				2	2	A-35	
1	0	A-13		2	3	A-36	
1	.5	A-14		2	4	A-37	
1	1	A-15					
1	2	A-16					
1	3	A-17					
1	4	A-18					
2	0	A-19					
2	.5	A-20					
2	1	A-21					
2	2	A-22					
2	3	A-23					
2	4	A-24					

Figure 9-4 Footprint Catalogue. See Section 9.5 for Footprints.

2. Drawing the contours (this will be easiest if you have your plan on a paper you can trace through):

a. Sketch the openings on your plan.
Skylights: draw the outline of the skylights as for a reflected ceiling plan (as if they were dropped straight down on to working plane).
Windows: using the section, rotate the windows 90° on to the workplane as though they were hinged at the inter-section of the workplane and the vertical plane. This locates them so you can then sketch the outline on the plan.
b. Sketch the contours for each opening. Position the plan with openings sketched on it over the appropriate footprint for each opening and trace the contours for the opening. (To use the pattern, the <u>solid rectangle should</u> <u>be lined up with the outline of the opening on the plan.</u> The <u>solid line at the bottom of the pattern</u> indicates the position of the window plane, so you don't get them up-side down.) Be sure to note values of DF contours on your plan. The values of the contour lines are % avail-able daylight, or Daylight Factors. The dotted contour is = 0.5% DF and they increase by 1% each contour. It will probably be most useful in the evaluation if you sketch all the side lighting contours on one sheet, and all the top lighting contours on a second sheet.
c. Combine the contours. With an overlay of tracing paper, mark all the points where the contours cross, and note the sum of the values of the crossing contours. Now draw the resulting contour lines by connecting points of equal value. Outside the area of overlap, trace the original contours directly. If you have side lighting and top lighting on separate sheets, combine contours on each sheet to get total side lighting and total top lighting, and then repeat the procedure to end up with contour lines of the overall, combined pattern. Be sure to note the DF values of the contours. This is the final isolux contour drawing for your space.

At this point, things get more complicated; so if you don't like the look of your DF contour distribution, change the design <u>now</u>, and start again.

----

for daylighting contours, enlarged plans: if 1/8"= 1'-0", 1 square = 10'.

9.0 DAYLIGHTING

9.4 DEVELOPING

for daylighting contours, enlarged plans: if 1/8"= 1'-0", 1 square = 10'.

for daylighting contours, enlarged plans: if 1/8"= 1'-0", 1 square = 10'.

---

for daylighting contours, enlarged plans: if 1/8"= 1'-0", 1 square = 10'.

for daylighting contours, enlarged plans: if 1/8" = 1'-0", 1 square = 10'.

for daylighting contours, enlarged plans: if 1/8"= 1'-0", 1 square = 10'.

---

for daylighting contours, enlarged plans: if 1/8" = 1'-0", 1 square = 10'.

for daylighting contours, enlarged plans: if 1/8"= 1'-0", 1 square = 10'.

----

for daylighting contours, enlarged plans:			if 1/8"= 1'-0",	1 square = 10'.	

3. Correcting the Daylight Factors
a. Increases due to IRC are the product of the internally reflected component and the interior maintenance factor (IMF).

$$\text{adjusted DF} = \text{DF} + (\text{IRC} \times \text{IMF})$$

Values of IMF for clear work in non-industrial areas; clean work in dirty-industrial areas; dirty-industrial work in non-industrial areas; and dirty-industrial work in dirty-industrial areas are 0.9, 0.8, 0.7, and 0.6 respectively (Millet and Bedrick, 1978).

Document the characteristics of your room by completing the table below (if you made assumptions about your design for the footprints use them for this section, too).
Reflectances: Flynn & Segil, Table 2-17, p. 126. MEEB, Table 20.3, p. 864.
Glass Transmission & Reflectance: MEEB, Table 19.5, p. 797. Flynn & Segil, Table 2-18, p. 127. Time Saver Standards, p. 925 (1974). ASHRAE Fundamentals. IES-Recommended Practice of Daylighting.

Surface	Material or Finish	Surface Area (sf)	Reflectance	Transmission of Glazing (and shading devices)

Total Room
Surface Area:

9.0 DAYLIGHTING

9.4 DEVELOPING

Figure your net glazed areas for horizontal and vertical
openings separately. Measure the actual <u>glazed</u> area
(omitting mullions, glazing bars, etc.) in your design,
or multiply the total opening size by a typical factor.

Window type	Net glazed factor
all metal windows	.80
metal windows in wood frames	.75
all wood windows	.65-.70

<u>net</u> glazed area, side lighting   _____ Trial 1   _____ Trial 2

<u>net</u> glazed area, top lighting   _____ Trial 1   _____ Trial 2

<u>net glazed area in vertical surfaces</u> (side lighting) =   _____ Trial 1   _____ Trial 2
    total room surface area

<u>net glazed area in horizontal surfaces</u> (top lighting) =   _____ Trial 1   _____ Trial 2
    total room surface area

Figure your average room reflectance.
(If you are working with one typical bay or a large space
such as the factory, assume 50% reflectance from the
"sidewalls" that represent adjacent bays, and include
their area in the total surface area for a typical bay.
It is not necessary to calculate the entire factory.)

Average room reflectance =
$$\frac{[(surface^1)\,(\%\ reflectance^1)] + [(surface^2)\,(\%\ reflectance^2)]}{total\ room\ surface\ area}$$

Trial 1 _____ %   Trial 2 _____ %

Find the TOP-LIT IRC values (if you have top lighting)
from figures 9-5 and 9-6.

Find the SIDE-LIT IRC values (if you have side lighting)
from figure 9-7.

Net Glazing Area	Average Reflectance (I.R.C.) of Room Surfaces								
Total Room Surface Area	20%	25%	30%	35%	40%	45%	50%	55%	60%
.01	.22	.32	.40	.49	.60	.72	.88	1.09	1.40
.02	.46	.62	.78	.90	1.20	1.50	1.80	2.20	2.60
.03	.60	.90	1.15	1.40	1.80	2.15	2.70	3.40	4.10
.04	.90	1.20	1.65	1.90	2.40	3.00	3.70	4.40	5.30
.05	1.15	1.50	1.90	2.40	3.00	3.70	4.30	5.40	6.60
.06	1.40	1.85	2.30	2.90	3.70	4.40	5.40	6.60	8.00
.07	1.65	2.10	2.30	3.80	4.20	5.00	6.20	7.40	8.60
.08	1.35	2.90	3.20	3.50	4.70	5.90	7.20	8.60	10.50
.09	2.00	2.80	3.60	4.40	5.40	6.60	8.00	9.60	11.70
.10	2.20	3.10	4.10	4.50	5.90	7.30	8.80	10.70	13.00

Figure 9-5   Internally Reflected Component for TOP-LIT Rooms (glazing from horizontal up to 60°) (Millet and Bedrick, adapted from Hopkinson <u>Daylighting</u>)

Net Glazing Area	Average Reflectance (I.R.C.) of Room Surfaces								
Total Room Surface Area	20%	25%	30%	35%	40%	45%	50%	55%	60%
.01	.09	.13	.16	.20	.25	.31	.37	.45	.54
.02	.19	.25	.32	.41	.50	.62	.76	.90	1.10
.03	.27	.38	.48	.60	.75	.90	1.20	1.30	1.65
.04	.38	.50	.69	.80	1.00	1.20	1.48	1.80	2.20
.05	.47	.63	.80	.99	1.22	1.60	1.85	2.20	2.70
.06	.56	.77	.95	1.20	1.50	1.80	2.20	2.70	3.40
.07	.66	.88	.96	1.40	1.72	2.10	2.55	3.20	3.70
.08	.77	.99	1.25	1.60	1.95	2.40	3.00	3.70	4.40
.09	.86	1.15	1.42	1.80	2.20	2.70	3.45	4.10	3.00
.10	.94	1.20	1.65	2.00	2.40	3.10	3.80	4.50	5.50

Figure 9-6   Internally Reflected Component for TOP-LIT Rooms (glazing vertical) (Millet and Bedrick, adapted from Hopkinson <u>Daylighting</u>)

Net Glazing Area	Average Reflectance (I.R.C.) of Room Surfaces								
Total Room Surface Area	20%	25%	30%	35%	40%	45%	50%	55%	60%
.01	.05	.07	.10	.13	.18	.25	.32	.40	.50
	.02	.03	.05	.08	.12	.17	.25	.33	.43
.02	.09	.13	.20	.27	.36	.48	.62	.78	.96
	.03	.06	.10	.16	.24	.34	.47	.64	.84
.03	.13	.20	.29	.40	.54	.74	.94	1.17	1.45
	.03	.09	.15	.24	.36	.52	.72	.98	1.24
.04	.18	.27	.39	.53	.72	.96	1.25	1.60	1.95
	.07	.12	.20	.31	.48	.70	.96	1.27	1.66
.05	.23	.34	.48	.66	.85	1.18	1.55	1.95	2.40
	.08	.15	.26	.39	.60	.86	1.18	1.60	2.08
.06	.27	.41	.58	.80	1.07	1.43	1.80	2.35	2.92
	.10	.18	.30	.46	.72	1.03	1.43	1.95	2.50
.07	.32	.47	.68	.92	1.25	1.68	2.15	2.70	3.40
	.11	.21	.35	.56	.84	1.20	1.67	2.22	2.90
.08	.36	.54	.78	1.07	1.42	1.95	2.50	3.15	3.90
	.13	.24	.41	.64	.96	1.38	1.85	2.60	3.33
.09	.41	.60	.87	1.18	1.61	2.15	2.80	3.52	4.39
	.15	.27	.45	.72	1.06	1.55	2.12	2.92	3.70
.10	.45	.68	.96	1.30	1.74	2.40	3.10	3.90	4.80
	.16	.29	.50	.79	1.18	1.71	2.39	3.20	4.15
.12	.54	.80	1.15	1.58	2.10	2.85	3.70	4.65	5.80
	.20	.35	.60	.94	1.40	2.00	2.70	3.80	4.90
.14	.64	.94	1.35	1.85	2.50	3.33	4.30	5.40	6.90
	.24	.42	.71	1.10	1.65	2.40	3.32	4.50	5.80
.16	.72	1.07	1.55	2.10	2.80	3.80	4.90	6.55	7.70
	.27	.47	.82	1.25	1.90	2.72	3.80	5.10	6.60
.18	.81	1.20	1.70	2.35	3.20	4.30	5.50	6.90	8.30
	.29	.53	.92	1.38	2.10	3.10	4.25	5.70	7.40
.20	.90	1.30	1.80	2.60	3.50	4.70	6.00	7.70	9.70
	.33	.60	1.00	1.55	2.35	3.80	4.70	6.40	8.10

Figure 9-7   Internally Reflected Component of SIDE-LIT Rooms
(Millet and Bedrick, adapted from Hopkinson, Daylighting)

Note: Values in each box: top = average IRC
                           bottom = minimum IRC

b. Reductions in Daylight Factors.
Reductions due to glass transmission loss and shading
devices. Glazing-shading transmission factor _____ %
Figure your reduction or maintenance factor for dirt from
the table below or MEEB Table 19.5 (b), p. 797.

Building Location	Inclination of Glazing	Interior Conditions	
		No industry Clean	Dirty Industry
Non-industrial area	Vertical	.9	.8
	Sloping	.8	.7
	Horizontal	.7	.6
Dirty industrial area	Vertical	.8	.7
	Sloping	.7	.6
	Horizontal	.6	.5

Figure 9-8 Maintenance Factors

Multiply all reduction factors to get total:

Total reduction factor = transmission   x   maintenance
                          loss factor   factor

Total reduction factor = _____ Trial 1   _____ Trial 2

c. Correct all Daylight Factors
For all contours with the lowest Daylight Factors and
those in dimmer parts of the space:

corrected DF = (DF + toplit IRC + min sidelit IRC) x (total reduction factor)
(show your complete work in space below)

_____

For all other contours:

corrected DF = (DF + toplit IRC + <u>avg</u> sidelit IRC) x (total reduction factor)

d. Show the final contours labelled with corrected Daylight Factors on your plan, and graph them in section
.
From the Climate Exercise, section 3.5 (Criterion A Analysis tools, and summary questions), the daylight factors required in my climate for this space are:_____

From Section 9.3, Criterion C, the recommended minimum D.F. for this space is:_____%

Does your first trial design as evaluated by footprints meet these criteria?

If not, what will you change for your second trial?

Does your first trial design as evaluated by footprints meet criterion B?

If not, what will you change for your second trial?

Alternate Evaluation Tool for Criteria A & B - "Daylight Model"
The great advantages of this evaluation tool are its combination of accuracy and the variety of resulting design information. No other tool presents such a wide range of evaluation possibilities. It also allows quick evaluation of minor design changes, such as the substitution of one set of shading devices for another.

1. Build a model of the space you are investigating. Your model must be at least 1/2"=1'-0" to 3/4"=1'-0" scale to accurately use a standard light meter. Materials should reasonably match the opacity, surface texture and % reflectance of the floors, walls and ceilings of your space. For large spaces with repetitive bays, such as the factory, build just one bay and insert opaque "side walls" where adjacent bays would otherwise occur. The reflectance of these "walls" should be 50%.

The most important model details are those around the daylight openings; the size and depth of the mullions, the depth and reflectivity of the sill, louvres or other shading devices, and the reflectivity of surfaces just outside the daylight openings.

Louvres can be constructed at a larger scale than the windows, as long as they accurately show their own proportion of louvre-to-opening. Thus, a set of 1 1/2" scale louvres made to cover a 3/4" scale window will give reasonably accurate results - with only half as much louvre cutting/glueing as a 3/4" set of louvres would have required.

Glazing should be placed over openings; choose model materials that match the transmission characteristics of your actual glazing. Alternatively, you could omit model glazing and decrease your internal measurement to account for transmission loss through the glazing. Transmission loss factors can be found in MEEB, Table 19.5, p. 797 and Flynn and Segil, Table 2-18, p. 127.

Joints between walls/floors/ceilings/etc., should be taped to avoid cracks of daylight that would hurt the accuracy of the test. <u>However</u>, consider how the probe of the light meter will be inserted, <u>and</u> how it will accurately be positioned at various measurement points. You may need an openable flap of some sort.

Note: Refer to Benjamin Evans, Daylight in Architecture, Chapter 6 for more complete information on daylighting models.

2. Preparing your documentation
a. Set up a grid on your plan with intersections at which the daylight measurements will be taken. Unless a space is very small, at least 15 measurements (spaced equally throughout the floor area) should be taken or the grid should be a 5' or 10' grid. The surface of the light meter's probe should be at the approximate level at which DF are desired - the "workplane", which is usually about 30" above the floor (to model scale, of course!). This may require adding some material below the probe to bring it to the proper height.

If you know the size of the light meter probe in advance, it could be advantageous to put stops on your model floor at each measurement point, to quickly and accurately position the probe for each measurement.

b. Prepare several floor plans, clearly showing the grid, and allowing space to record daylight measurements. You may need to make more than one set of measurements; one set should be made for each design alternative you wish to explore.

3. Taking measurements.
a. The most critical factor is the sky condition, which should match as closely as possible the condition for which you are daylighting:
Overcast skies should be truly overcast - the actual position of the sun should not be discernable. Avoid "partly cloudy" skies.
Clear skies can be a problem, because the sun position should be at an altitude/azimuth similar to the time of day in the climate for which you are designing. Do not tilt the model a la "sun peg" approach to achieve this altitude/azimuth, because your building's openings will "see" a wildly different portion of sky vs. ground as your model leaves a horizontal position.

Another critical factor is the constancy of the outside light, against which interior measurements are compared to obtain daylight factor. If you are fortunate to have a "ratiometric" meter, which has at least two probes (one outside, one inside) and gives you the D.F. ratio directly, you can simultaneously measure exterior levels as well as interior ones, to assure that your DF will be based on accurate comparative measurements. This can also be accomplished by using two light meters simultaneously, one inside the model and one in an open area for the exterior measurement.

A somewhat less critical factor is the ground reflectance. Set your model on a surface with texture and reflectance matching the "real" condition. For the first 10' height of opening above grade, the 40' of horizontal surface in front of your model is particularly critical in this regard. Likewise, each 10' height increase should be matched by another 40' width of controlled ground reflectance.

b. For each point, measure and record the daylight both inside the model at that point and simultaneously outside the model in a large open area (be careful you don't obstruct the measurement by leaning over the meter).

4. When you have completed measurements, calculate your daylight factor for each point

DF = (interior fc/exterior fc)x(total reduction factor)

Plot the daylight factor contours on your plan, and graph them in section.

For criterion A (Analysis tools and summary questions), from the Climate Exercise section 3.5, the daylight factors required in my climate for this space are: _____ %

From Section 9.3, Criterion C, the minimum recommended D.F. for this space is: _____ %

Does your first trial as evaluated by model measurements meet this criteria?

If not, what will you change for your second trial?

Does your first trial design meet criterion B?

If not, what will you change for your second trial?

A-7

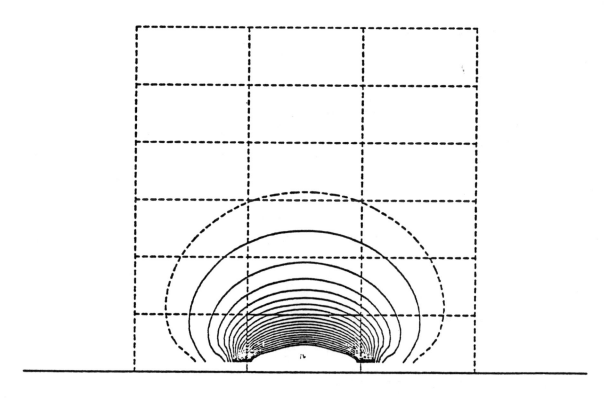

H/W = .5   S/H = 0

A-8

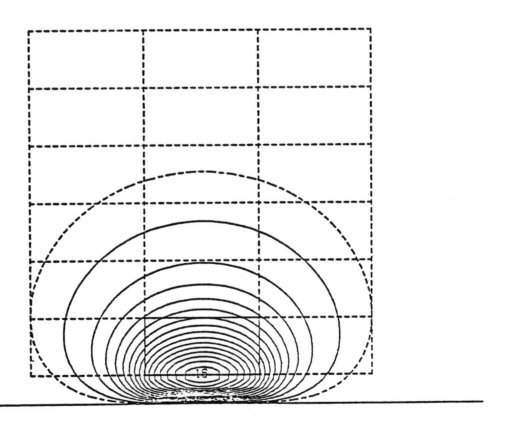

H/W = .5   S/H = .5

A-9

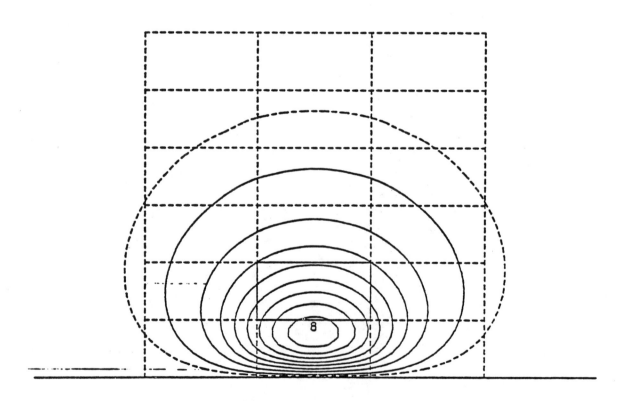

H/W = .5   S/H = 1

A-10

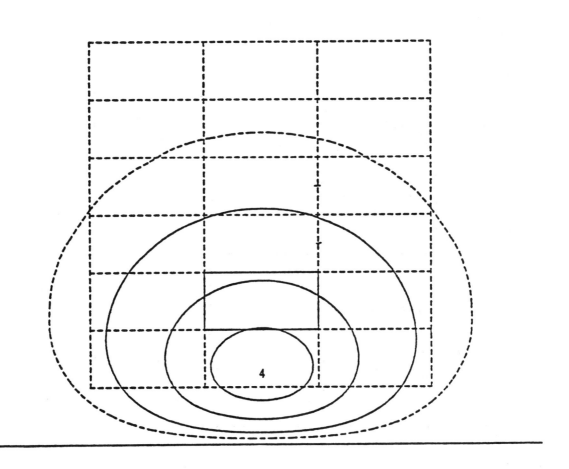

H/W = .5   S/H = 2

A-11

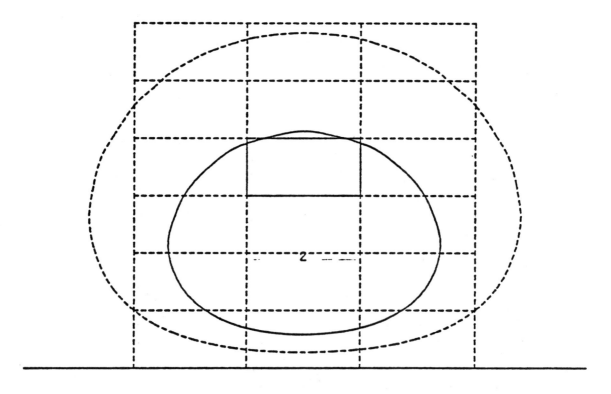

H/W = .5   S/H = 3

A-12

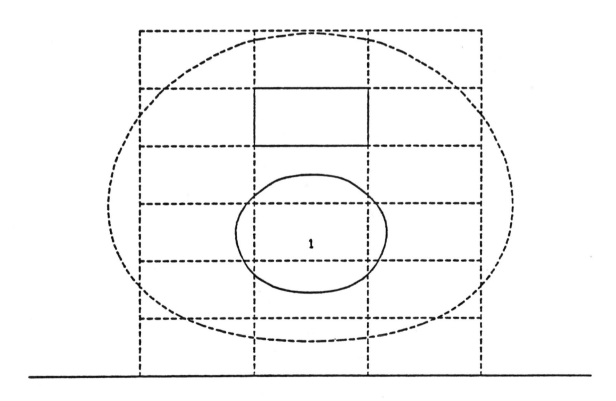

H/W = .5   S/H = 4

A-13

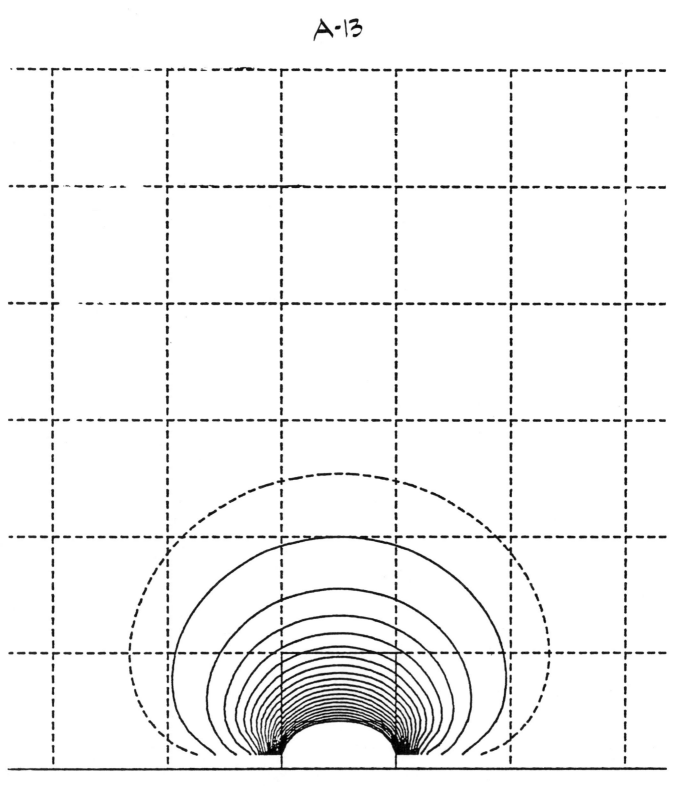

H/W = 1   S/H = 0

A-14

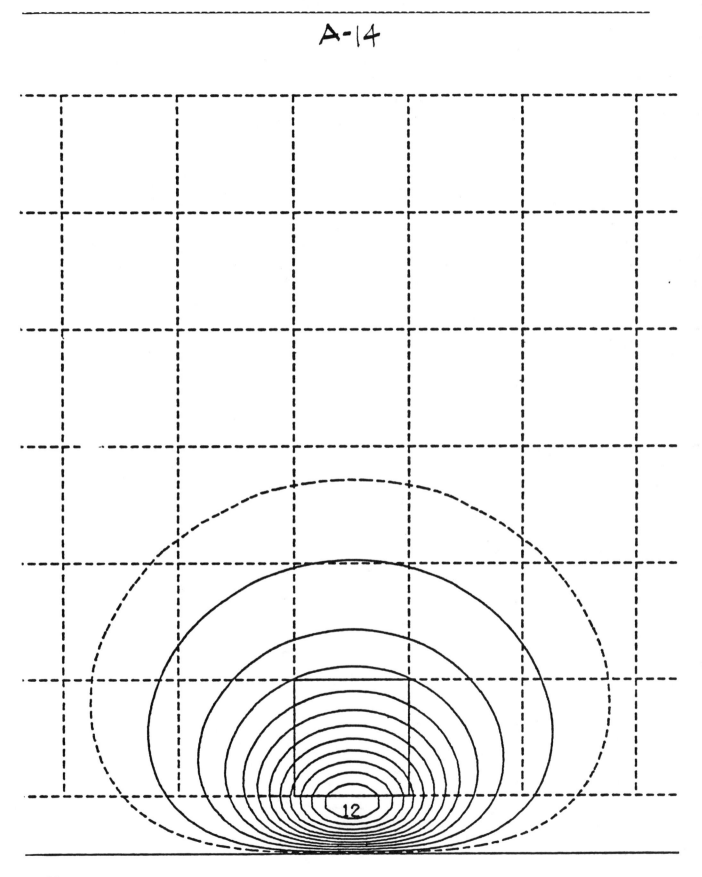

H/W = 1   S/H = .5

INSIDEOUT

A·15

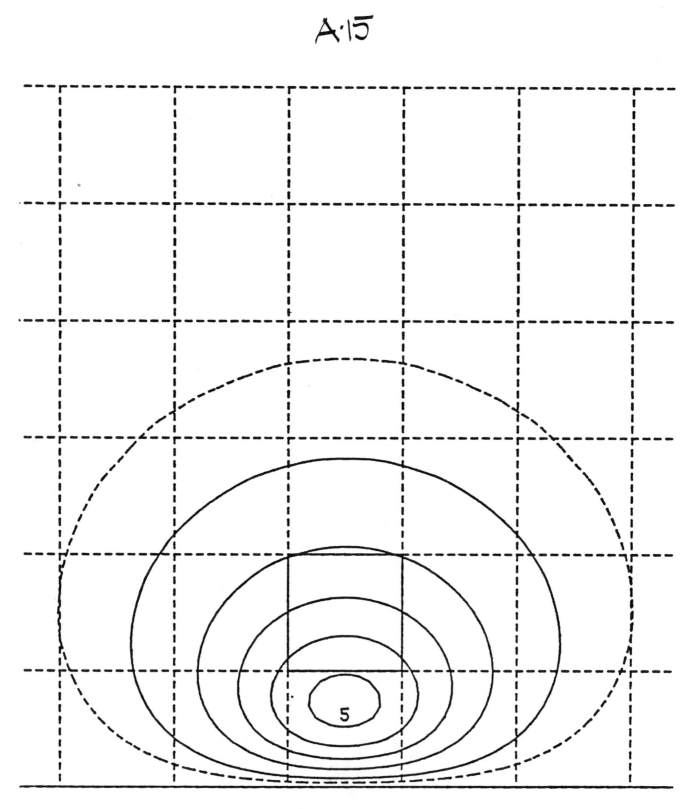

H/W= 1   S/H = 1

A.16

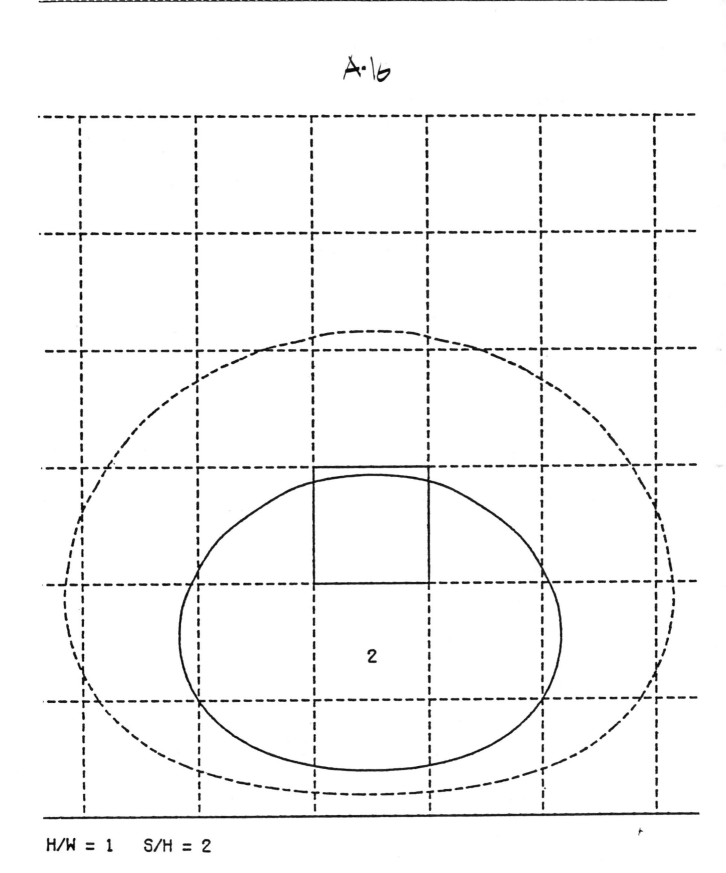

H/W = 1   S/H = 2

A·17

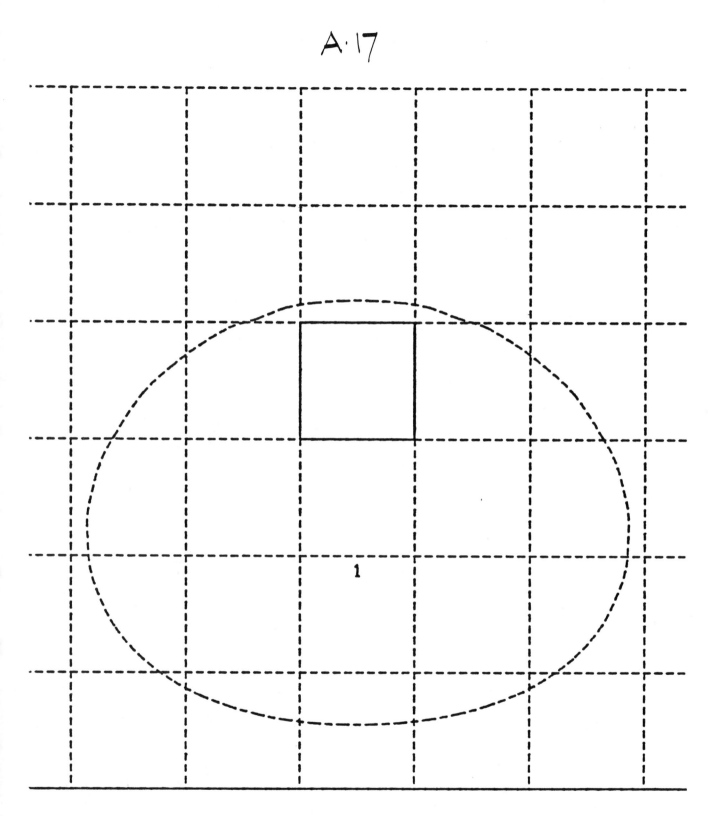

H/W = 1   S/H = 3

A-18

H/W = 1   S/H = 4

A-19

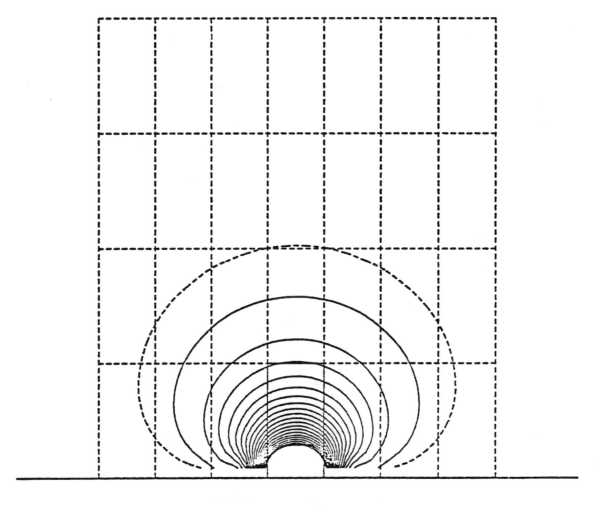

H/W = 2   S/H = 0

A-20

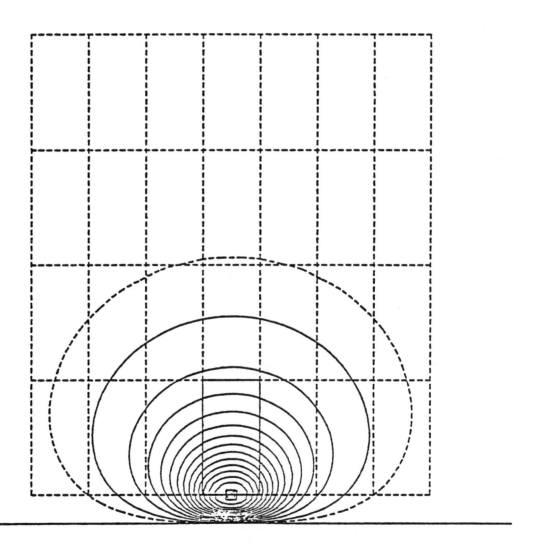

H/W = 2   S/H = .25

A-21

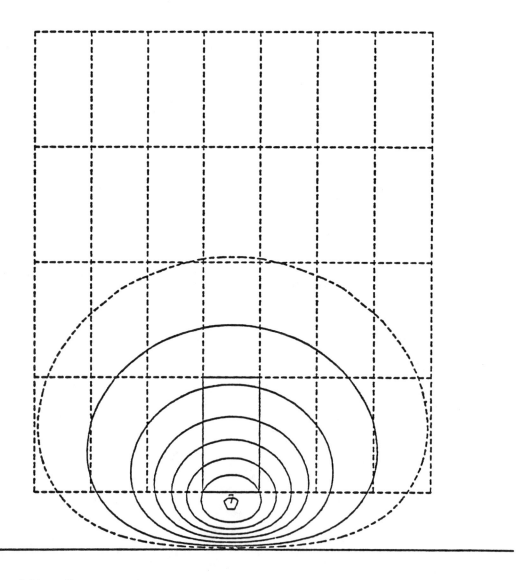

H/W = 2   S/H = .5

A-22

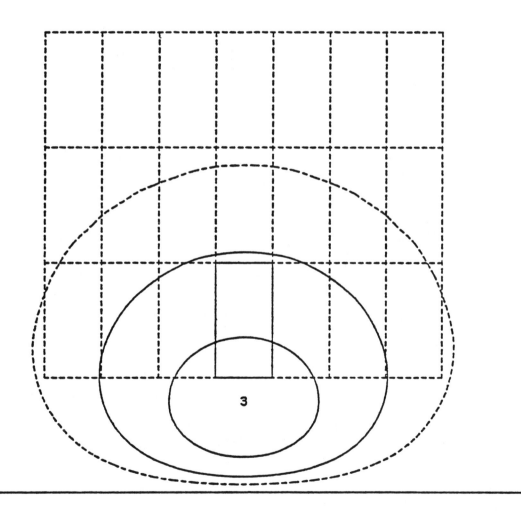

H/W = 2   S/H = 1

A-23

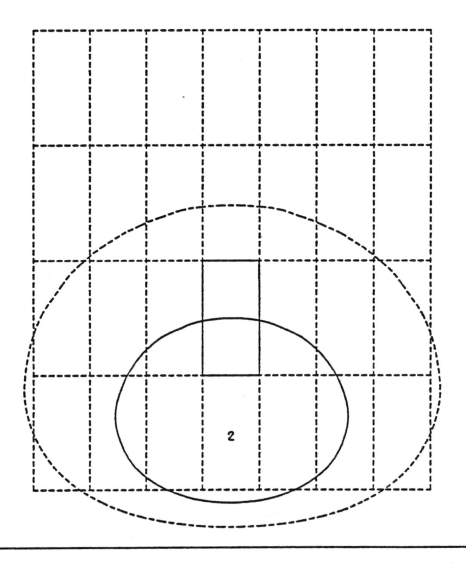

H/W = 2   S/H = 1.5

A-24

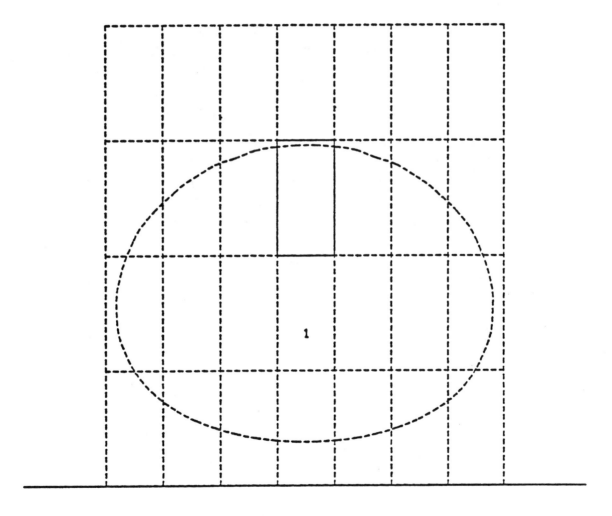

H/W = 2   S/H = 2

A·30

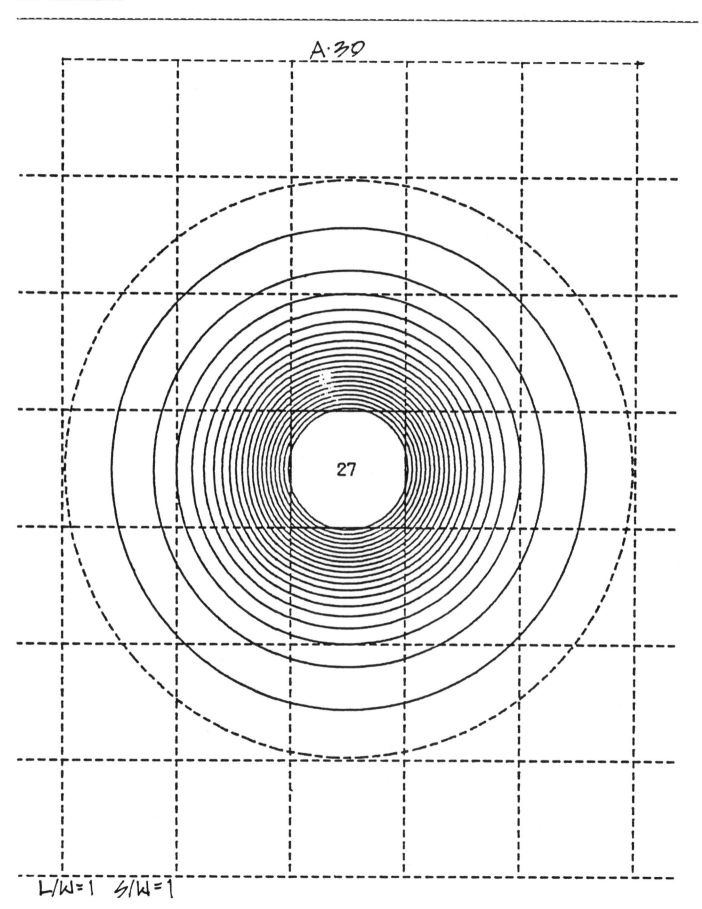

27

L/W=1 S/W=1

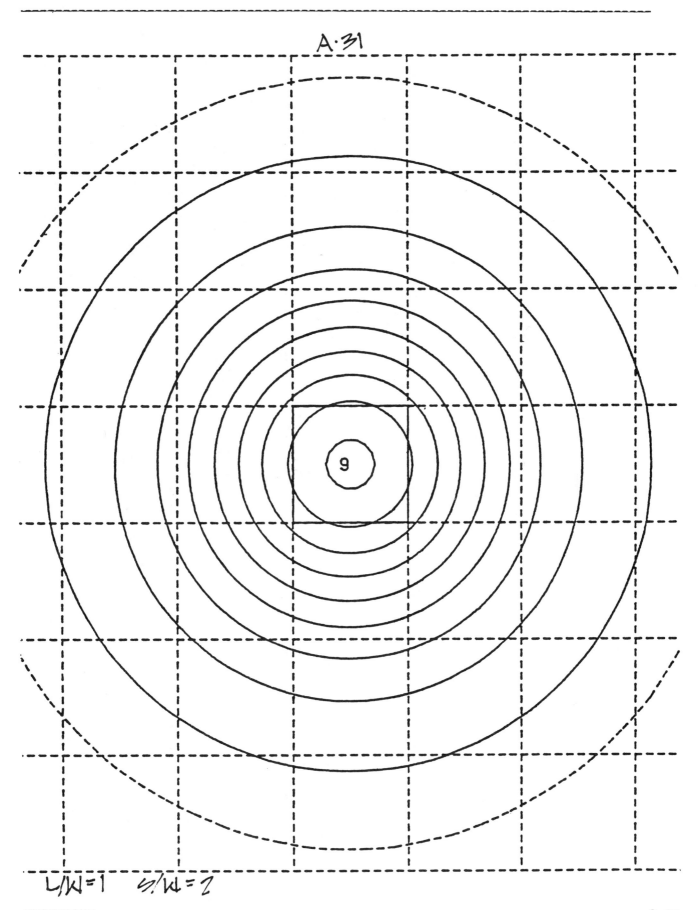

A·31

L/W=1 S/W=2

A-32

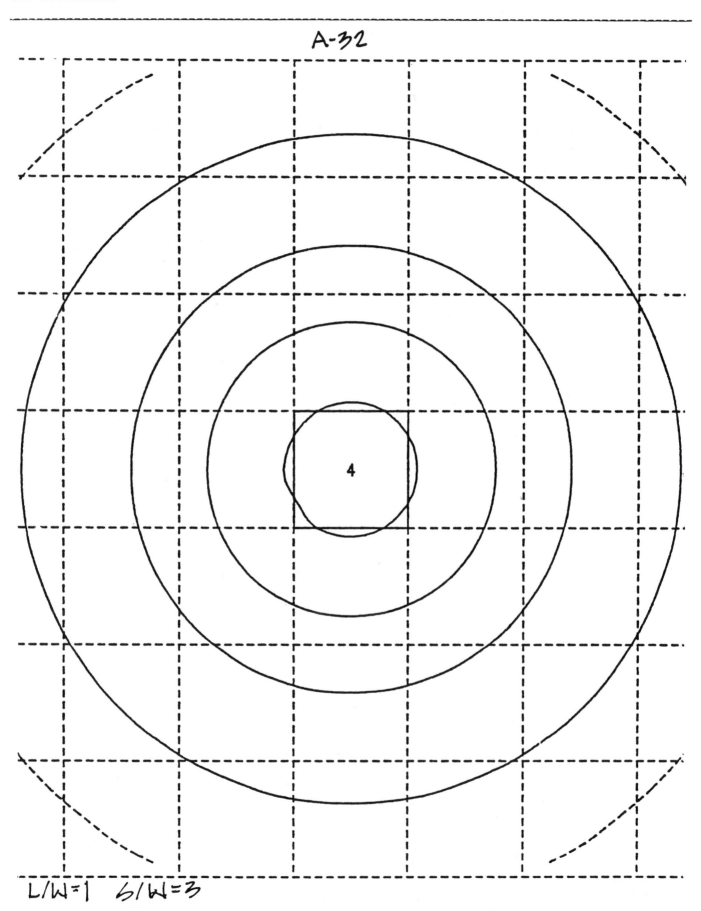

L/W=1 S/W=3

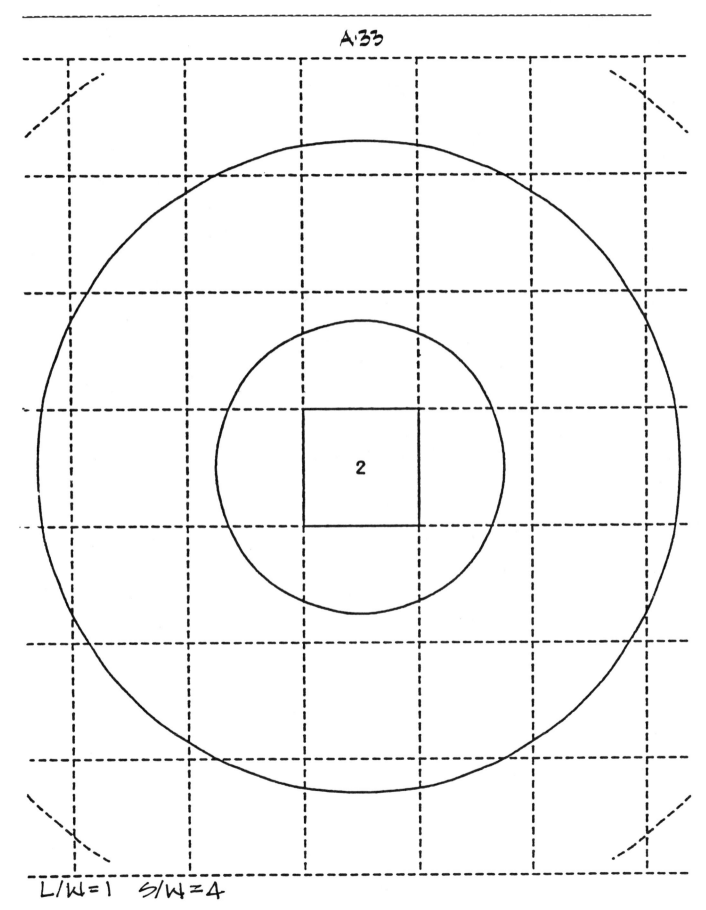

A·33

L/W=1 S/W=4

INSIDEOUT

A-34

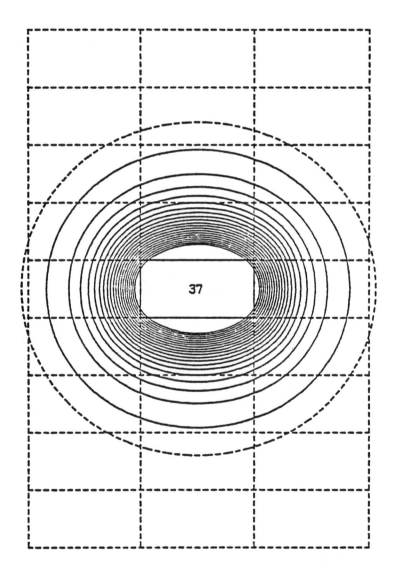

L/W = 2   S/W = 1

A-35

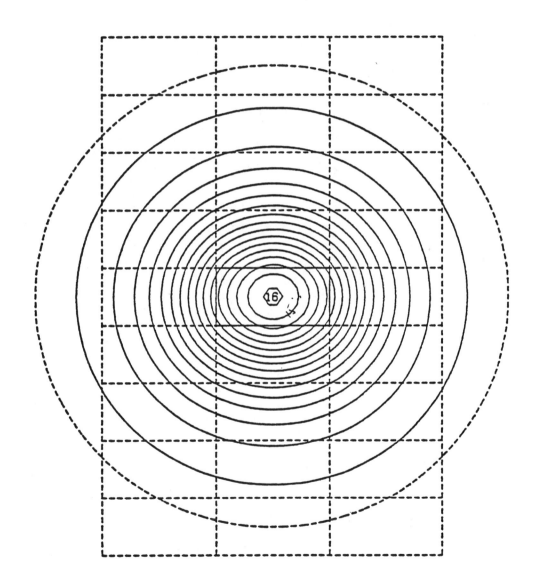

L/W = 2   S/W = 2

A-36

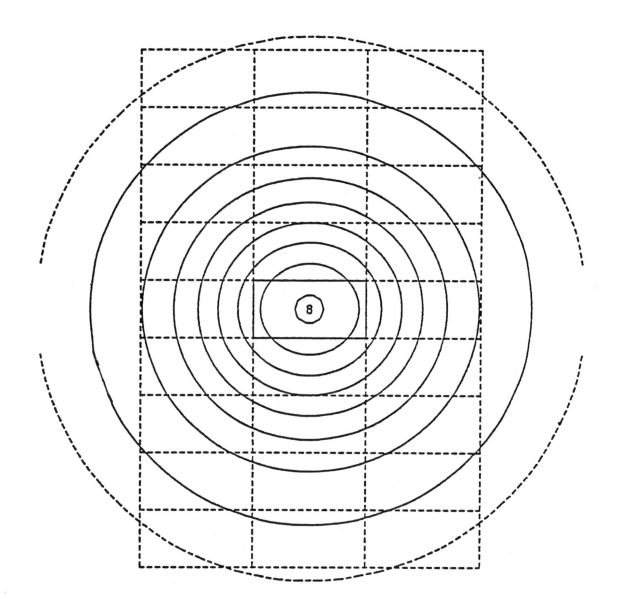

L/H = 2   S/H = 3

A-37

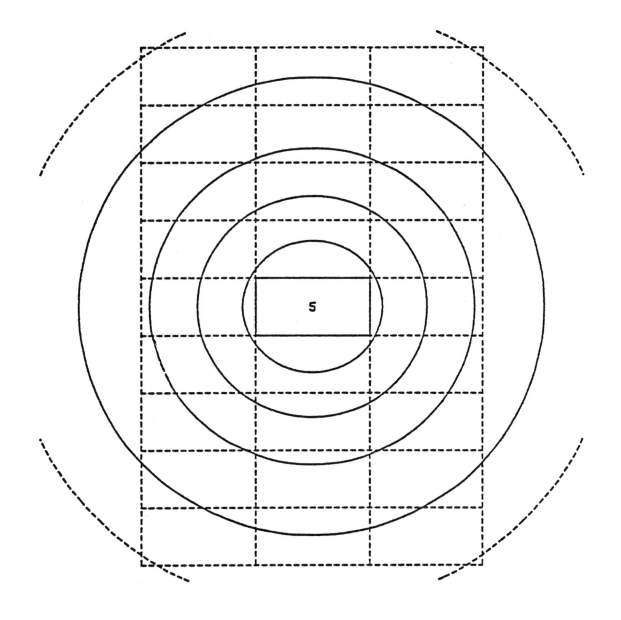

L/W = 2   S/W = 4

10.1 INTRODUCTION

From the approximate window sizes of the Climate exercise
and the more precise daylight factor predictions of
Section 9.4, you know how daylighting serves the tasks in
a particular space, and how that varies by the time of
day and by season. Now you propose a scheme for supplementary electric lighting which responds to these
changes. This enables you to find the energy requirements (watts/sq.ft.) of your overall lighting design and
understand the contribution daylighting is making in
reducing electric lighting's energy requirements.

Contents

10.2 DAYLIGHT PERFORMANCE

GOAL

Minimize energy use for electric lighting.

CRITERIA

A. Recommended minimum foot candles of illumination are
provided for tasks using daytime electric lighting equal
to or less than 1 watt per sq.ft.

DESIGN

Present full documentation of the design which you are
evaluating in this exercise.   Include at least the
following:
A. Floor plans of your space (display/conference or
factory), with the significant task areas labelled.
B. Sections of your space.
C. Partial roof plan and elevations (or paralines)
clearly showing all sides and the top of your space.
D. Reflected ceiling plan(s) of your space showing positions of lighting fixtures and exposed structure (if
any).

for plans, sections, & elevations: if   1/16"=1'-0",   1 square = 10'.

---

for plans, sections, & elevations: if   1/16"=1'-0",   1 square = 10'.

for plans, sections, & elevations :   if   1/16"=1'-0",   1 square = 10'.

for plans, sections, & elevations: if   1/16"=1'-0",   1 square = 10'.

for plans, sections, & elevations: if   1/16"=1'-0",   1 square = 10'.

for plans, sections, & elevations :   if   1/16"=1'-0",   1 square = 10'.

EVALUATION

A. Criterion A: Provide recommended minimum footcandles of illumination for your tasks using daytime electric lighting equal to or less than 1 watt per s.f.

Evaluation Tool For Criterion A - "Specifying 100% Electric System"
1. Show on your plans how you would employ a strategy of task and ambient lighting for night conditions.
2. Complete this table to relate electric lighting footcandle levels to watts/sq.ft. of installed lighting, refer to MEEB Figure 20.2, p. 867.

Design Condition	Type of electric light source installed	Approx. fc to be provided by source	Task area to be lit, X sq.ft.	Watts/sq.ft. required for this fc	Total watts = of electric light
Night lighting	_____	_____	_____	_____	_____

Evaluation Tool for Criterion A - "Sizing the Daytime Backup System"
1. To size the supplementary daytime electric light system, determine when and where daylight is less than adequate by itself. Because the daylight factor is the ratio between interior illumination and available light outside, you can use it to determine actual illumination levels (fc on working plane) from available daylight:

interior fc from daylight = DF x exterior fc available daylight

Convert the daylight factors you found in Lighting Exercise 9.4 p. 9-46 and 9-49 to footcandle levels for the sky condition you are investigating. (Note that 1 footcandle = 1 foot lambert.)

Use the Equivalent Sky Brightness tables (MEEB Table 19.2a for overcast skies, 19.2c for clear skies where direct sun is excluded from the window opening, p. 774-776), to find the interior illumination of your space in fc for minimum and maximum DF for seasons when sky conditions you tested are applicable in your climate.

10.0 LIGHTING FINALIZATION

10.2 DAYLIGHTING PERFORMANCE

Note: For clear sky conditions, list the maximum and minimum fc for each hour's entry, for each window orientation (north, south, east, west).

	Dec 21 Sky Condition _____		Mar/Sept 21 (1) Sky Condition _____		June 21 Sky Condition _____	
Trial 1	(Max % DF)	(Min % DF)	(Max % DF)	(Min % DF)	(Max % DF)	(Min % D
8 am	_____ fc	_____ fc	_____ fc	_____ fc	_____ fc	_____ fc
10 am	_____	_____	_____	_____	_____	_____
noon	_____	_____	_____	_____	_____	_____
2 pm	_____	_____	_____	_____	_____	_____
4 pm	_____	_____	_____	_____	_____	_____

	Dec 21 Sky Condition _____		Mar/Sept 21 Sky Condition _____		June 21 Sky Condition _____	
Trial 2	(Max & DF)	(Min % DF)	(Max % DF)	(Min %DF)	(Max % DF)	Min % DF
8 am	_____ fc	_____ fc	_____ fc	_____ fc	_____ fc	_____ fc
10 am	_____	_____	_____	_____	_____	_____
noon	_____	_____	_____	_____	_____	_____
2 pm	_____	_____	_____	_____	_____	_____
4 pm	_____	_____	_____	_____	_____	_____

2. Compare these fc levels with the minimum recommended ones   (See   Climate   Exercise   section   3.4).

Task Area	Recommended Footcandles	Hours Occupied Begin: Finish:

----

3. Graph and label the following information on the chart below, to understand how the daylight you have compares to illumination required by the task.
a. Band of interior fc (min to max) for the chosen sky condition in all seasons applicable.
b. Line of recommended fc for each task area (for the hours required only)

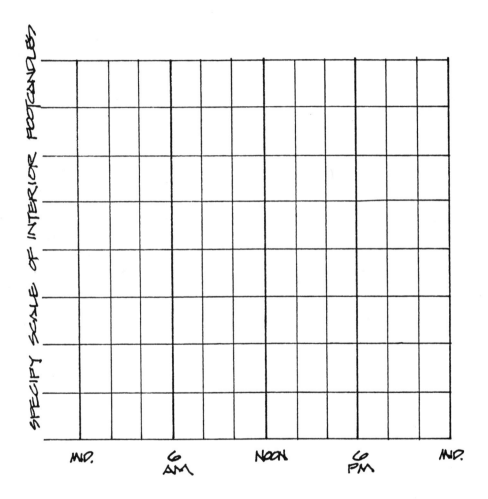

4. From the graph, determine <u>when and where</u> daylight, by itself, is less than adequate for each task. List the supplementary fc needed from electric lighting in order to meet the minimum light required for this task. Then compare this supplementary electric lighting level with the total electric lighting provided for night conditions.

Season & task	Time(s) that Daylight is available, but inadequate	Supplementary fc of electric light required	% of total installed electric lights needed to provide supplementary fc

Trial 1:

Trial 2:

5. Finally, show on your plans which of your installed electric lights will be used under the various supplementary conditions.

Evaluation Tool for Criterion A - "% Savings from Daylighting"

1. From the information you listed in the two previous Evaluation Tools, complete this table:

Trial 1	Column 1 sf of elec light needed	X	Column 2 Watts/sq.ft. of electric light	X	Column 3 Hours per day of operation under these conditions	=	Column 4 Total watt hours electric light

Night: summer:

winter:

Least Daylight:

      summer:

      winter:

Intermediate
Daylight:

      summer:

      winter:

Most Daylight:

      summer:

      winter:

Total   Total Watt hours Summer:
hrs/day operation _____   per 24 hrs   Winter:

10.0 LIGHTING FINALIZATION

10.2 DAYLIGHTING PERFORMANCE

Trial 2	Column 1 sf of elec light needed X	Column 2 Watts/sq.ft. of electric X light	Column 3 Hours per day of operation under these conditions	Column 4 Total watt hours = electric light

Night: summer:

    winter:

Least Daylight:

    summer:

    winter:

Intermediate
Daylight:

    summer:

    winter:

Most Daylight:

    summer:

    winter:

Total
hrs/day operation _____

Total Watt hours summer:
per 24 hrs   winter:

2. Compare the actual internal gains that will result
from electric lights with your estimate in the Climate
exercise, section 3.6.

$$\frac{\text{total summer watt hours per 24 hours (Col. 4)}}{\text{(sq.ft.) (24 hours)}} = \text{watts per sq.ft.}$$

Watts per sq. ft. = _____

10.2 DAYLIGHTING PERFORMANCE

What did you assume in the Cooling Exercise as the watts/sq.ft. due to electric lighting?

3. To estimate the energy saved by daylighting compare your total watts per 24 hours with what that would be if <u>no</u> daylighting were admitted:

Total electric = Night sq.ft. x watts/sq.ft. x total hrs/day operation = _____ watt-hrs.
          (Col. 1)   (Col. 2)

% energy = 100-100 x $\frac{\text{(total watt hours per 24 hours, Col. 4)}}{\text{total electric watt hours per 24 hours}}$ =
  savings

                           summer _____ % Trial 1
                           winter _____ %
                           summer _____ % Trial 2
                           winter _____ %

Does your first trial design meet Criterion A for Daylighting Performance?

Have you provided night time lighting not exceeding 2 watts/sq.ft.? Have you provided daytime electric lighting of less than 1 watt/sq.ft.?

If not, what will you change for your second trial?

10.3 BACKUP SYSTEM INFORMATION

In finalizing you calculated the % electrical energy savings due to daylighting for a given sky condition and season. This information can be expanded in several ways.

You can model the predominant sky conditions for all seasons in your climate, thereby being able to calculate the electrical energy savings on an annual basis rather than a single condition.

Reducing the use of electric lighting also means a reduced cooling load (1 watt = 3.41 Btu's), which will affect the sizing of your passive cooling design components, as well as any mechanical cooling system.

Design of electric lighting systems involves many more issues than watts per sq.ft. We have included some

---

Electric light sources have widely varying color rendition characteristics. See MEEB Table 19.26, p. 853; and Fig. 18.56 p. 768 and Fig. 18.52 p. 764.

Lamp Type	Application
Incandescent	1. Decorative display lighting 2. Religious worship halls 3. Work closets or other very confined spaces 4. Stage spotlighting 5. Tasks which require a small light source
Fluorescent	1. Office and other relatively low-ceiling applications 2. Flashing advertising signs 3. Islands at service stations 4. Display cases in stores 5. Desk lamps 6. Classrooms or training centers 7. Cafeterias
High-intensity discharge	1. Stores and some office areas 2. Auditoriums 3. Outdoor area lighting 4. Outdoor floodlighting 5. Outdoor building security lighting 6. Marking of obstructions

Fig. 10-1 Examples of Energy-Conserving Lamp Applications from Energy Conservation Standards by Dubin and Long. Copyright (c) 1978. Used with the permission of McGraw-Hill Book Company.

See also William Lam, Perception and Lighting as Formgivers for Architecture, 1977.

Finally, a valuable graphic tool for electric lighting design is to render a night perspective of your design (interior and exterior).

----

for daylighting contours, enlarged plans: if 1/8" = 1'-0", 1 square = 10'.

The fact that we probably have little idea of the amount of water that we use during the course of a day is some indication of its value to us. This is understandable in part because water is readily available and inexpensive. On the other hand, it is absolutely indispensable to us and the ecosystems which support us. In addition, the level of consumption in most of the U.S. seems somewhat irresponsible in a world where most people don't have access to piped water.[1]

Although water itself is inexpensive and readily available, the path it travels is often long, expensive, and environmentally damaging. Murray Milne in Residential Water Consumption describes the process. "Imagine how long it took to move through the stream, lake, river reservoir, spillway, aqueduct, pump, main, lateral meter, pipe, valve, tube and faucet before it finally flows out to fill your glass. Now think about the rest of the trip; down the drain, through the trap, out and down into a network of merging pipes and tunnels, and finally through a treatment plant and back into the water course."

Small amounts of our direct consumption of water are actually used for drinking or cooking (approximately 5% of residential consumption). Most of it is used as a transport medium for waste. One person, on the average, in the U.S. contaminates 13,000 gallons of fresh water too flush 165 gallons of human body waste a year.[2]

Indirectly we also use tremendous amounts of water. Every time an architect specifies a ton of structural steel for a building he/she also specifies the consumption of 37 tons of water.[3]

With the exception of some reduced costs, because of reduced volumes, most reasons for conserving water involve considerations that are outside the bulding and frequently off the site. Even though most places in the U.S. have abundant water supplies there is ample evidence to support the need for conservation. Most of these reasons are related to the costs, other than water, of the water and waste systems. for example:

A) energy use for pumping is the fastest growing item in water system budgets. Some experts believe that within ten years energy shortages will force the institution of strict water conservation program in many parts of the U.S.

B) It is likely that most homeowners spend more to heat water than they do to acquire the water. Because 4% of our national energy use is for domestic water heating, conserving water means conserving energy.

C) The costs (economic, materials, energy, etc.) of building additional waste treatment plants is directly related to the volume of contaminated water to be treated.[4]

---

1. 70% in 1972 according to the World Health Organization estimates.

2. Carol Stoner, ed., Goodbye to the Flush Toilet, p. vii.

3. B. Vale, The Autonomous House, p. 139.

4. These examples are from Murray Milne, Residential Water Consumption.

12.0 WATER AND WASTE

12.1 INTRODUCTION

12.1 INTRODUCTION --

Just as you set the context for your design with climate analysis for the thermal exercises, some preliminary analysis of the water and waste context must be done.

You start by exploring the water supply and the soil conditions for your location, the hot water usage and minimum number of fixtures for your building program, and resulting total water use for conventional fixtures.

You then select the one most appropriate combination of fixtures for the conditions you have analyzed.

Contents

12.2 ANALYZING ---

GOAL

Determine the importance and feasibility of conserving
water in your site/building design.

A. Analysis Tool - "Annual Rainfall Comparisons"
Your annual rainfall (from climate data) is ___ inches/
year. From MEEB, Fig. 2.20, p. 51, determine whether
your annual rainfall seems unusually low, about average,
or unusually heavy.

B. Analysis Tool - "Soil Characteristics"
Soil conditions are a major influence on the required
size of drainage fields. Based on generalizations about
regional soil types assume these site soil
characteristics:

Location	Soil Type	Time Required for Water Level to Drop 1" in a Test Hole:
Madison Dodge City Salem	Medium to good drainage; Sandy loam or sandy clay	5 minutes
Phoenix	Excellent drainage; Coarse sand or gravel	1 minute
Charleston	Medium to poor drainage; Clay with large amounts of sand or gravel	10 minutes

Figure 12-1 Soil Characteristics

C. Analysis Tool - "Required Minimum Plumbing Fixtures"
The population of your buildings, combined with their functions, determines the <u>minimum</u> number of plumbing fixtures required in toilet rooms (see 4.6 Analysis Tool A for population).

Factory population _____

Office population _____

Total   _____

Men and women are equally likely to be hired as office and factory workers here, so up to two-thirds of this total should be able to be accomodated by either sex's toilet room. Use MEEB, table 11.8, p. 459-461 to determine minimum number of fixtures.

	men	women	both
water closets	____	____	____
urinals	____	____	____
lavatories	____	____	____
drinking fountains	____	____	____
vending machines	____	____	____
showers	____	____	____
other: slop sink	____	____	____

D. Analysis Tool - "Conventional Fixture Water Usage Estimate"
Determine the average consumption per person assuming conventional fixtures. Use the population determined above, multiplied by the average gallons per person. If your building has use characteristics which indicate a greater need for fixtures than this procedure would lead you to believe then you should design for a larger number. Use MEEB, table 11.9, p. 466 to determine gallons per person.

	Gallons per person per day	Population	Total Gallons per day
Office	_____	_____	_____
Factory	_____	_____	_____
		TOTAL	_____

QUESTIONS

You now have an idea how much water is available in your location (input), how much water might be used, and how effectively waste may be processed on your site (output). Use this information to answer the following questions.

A. Viewed from the standpoint of availability, is water conservation important in your location?

B. Viewed from the standpoint of conventional waste treatment on site (septic tank and drainfield), is water conservation important in your location?

C. How does your need to conserve water compare to the other locations?

E. Analysis Tool - "Water and Waste Flow Diagrams"
The flows in a conventional system using 1000 gal. per day might look like this:

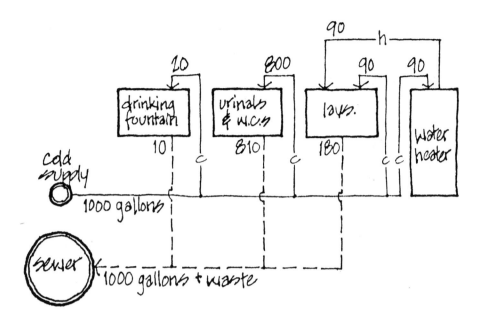

Figure 12-2 Typical Conventional System Flows

A septic tank system should conserve water compared to a
conventional system. Assemble a system which achieves
what you feel is an appropriate unit of conservation for
your area by selecting fixtures from the following list,
and putting them together in a diagramatic system for
your building:

	Supply fixture units (MEEB p. 415)	Gals/day* For this building group
Conventional:		
flush valve water closet	10	200
flush tank water closet	5	100
flush valve urinal	5	100
flush tank urinal	3	60
lavatory	2	40
shower	4	80
service sink	3	60
Conserving:		
conserving wc, valve	--	120
conserving wc, tank	--	60
conserving urinal, valve	--	40
conserving urinal, tank	--	24
compost toilet	--	0
conserving lavatory	--	16

Figure 12-3 Typical Plumbing Fixtures

---12-—

For examples of alternative waste and water systems see:
MEEB Chapter 2(p. 50-52) and Chapter 9.
B. Vale, The Autonomous House
C. Stoner, ed., Goodbye to the Flush Toilet
M. Milne, Residential Water Conservation (this is a good
general review of water conservation components)
J. Lackie, et.al., Other Homes & Garbage, Chapter 5 -
Waste; Chapter 6 - Water.

*Note: The total gallons per day (1,000 in the above
diagram) was divided by the total supply fixture units to
get a general indication of the gallons per day per
fixture unit. This assumes that water supply fixture
units are a reasonable approximation of water demand
(they consider use patterns and flow rates) for each
fixture. Conserving fixture gallonages were determined
as follows:

$$\text{WC: } \frac{\text{conserving model 3 gal/flush}}{\text{conventional} \quad \text{5 gal/flush}} = 0.6 \text{ conventional usage}$$

$$\text{Urinals: } \frac{\text{conserving model} \quad \text{1 gal flush}}{\text{conventional} \quad \text{2.5 gal flush}} = 0.4 \text{ conventional usage}$$

$$\text{Lavatories: } \frac{\text{conserving model 2 gpm flow rate}}{\text{conventional} \quad \text{4.5 gpm flow rate}} = 0.4 \text{ conventional usage}$$

QUESTIONS

A. What percent reduction of water use did you achieve compared to a conventional system?

B. Compared to the conventional system what percentage of your usage was the following:

conventional inputs		your system
rain	0	_____
city main	100%	_____
other	0	_____
outputs		
grey water	0	_____
black water	100%	_____
city sewer	100%	_____
other	0	_____

12.3 CONCEPTUALIZING _____

GOAL

A. Use rainwater where possible.
B. Decrease the use of fresh (potable) water and the production of waste water.

DESIGN STRATEGIES

A. Site Scale
Minimize rain run off to encourage soil recharging.
B. Cluster Scale
Water moving downhill requires no pumping energy or pressurized piping; store water at higher elevations than it is used, and treat waste at still lower elevation.

C. Building Scale
"Service" spaces, which include toilet and mechanical rooms, can help organize the clustering of buildings and the other spaces they serve.

Toilet spaces have special environmental characteristics, which usually suggest acoustic isolation from other spaces, and a ventilation pattern unshared by most other spaces.

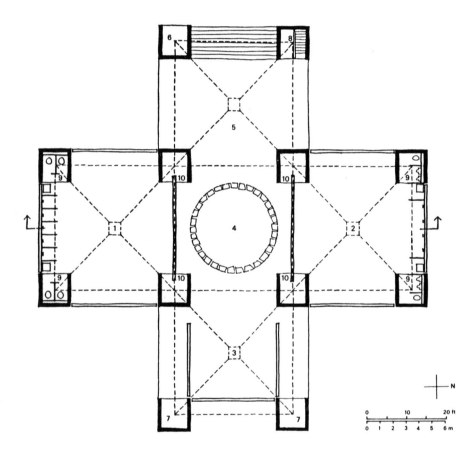

Kahn, Trenton Bath House. (Ronner, 1977, p. 94)

D. Component Scale
Plumbing fixtures are often grouped together within buildings for cost savings in plumbing, and for convenience in clustering environments containing water usage.

DESIGN

Choose an existing building/site which has a clear conceptual approach, at any scale, to achieving these goals. Document your choice as follows:

12.4 SCHEMING

A. identify the location, program, architect (if any) and the source of your information.
B. Include xerox copies or drawings (whatever is quick and easy for you) to explain the design.

EVALUATION

Evaluation Tool - "Building Response Diagrams"
Diagram how this design is organized to achieve these goals.

12.4 SCHEMING

GOALS

A. Use the site and buildings to the degree possible to heat water, collect rainwater and dispose of wastes.
B. Arrange the elements of the water/waste system within the cluster and buildings to minimize plumbing.

CRITERIA

A. Provide enough solar collectors to heat at least 50% of the hot water used in your buildings.
B. Provide adequate collection area and water storage for your buildings' needs. (If you elect to use this method)
C. Size the septic tank and drainfield to adequately treat the waste generated in your building considering your soil conditions.
D. Provide enough composting toilets to meet your buildings' needs.

DESIGN STRATEGIES

A. Site Scale
Locate the drainfield and septic tank at a lower elevation than the building.

Utilize the naturally irrigated area over the drainfield as a flat unpaved open area for view or active use (excluding cars, etc.).

Provide adequate clearances to other site elements.

Minimum Horizontal Distance	Building Sewer	Septic Tank	Disposal Field	Seepage Pit or Cesspool
Buildings or Structures	2 ft (.6 m)	5 ft (1.5 m)	8 ft (2.4 m)	8 ft (2.4 m)
Property line adjoining private property	Clear	5 ft (1.5 m)	5 ft (1.5 m)	8 ft (2.4 m)
Water supply wells	50 ft (15.2 m)	50 ft (15.2 m)	100 ft (30.5 m)	150 ft (45.7 m)
Streams	50 ft (15.2 m)	50 ft (15.2 m)	50 ft (15.2 m)	100 ft (30.5 m)
Large trees	---	10 ft (3 m)	---	10 ft (3 m)
Seepage pits or cesspools	---	5 ft (1.5 m)	5 ft (1.5 m)	12 ft (3.7 m)
Disposal field	---	5 ft (1.5 m)	4 ft (1.2 m)	5 ft (1.5 m)
Domestic water line	1 foot (.3 m)	5 ft (1.5 m)	5 ft (1.5 m)	5 ft (1.5 m)
Distribution box	---	---	5 ft (1.5 m)	5 ft (1.5 m)

Figure 12-4 Location of Sewage Disposal System, table I-1, p. 186, Uniform Plumbing Code.

Drainfields may not be used for parking, and they should be free from trees and shrubs because roots may block the lines.

B. Cluster Scale
Where rainfall is to be collected and stored for use, consider arranging roof slopes so that the rain water falling on the roof readily converges upon a single location, above the cistern inlet.

Celebrate the flow of water from collection area to storage area.

Group the things that use water in the same area.

C. Building Scale
If the domestic hot water solar collectors aren't integrated with the roof, consider using them to provide the roof with an exterior shaded area.

D. Component Scale
Provide the collectors with an unshaded south facing area. Set them at an angle above horizontal equal to or less than your latitude.

DESIGN

Document your design as follows:

A. Select and circle the water and waste system you will use in your buildings/site.

----

supply $<$ $\begin{array}{l} \text{city} \\ \text{site} < \end{array}$ $<$well

cistern & collection area

waste $<$ $\begin{array}{l} \text{city} \\ \text{site} < \end{array}$ septic tank & drainfield

composting toilets

B. Site or cluster plan showing parking, drives and depending on what system you are using septic tank, drainfield, cistern, well, supply and waste line from city lines, rain collection areas.
C. Floor plans showing location of hot water storage toilet rooms, and location of plumbing lines.
D. Building section(s) showing solar collectors, hot water storage, location of plumbing lines, and if you have them, cisterns and composting toilets.

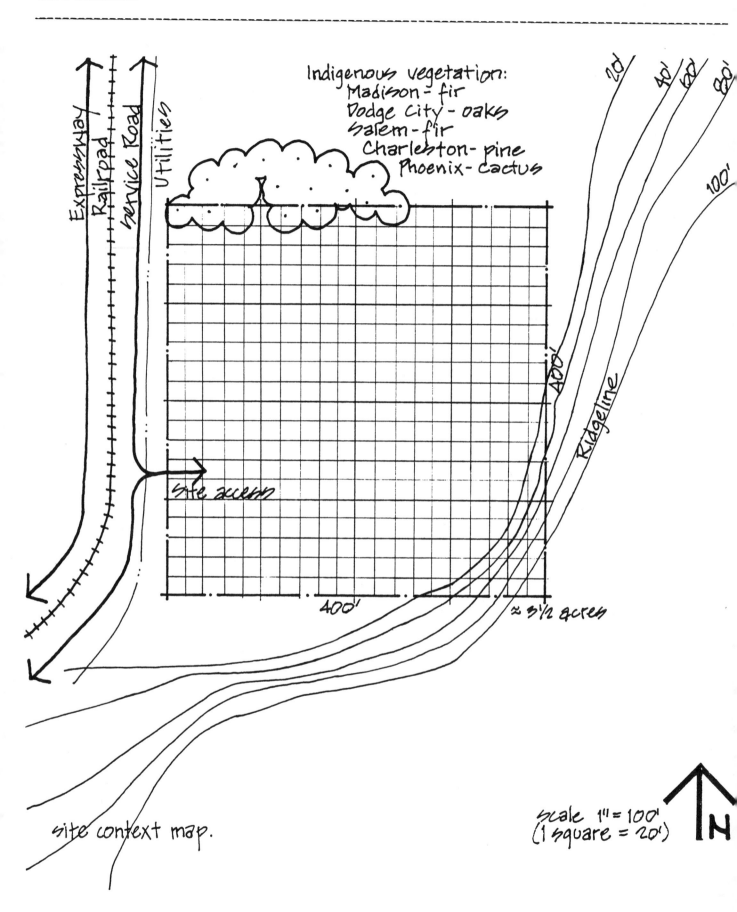

Indigenous vegetation:
Madison - fir
Dodge City - oaks
Salem - fir
Charleston - pine
Phoenix - cactus

Expressway
Railroad
Service Road
Utilities

Ridgeline

site access

400'

400'

≈ 3½ acres

scale 1" = 100'
(1 square = 20')

N

site context map.

----

cluster site plan: 1" = 40'-0".

for plans, sections, & elevations: if 1/16"=1'-0", 1 square = 10'.

----

for plans, sections, & elevations: if   1/16"=1'-0",   1 square = 10'.

---12--

for plans, sections, & elevations: if   1/16"=1'-0",   1 square = 10'.

12.0 WATER AND WASTE

12.4 SCHEMING

----

for plans, sections, & elevations: if   1/16"=1'-0",   1 square = 10'.

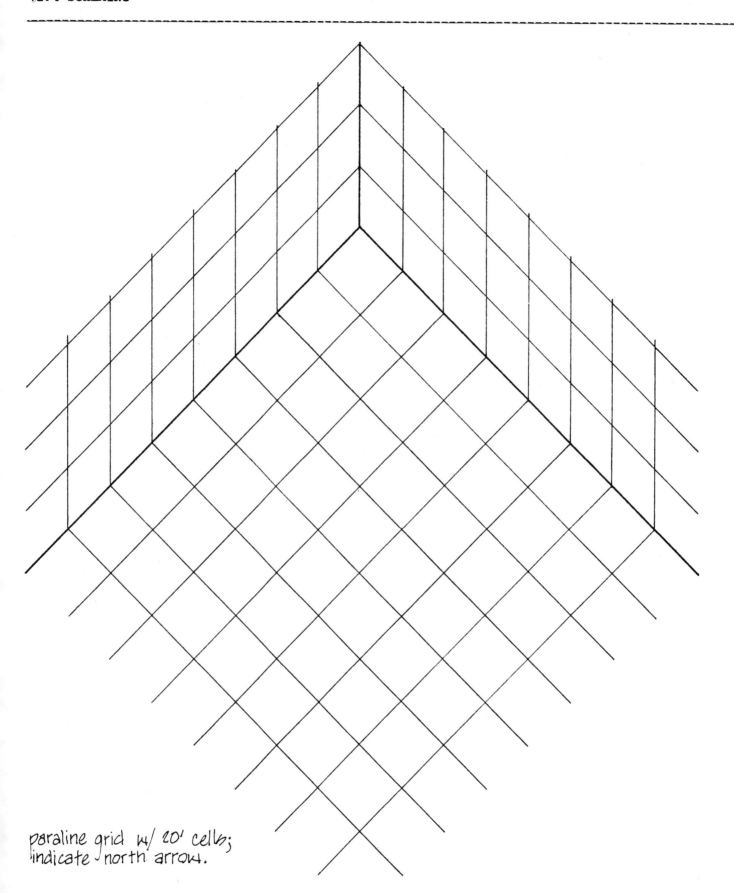

paraline grid w/ 20' cells;
indicate north arrow.

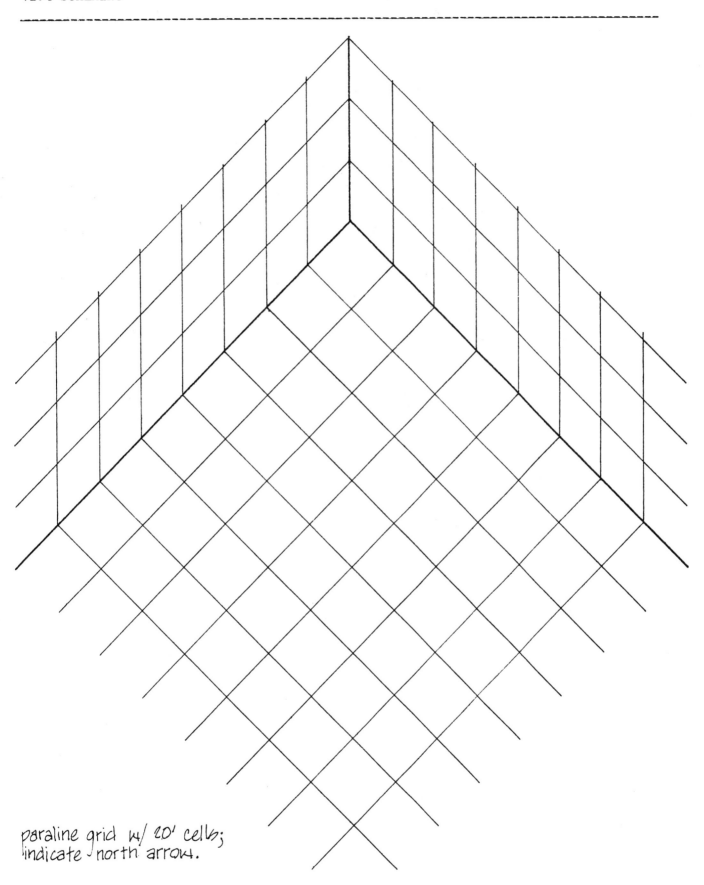

paraline grid w/ 20' cells;
indicate north arrow.

---12---

EVALUATION

A. Criterion A: Provide enough solar collectors to heat at least 50% of the hot water used in your buildings in the month with the most solar radiation.

Evaluation Tool for Criterion A - "Solar Collector Sizing"

sf of collector per 100 employees*

Phoenix	30 to 45
Charleston	42 to 63
Salem	50 to 75
Madison	50 to 75
Dodge City	36 to 54

Figure 12-5 Estimated Solar Collector Sizing
*Based on 50 to 75 sf for Salem; adjusted collector area by a percent equal to the increase in the annual radiation ona horizontal surface for the other locations. This neglects ambient temperature and actual collector tilt. Thus, a very rough rule of thumb!

Does your first trial meet the design criteria?

If not, what will you change for your second trial?

B. Criterion B: Provide adequate collection area and water storage for your buildings' needs.

Evaluation Tool for Criteria B - "Cistern Sizing"
Monthly usage = _____gallons/day x 31 days = _____gallons/month.

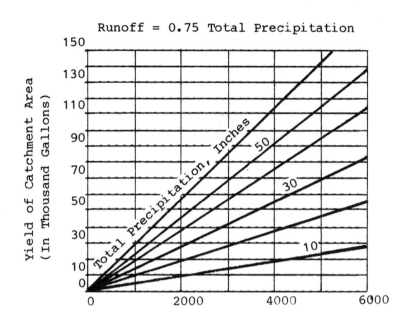

Runoff = 0.75 Total Precipitation

Horizontal Area of Catchment (In Square Feet)

Figure 12-6 Rainfall Catchment Area Yield (Lehr, et al
Domestic Water Treatment, p. 46)

Determine the average yield of the catchment area
(= approximate ground floor area) for each month.

Aver. inches rain		Gallons collected
Jan	_____	_____
Feb	_____	_____
March	_____	_____
April	_____	_____
May	_____	_____
June	_____	_____
July	_____	_____
August	_____	_____
Sept	_____	_____
Oct	_____	_____
Nov	_____	_____
Dec	_____	_____

---

cistern size $=$ (sum of average consumption for months $-$ (yield from catchment
(in gallons)   in which consumption exceeds catchment yield)   area in those months)

When the collectable rainfall from the roof <u>exceeds</u> the
consumption on a monthly basis store water in a cistern
from the excess capacity months for the below capacity
months.

When collectable rainfall from the roofs doesn't usually
exceed consumption for most months, store water in a
cistern which is typical of a maximum one month rainfall.
Maximum one month rainfall catchment yield = cistern size
in gal.

Does your first trial meet the criteria?

If not, what will you change for your second trial?

C. Criterion C: Considering your soil conditions, size
the septic tank drainfield to adequately treat the waste
generated in your building.

**Evaluation Tool for Criterion C - "Septic Tank and Drain
Field Size"**
To determine the square footage of drainfield required,
multiply the number of fixtures* determined in analysis
by:

34 for coarse sand and gravel
43 fine sand
68 sandy loam or sandy clay
102 clay with considerable sand or gravel
153 clay with small amount of sand or gravel

(In poor soils, a stand-by drainage field "reserve" is
sometimes required, equal in size to the original field.)

Tne number of fixtures multiplied by 170 equals the
minimum septic tank capacity in gallons.

Does your first trial design meet the criteria?

If not, what will you change for your second trial?

----

*Note: Derivation of the Septic Field Rule of Thumb is based on the Uniform Plumbing Code 1976 edition, table I-2, p. 187, table I-4, p. 179, and table 4-1, p. 38. Fixture units were determined by multiplying the number of fixtures by 4, which is an average number between lavs at 2 and valve WC's at 6. Codes seem to require about an equal number of lavs and WC's for most buildings.

Fixture units were converted to minimum septic tank capacity in gallons by multiplying by 42.5. 42.5 is the average of 50 (which is the minimum septic tank capacity in gallons, 750, divided by the maximum fixture units served, 15, from table I-2) and 35 (which is the minimum septic tank capacity in gallons, 3500, divided by the maximum fixture units served 100).

.425 was divided by 100 to convert septic tank capacity into sf of leaching area per 100 gallons of septic capacity. See table I-4.

20 sf of leaching area per 100 gallons of tank capacity was used for porous soils and 90 sf of leaching area per 100 gallons of tank capacity for impervious soils. See table I-4.

Therefore:

Fixtures x 4 (av. fixture units) x 42.5 (conversion to septic tank capacity) ÷ 100 (conversion to leaching area in S.F.) x 20 for poor soils or 90 for good soils.

$$4 \times 42.5 \div 100 \times 20 = 34$$
$$4 \times 42.5 \div 100 \times 90 = 153$$

fixtures x 34 in good soil = sf leaching field
         x 153 in poor soil = sf leaching field

D. Criterion D: Provide enough composting toilets to meet your buildings' need.
Evaluation Tool for Criteria D - "Compost Toilet Capacity."
3 to 10 persons per unit in a residential application.
(Stoner, Goodbye to the Flush Toilet, p. 132)

---12---

Evaluation Tool for Criteria D - **"Removing Compost"**
Draw a diagramatic section through the toilet room
showing how compost is removed and how access to the
removal area is achieved:

Did your first trial meet the design criteria?

If not what will you change for the second trial?

12.5 DEVELOPING ---

GOAL

Design toilet facilities which are both adequately sized
and comfortable for all users, including the handicapped.

CRITERIA

A. Your design meets handicapped requirements of the 1979
Uniform Building Code.
B. Vision into either toilet room is blocked beyond the
entry way.

DESIGN STRATEGIES

A. Component Scale
Clear space within a toilet room should be sufficient to inscribe a circle of 5' diameter. Put fixtures back to back to reduce the length of pipes and increase ease of installation.

DESIGN

Develop the design of your toilet rooms to achieve the goal. Document your design as follows:

Floor plan of toilet rooms @ 1/4" = 1'-0"

for daylighting contours, enlarged plans: if 1/8" = 1'-0", 1 square = 10'.

EVALUATION

A. Criterion A: Meets handicapped requirements for the 1979 Uniform Building Codes.

Evaluation Tool for Criteria A - "Code Review"
The 1979 Uniform Building Code specifies (Section 1811):
<u>Doorways to toilet rooms</u>: 30" door width, with 44" clear beyond each side of door.
<u>Clear space within toilet room</u>: sufficient to inscribe a circle of 5' diameter. Doors may encroach this circle by a maximum of 1' when open.
<u>WC stall</u>: one for each sex must have 42" wide stall, with 48" clear in front of stool within stall. (Stall door cannot encroach on this space, so it usually swings outward.)
A minimum of 44" clear access is needed in front of the stall (into which door may swing).
<u>Lavatory</u>: under one lavatory, a space 26" wide, 27" high and 12" deep.

Does your first trial design meet the criteria?

If not, what will you change for your second trial?

B. Criterion B: Vision into either toilet room is blocked behind the entry way.

Evaluation Tool for Criterion B - "Sight Lines"
Draw sight lines into your toilet rooms as if the doors were not in place. Show the result on the 1/4"=1'-0 plan.

Does your first trial design meet criterion B?

If not, what will you change for your second trial?

--12-28

12.6 FINALIZING _____

GOAL

Insure that the major parts of your systems are
accurately sized so that they perform well and conserve
water.

CRITERIA

A. Supply and waste pipes are adequately sized to carry
their expected loads.
B. Drainfield and septic tank are adequately sized to
carry the expected loads of the conserving system.
C. Solar collectors should be sized to supply 100% of the
hot water requirement in the month with the most solar
radiation.
D. The cistern is sized to meet the needs of the
conserving system.

DESIGN

Draw a schematic diagram of your water and waste system
showing the major elements and the flow of water through
the system in gallons.

EVALUATIONS

A. Criterion A: Supply and waste pipes are adequately sized to carry their expected loads.

Evaluation Tool for Criterion A - "Supply Pipe Sizing"
For the water/waste system you have chosen, size the potable water supply system and the waste lines to the septic tank (or filter).

Supply: (MEEB, Example 10.4, p.413 may be helpful)

Given: main (or pump) pressure 40 psi
         main location: at street 6' below grade
         (pump location: submersible, in cistern)
         total equivalent length of pipe = 150% actual
         length

1. Estimate the diameter of the supply pipe _____ (to develop your ability to guesstimate)

2. Total actual length of pipe, main (or cistern) to furthest fixture:

	Trial 1	Trial 2
3. Determine the supply fixture units (MEEB, Table 10.10)	_____	_____
4. Determine flow required gal/min (MEEB, Fig. 10.27)	_____	_____
5. Pressure available to overcome friction in piping:		
main or pump pressure	_____	_____ psi
water meter friction (MEEB, Fig. 10.28)	_____	_____ psi
remainder	_____	_____ psi
fixture min. pessure (MEEB, Tab. 10.9)	_____	_____ psi
remainder	_____	_____ psi
static head (height x .434 constant)	_____	_____ psi
remainder, available to overcome friction	_____	_____ psi
unit friction loss, psi/100' of pipe	_____	_____ psi

= (pressure available to x 100' constant
   overcome friction)
   _____
      total equivalent length of piping

----

6. Size (cast iron) (MEEB, Fig. 10.26)_____"diam.

7. Velocity is _____ft/sec

Does your first trial design meet the criteria?
If not, what will you change for your second trial?

Evaluation Tool for Criterion A - "Waste Pipe Sizing"
(MEEB Examples 11.1 and 11.2 may be helpful.)
1. Locate and diagram waste stacks and list the number of
fixtures on each branch; size stacks and branches from
MEEB, Table 11.5.

Branch #1				Branch #2				Branch #3			
Men's lavatories				Women's lavatories				Common wall			
Fixt. type	#	Waste FU each	Total FU	Fixt. type	#	Waste FU each	Total FU	Fixt. type	#	Waste FU each	Total FU
Total		_____		Total		_____		Total		_____	
Size horiz branch		_____		Size horiz branch		_____		Size horiz branch		_____	
Capacity		_____		Capacity		_____		Capacity		_____	

Note: Capacity is the total possible FU that this size line can handle (MEEB, 11.5).

2. Minimum size vertical stack to serve all facilities (MEEB, Tab. 11.5) is determined by the sum of waste FU:

_____ total F.U.   _____ inch stack   capacity _____ FU

3. Minimum size of vent required to serve total toilet room fixtures (MEEB, Table 11.6):   _____ in

_____ ft max length

----

4. Minimum size building drain, using a <u>minimum</u> 1/8" fall/ft. (MEEB, Table 11.4): _____in

capacity _____FU

5. Given the horizontal distance between the toilet room and the septic tank, what **vertical drop** is required to maintain the minimum slope of this drain line?

_____ft horizontal distance x 1/8" fall/ft = _____inch vertical drop

Is this compatible with elevations on your site? Remember, trenches in drainage field will be even lower, to maintain 1/8" fall per foot.

B. Criterion B: Drainfield and septic tank are adequately sized to carry their expected loads.

Evaluation Tool for Criterion B - "Sizing Drainfield and Septic Tank"
The total gallons/day to be handled by your septic tank and drainfield combination, from your conservation potential analysis, or schematic:

_____gallons/day total _____waste FU total

<u>Septic tank size:</u> MEEB, Table 11.10, based on number of fixture units:

_____drainage FU, needs _____gallons septic tank

<u>Drainage field size</u> MEEB, Table 11.12 lists the length of tiles needed for various trench sizes, in various soil conditions. By assuming either 24" or 36" wide trenches, the length and spacing of the tile lines and their trenches can be determined from MEEB, Tables 11.12 and 11.13.

Determine the overall size of your drainage field:

Consider the possibility of building (and drainage field) expansion. Show on the site plan how your drainage field meets the restrictions placed on its location (see MEEB, Fig. 9.7, p. 358).

How does this final size compare with the estimate you made earlier?

C. Criterion C: Solar collectors are sized to supply 100% of the hot water requirement in the month with the most solar radiation.

Evaluation Tool for Criterion C - "Estimate Hot Water Usage"
In this building group, hot water is only used in the lavatories. Hot water for washing hands and face becomes uncomfortably hot at about 110°F. Assuming that energy-conscious management has resulted in the water heater thermostat set no higher than 120°F, that an average water usage temperature is about 100°F, and that cold water enters the buildings at GWT (see below for your climate's GWT, or ground temperature),

X gallons @ 120° + Y gallons @ GWT = Z gallons at 100°
(X+Y=Z=100%)

$$\% \text{ hot} = x = \frac{100-GWT}{120-GWT}$$

% cold = y = 100% - x

TOTAL hot water = _____ gallons/day in lavatories x _____ % hot = _____ gallons/day, hot (120°)

TOTAL Btu/day: ___ gallons/day hot x 8.33 lb/gal x(120° - ___ GWT = _____ Btu/day.

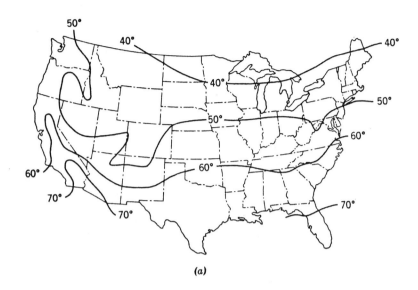

(a)

Figure 12-7   Ground Water Temperature, from MEEB, Fifth
Edition 1971.   Used by permission.

Evaluation Tool for Criterion C - "Sizing DHW Solar Collectors"

Listed below are each climate's average insolation values
for the worst and best solar months, at several tilt
angles (from SOLMET data, distributed by U. S.
Government)

	Worst Month				Best Month			
	Horiz.	30°	60°	Vert.	Horiz.	30°	60°	Vert.
Madison	389	429	525	477	1933	1364	1010	505
Dodge City	731	857	1047	952	2294	1565	1188	447
Salem	277	281	343	312	2141	1499	1166	555
Phoenix	931	1104	1214	1104	2737	1765	1073	346
Charleston	744	732	789	665	1859	1251	876	417

Figure 12-8   Solar Radiation Data

You need to choose a collector tilt angle that will:
 - be close to optimum in the best month (high insolation
and collector efficiency), and
 - be somewhat better than the lowest insolation in the
worst month, since hot water is also desirable in winter.

To do this, find the collector size required to provide 100% solar heating in the best month; then check its performance in the worst month (the one with the least radiation, that's the coldest).   Trial 1   Trial 2

Collector tilt angle chosen:

To size the collector:
1. Collector efficiency: this graph is a typical efficiency curve for selective surface, single glazed flat plate collectors of moderately low cost.

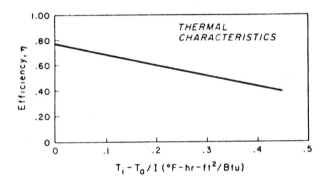

Figure 12-9 Collector Efficiency Curve

T_i = input temperature, a little lower than the thermostat setting in your water heater.
(summer: higher, since a hot sunny day will heat your tank well above this setting)
(winter: somewhat lower, since the setting will be the maximum temperature, and you are feeding your maximum collectors with colder water from the bottom of the tank.)

T_a = outdoor daytime temperature (a little below the maximum temperature for the month, from climate analysis)

I = insolation, in Btuh. Interpolate if necessary from the insolation table in this exercise, to find the Btu/day; the hourly insolation, during collection hours, is about 18% of the daily total.

	Trial 1	Trial 2
Best month collector efficiency	_____%	_____%
Worst month collector efficiency	_____%	_____%

----

2. System Efficiency: Your collector will not usually operate during every hour that the sun is above the horizon; it will only collect during hours when enough sun enters the collector to heat water above the temperature at the bottom of the water heater tank. Therefore, reduce your collector efficiency by about 20% to account for hours of too-low-insolation, as well as heat losses in the lines to and from your collector.

System efficiency = (0.8) x (collector efficiency) =

<u>Trial 1</u>:

____% Best month ____% Worst month

<u>Trial 2</u>:

____% Best month ____% Worst month

3. Collector size, <u>Trial 1</u>: Best month

$$\frac{\text{_____Btu/day needed for hot water}}{\text{_____Btu/sf-day (summer) x summer system effic.\_\_\_\_}} = \text{\_\_\_\_ sf (summer optimum) collector area}$$

Now <u>check</u> to see what % solar water heating this will produce in your worst month:

____Btu/sf-day (winter) x ____winter system effic. x ____sf collector area = ____Btu/day

What % of total daily Btu need is this? ____%

<u>Trial 2</u>: Best month,

$$\frac{\text{_____Btu/day needed for hot water}}{\text{\_\_\_\_\_Btu/sf-day (summer) x summer system effic.\_\_\_}} = \text{\_\_\_ sf (summer optimum) collector area}$$

Now <u>check</u> to see what % solar water heating this will produce in your worst month:

____Btu/sf-day (winter) x ___winter system effic. x ____sf collector area = ____Btu/day

What % of total daily Btu need is this ____%

For your collector size and tilt, what average yearly percent solar water heating might you expect, based on simple interpolation between the best and worst months?

How does this collector area compare to your earlier estimate?

D. Criterion D: The cistern is sized to supply the needs of your conserving system design.

Evaluation Tool for Criterion D - "Cistern Sizing"
(See MEEB, p. 50-52)
1. Calculate catchment (roof) area: _____sf Trial 1
                                         _____sf Trial 2

2. Trial run on sizing - add monthly rainfall in gallons and identify your 'rainy season'. 1 cf = 7.48 gal

	Avg. Monthly Rainfall (in) ÷ 12 = (ft.)	Roof Area (sf)	Trial 1 Rain Collected (cf) x 7.48 = (Gal)	Roof Area (sf)	Trial 2 Rain Collected (cf) x 7.48 = (Gal)
Jan					
Feb					
Mar					
Apr					
May					
Jun					
Jul					
Aug					
Sep					
Oct					
Nov					
Dec					

                          TOTAL _____(Gal)   TOTAL _____(Gal)

What is your rainy season?_____

3. Cistern size should accomodate at least minimum capacity (See MEEB, p. 52) plus the average rainy season rainfall. (Use and bypass should mean this size will accomodate above average seasons.)

Minimum capacity _____gal _____cf

Average rainy season _____gal _____cf

TOTAL _____gal _____cf

What are dimensions of your cistern? ___ x ___ x ___

4. Compare yearly supply to demand

Supply _____gal Demand _____ gal

Is this ok? If supply is small, how can you increase supply, or reduce demand?

5. Compare use to supply on a month to month basis. Start with the month after the "rainy season" and assume a 90% full cistern at that time.

			Trial 1		Trial 2	
Month	Use (Gal)	Rainfall (Gal)	Balance* (Gal)	Backup Source (Gal)	Balance* (Gal)	Backup Source (Gal)

*Must not go below minimum capacity.

12.0 WATER AND WASTE

12.6 FINALIZING

Did you run dry in your first trial design?   Why?

If so, what will you change for your second trial?

At first glance, acoustic influences on building and site design seem unrelated to energy considerations. Heating, cooling, daylighting and electric lighting have energy consumption as a central theme; hot water energy is a consideration in water/waste system design. Where are the energy implications in acoustics?

A closer look at some basic acoustical characteristics, however, recalls some familiar terms. Absorptive materials - such as thermal insulation - soak up sound, producing "dead" spaces by reducing the length of time that a sound remains available in a space. Massive and dense materials - such as thermal mass - reflect sound, producing acoustically "live" spaces as sounds stay available longer as they are bounced rather than soaked up. Because they reflect sound, massive materials make good barriers to sound between spaces. Thus a distribution of hard and soft surfaces can have both thermal and acoustical implications; their color and texture involves lighting considerations as well.

More obviously, openings that admit ventilation will also admit sounds; cooling strategies that depend heavily on ventilation will often need close acoustical scrutiny.

The principle of shadowing reappears. One building's acoustic shadow can aid another building's quiet environment. The issue of comfort is also familiar. Acoustics has its comfort criteria and its evaluation procedures to predict success, just as did heating, cooling and lighting.

In this exercise, two basic acoustical situations are examined: the acoustical environment within spaces and sound transfer between spaces. Sound within a space will involve considerations of both quality and quantity of "background sound", liveness or deadness, and be heavily influenced by room volume and surface materials. Sound transfer between spaces also involves acceptable quantities of background sound and compatibility of functions in adjacent spaces. It will be heavily influenced by openings and the total construction (rather than just surfaces) of walls, ceilings and floors.

14.0 ACOUSTICS

14.1 INTRODUCTION

14.1 INTRODUCTION

Your design cycles through heating, cooling and lighting
have produced a combination of shapes and surfaces that
are optimized for energy conservation and the use of
passive solar heating and natural cooling sources, as
well as other design objectives you have brought to this
project. Specifically, combinations of non-porous surfaces (for thermal mass and for the admission of daylight) and highly porous materials (insulation) have been
distributed for heat, cold and light utilization.

The acoustics of space is highly influenced by the shape,
the surface materials, and enclosing wall and floor
construction.

Natural ventilation is a further major influence on
acoustics. Openings that freely admit cooling breezes
admit exterior noises just as freely. They also freely
allow interior noises to escape. The open plans that
facilitate natural ventilation also allow noises from one
area to move freely to other, perhaps quieter, spaces.

This exercise examines the often-conflicting goals of
acoustics as compared with your decisions in the preceeding energy-conscious exercises. Since acoustic
analysis is so often postponed until after most other
decisions are made, this conflict is both frequent and
visible.

This exercise will use dB at 500 Hertz (Hz), which is one
of the important frequencies in human speech. See MEEB,
figures 26.14 and 26.19.

Contents

14.0 ACOUSTICS

14.1 INTRODUCTION

-- 14-2

14.2 ANALYZING

GOAL

A. Determine the the acoustical conflicts and compatibilities in the site and program.

A. Analysis Tool - "Site Noise Contours"

Three major sources of sound are present here: the freeway, the railroad, and the factory itself. All three may be considered "linear" sources rather than "point" sources: their sound is generated along a line (such as windows along the side of the factory) rather than at a point (such as a siren would produce).

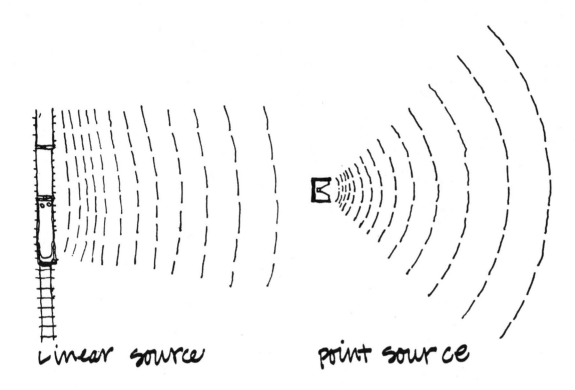

linear source   point source

From Environmental Acoustics by Doelle. Copyright (c)
1972.   Used with the permission of McGraw-Hill Book
Company.

For linear sources, the sound level outdoors falls off
3 dB with each doubling of the distance from the line
source. (Egan, 1972, p. 91). Given:

Freeway (constant) average noise level is 60dB (500 Hz)
at 50 feet away.
Train (occasional) average noise level is 70 dB (500 Hz)
at 100 feet away.
Factory's noise level will be more accurately determined
later. For now, assume: factory (constant) average noise
level is 60 dB (500 Hz) at 1 foot away.

1. Plot the noise contours for each source separately on tracing paper.

2. Combine the contours for the fixed sources (freeway and train). Decibels (dB) are not added arithmetically. Use the following chart to add the sources.

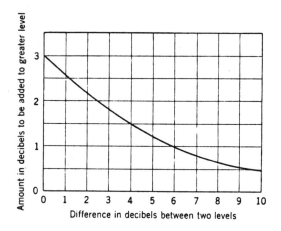

Figure 14.1 Chart for adding two noise levels or sound pressure levels (SPL). From MEEB, Fifth Edition 1971. Used by permission.

3. Plot the resulting combined "noise contours" at 10 dB intervals on the site plan, from the fixed noise sources (freeway and train).

4. On a tracing paper overlay, plot the noise contours at 10 dB intervals from all sides of the factory. (Since this is a "moveable" source in the design process, the overlay can be used to assess the acoustical merits of various factory locations on the site.)

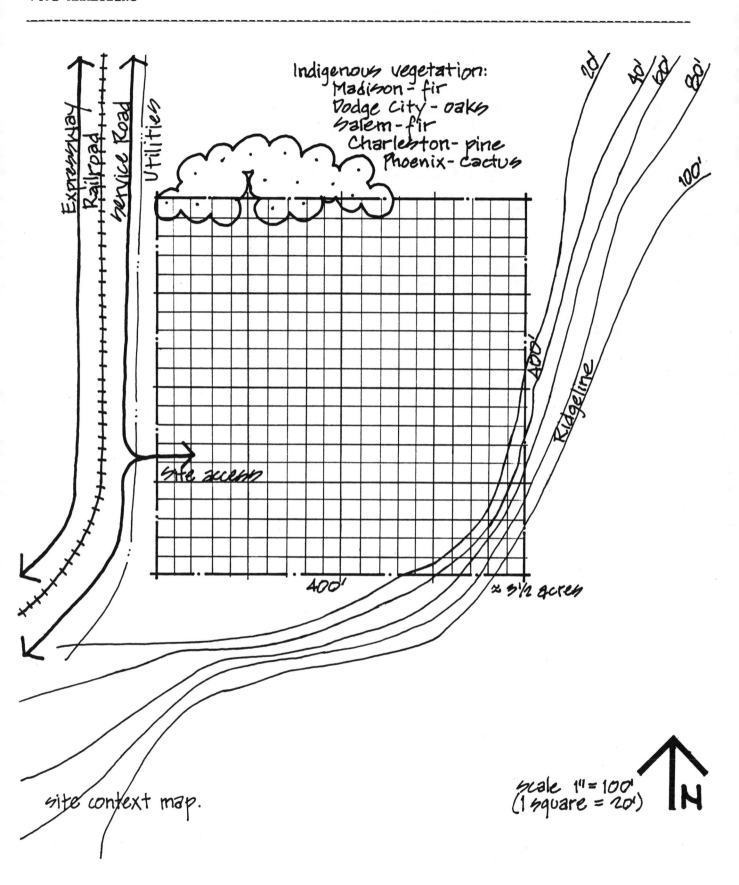

Indigenous vegetation:
Madison - fir
Dodge City - oaks
Salem - fir
Charleston - pine
Phoenix - cactus

Expressway
Railroad
Service Road
Utilities

20' 40' 60' 80' 100'

400'

Ridgeline

Site access

400'

≈ 5½ acres

site context map.

scale 1" = 100'
(1 square = 20')

N

B. Analysis Tool - "Noise Criteria-NC"
Noise Criteria are used to establish a maximum sound level for background noise within a space, based on the activity within the space. As such they begin to quantify which spaces need more quiet than others and give you a preliminary means for assessing your design decisions.

Spaces with low NC requirements should be placed away from noisy (high NC) spaces. Where adjacent spaces have differing NC, an acoustical barrier between them is necessary. The greater the difference in adjacent NC, the more impenetrable an acoustical barrier is needed.

Silence-goal spaces are those in which attention is focused on one sound source, such as a lecturer. Background sound is excluded from silence-goal spaces. Background sound is admitted to quiet-goal spaces, so that no one sound source becomes predominant.

Find the NC appropriate for each space. (See MEEB, Table 27.11, p. 1255, for Noise Criteria (NC) data.)

Factory (speech or telephone communication not required, yet desirable) NC = _____

Circulation and toilet rooms   NC = _____

Display and conference space   NC = _____

Private office spaces   NC = _____

Questions
A. Which spaces are silence goal spaces and which are quiet goal spaces?

B. Which are the acoustically similar program spaces?

C. Which are the acoustically incompatible spaces?

D. Are there potential acoustical conflicts between inside and outside?

14.3 CONCEPTUALIZING --

GOAL

A. Utilize wanted sounds on site, while minimizing noise (unwanted sound).

DESIGN STRATEGIES

A. Cluster Scale
Use buildings or other barriers to create quieter spaces, sheltered from noise sources.

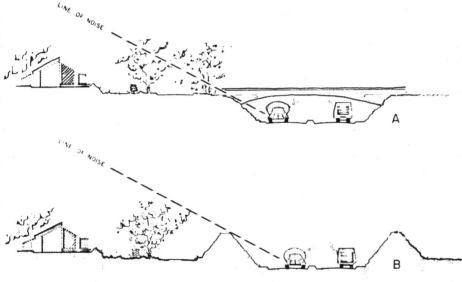

From <u>Environmental Acoustics</u> by Doelle.   Copyright (c) 1972.   Used with the permission of McGraw-Hill Book Company.

Quiet spaces for quiet uses.

B. Building Scale
Separate the quiet from the noisy spaces.
C. Component Scale
Noise sources are usually most economically quieted <u>at source</u>.

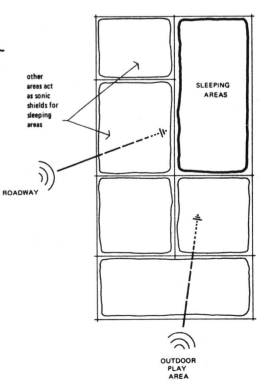

From <u>Concepts in Architectural Acoustics</u> by Egan. Copyright (c) 1972. Used with the permission of McGraw-Hill Book Company.

DESIGN

Choose an existing building/site which has a clear conceptual approach, at any scale, to achieving these goals. Document your choice as follows:
A. Identify the location, program, architect (if any) of the design and the source of your information.
B. Include photocopies or drawings (whatever is quick and easy for you) to explain the design.

EVALUATION

A. Evaluation Tool - "Building Response Diagrams"
Diagram how this design is organized to achieve these goals.

14.4 SCHEMING

GOALS

A. Separate or isolate acoustically incompatible spaces.
B. Exclude background sound from silence-goal spaces and admit background sound to quiet-goal spaces.

CRITERIA

A. Potential acoustical conflicts have been avoided.
B. Heating, Cooling, Daylighting performance has not been impaired in meeting the acoustic criteria.

DESIGN STRATEGIES

A. Cluster Scale
Consider introducing a sound source to make noisy sources (such as a courtyard fountain to mask traffic). When sound sources generated within buildings are preferable to noise sources offsite, these sounds can be similarily used for masking.

B. Building Scale
When openings for ventilation also admit noise, consider placing masking sound sources near air vents, or placing the most noisy spaces nearest to noisy ventilating edges of building.

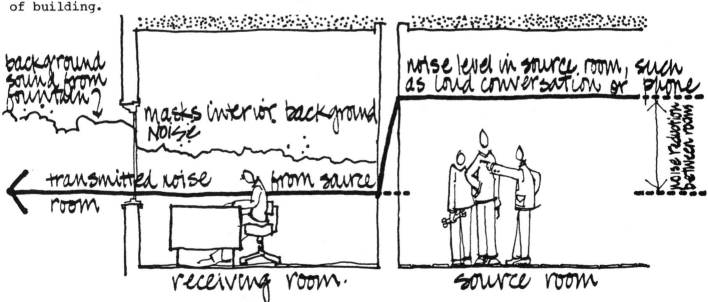

From Concepts in Architectural Acoustics by Egan. Copyright (c) 1972. Used with the permission of McGraw-Hill Book Company.

C. Component Scale

Thermally massive walls can be excellent sound barriers, such as between noisy and quiet spaces.

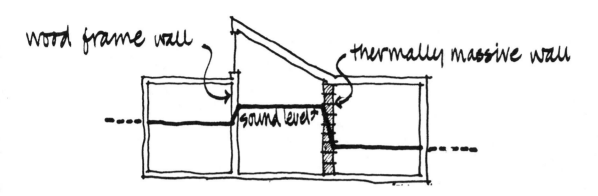

Sound barriers should have a minimum of openings, and these openings must be fully closable when sound isolation is desired. When a "weaker" element, such as a window or door, is used in a construction the composite TL for the combination is usually closer to the TL of the weaker element.

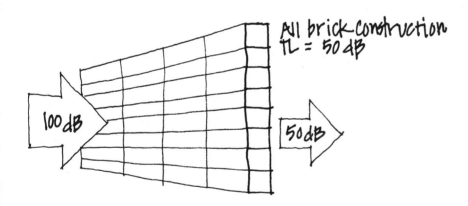

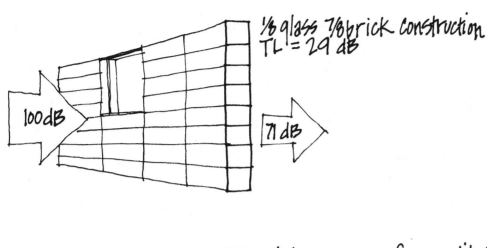

1/8 glass 7/8 brick construction
TL = 29 dB

100dB

71 dB

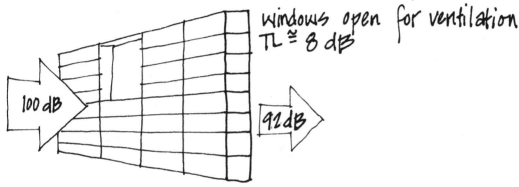

windows open for ventilation
TL ≅ 8 dB

100 dB

92 dB

Acoustically absorptive surfaces are rarely compatible with thermal storage, so such surfaces should either receive no winter sun, or be in a position to reflect it toward thermal mass surfaces. Similarly, such surfaces need no exposure to cooling winds.

DESIGN

Design your office building, factory and site to achieve the Scheming goals. Document your design:

A. Site Plan
B. Cluster Plan
C. Floor Plans
D. Sections
E. Roof plan and elevations or paralines showing all sides and top of your buildings.

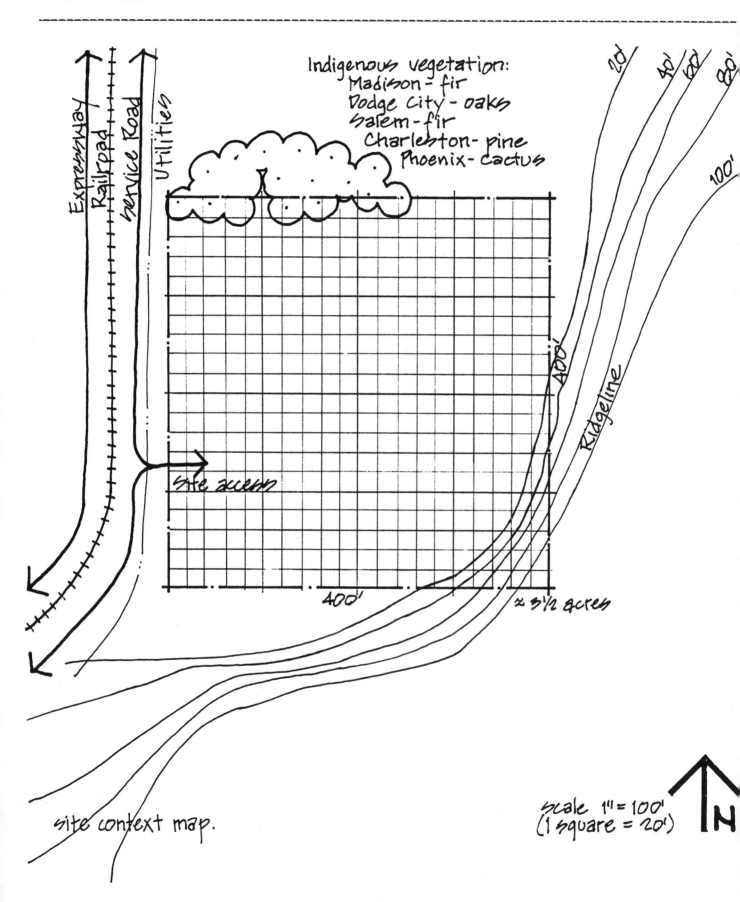

Indigenous vegetation:
Madison - fir
Dodge City - oaks
Salem - fir
Charleston - pine
Phoenix - cactus

Expressway
Railroad
Service Road
Utilities

20' 40' 60' 80'
100'

400'

Ridgeline

site access

400'   ≈ 3½ acres

site context map.

scale 1" = 100'
(1 square = 20')

N

cluster site plan: 1" = 40'-0".

----

for plans, sections, & elevations: if   1/16"=1'-0",   1 square = 10'.

for plans, sections, & elevations: if   1/16"=1'-0",   1 square = 10'.

---

for plans, sections, & elevations :   if   1/16"=1'-0",   1 square = 10'.

for plans, sections, & elevations: if 1/16"=1'-0",   1 square = 10'.

for plans, sections, & elevations: if   1/16"=1'-0",   1 square = 10'.

---14--

for plans, sections, & elevations: if   1/16"=1'-0",   1 square = 10'.

paraline grid w/ 20' cells;
indicate north arrow.

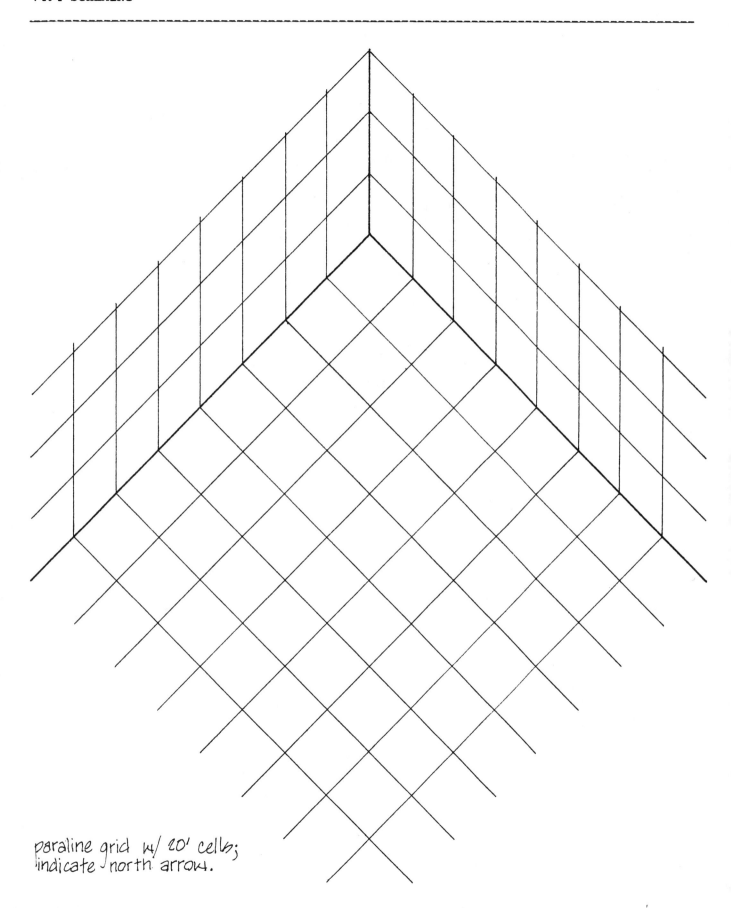

paraline grid w/ 20' cells;
indicate north arrow.

EVALUATION

A. Criterion A: The potential acoustical conflicts identified in 14.2 Analyzing Summary, Question D, have been
avoided on the site and within the buildings.

Evaluation Tool for Criterion A - "Site Noise Contours"
1. Locate your buildings on a site plan.
2. Use MEEB, Fig. 26.12, p. 1171 to add the dB (at 500
Hz) from the three noise sources (freeway, train and
factory) to achieve overall noise contours for the site.
(The total dB from more than one source is not the <u>sum</u> of
the individual dB measurements.) We are assuming that
our factory equipment has a peak noise level at 500 Hz;
remember that this is only one frequency, that noise
levels vary with frequency, and that factory equipment in
particular might peak at other frequencies.

**Evaluation Tool for Criterion A - "Acoustical Renderings
or Noise Drawings"**
1. For each space identify the NC and whether the goal is
silence or quiet (from the Analysis Section.)

2. In your cluster plan clearly show all barriers and all
openings.
3. Draw an acoustical picture (in plan, section, perspective, model, whatever) which shows sounds at their
sources and how they are used as background sound or
controlled by isolation or separation.

Does your first trial design meet Criterion A?

If not, what will you change for your second trial?

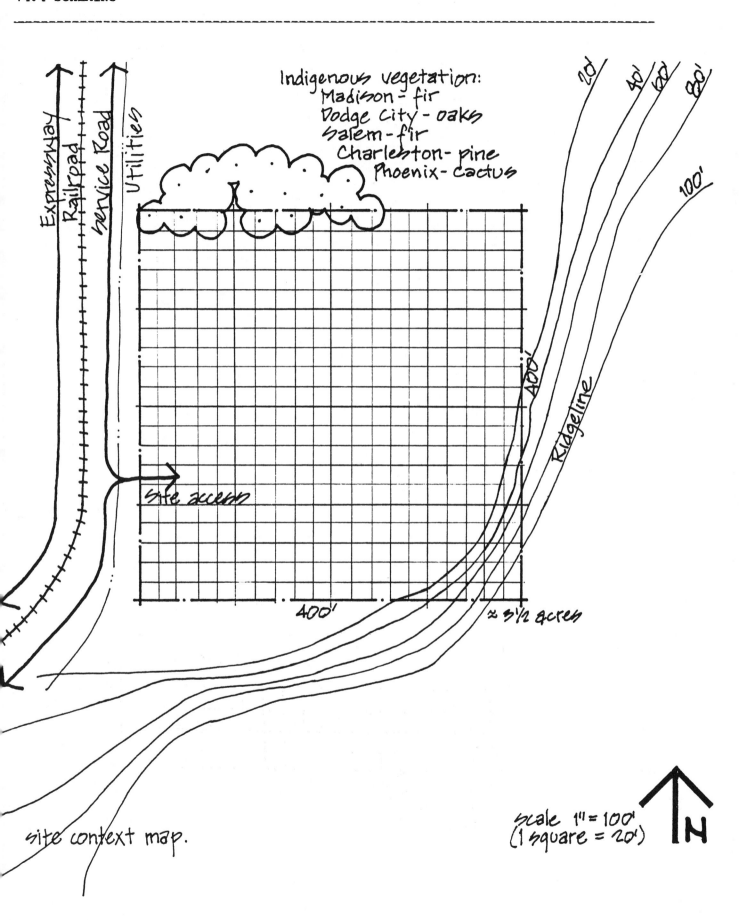

Indigenous vegetation:
Madison = fir
Dodge City - oaks
Salem - fir
Charleston - pine
Phoenix - cactus

Expressway
Railroad
Service Road
Utilities

20'
40'
60'
80'
100'

400'

Ridgeline

400'

≈ 3½ acres

site access

site context map.

scale 1" = 100'
(1 square = 20')

N

cluster site plan: 1"= 40'-0".

for plans, sections, & elevations: if   1/16"=1'-0",   1 square = 10'.

B. Criterion B: Heating, Cooling, Daylighting Check:
Re-check the criteria that produced your site and building organization in the energy-conscious exercises of
heating, cooling and daylighting. If acoustic goals are
incompatible with other considerations, decide on your
priorities and explain your decisions.

Does your first trial design preserve the heating, cooling and daylighting performance achieved in those exercises?

If not, what will you change for your second trial?

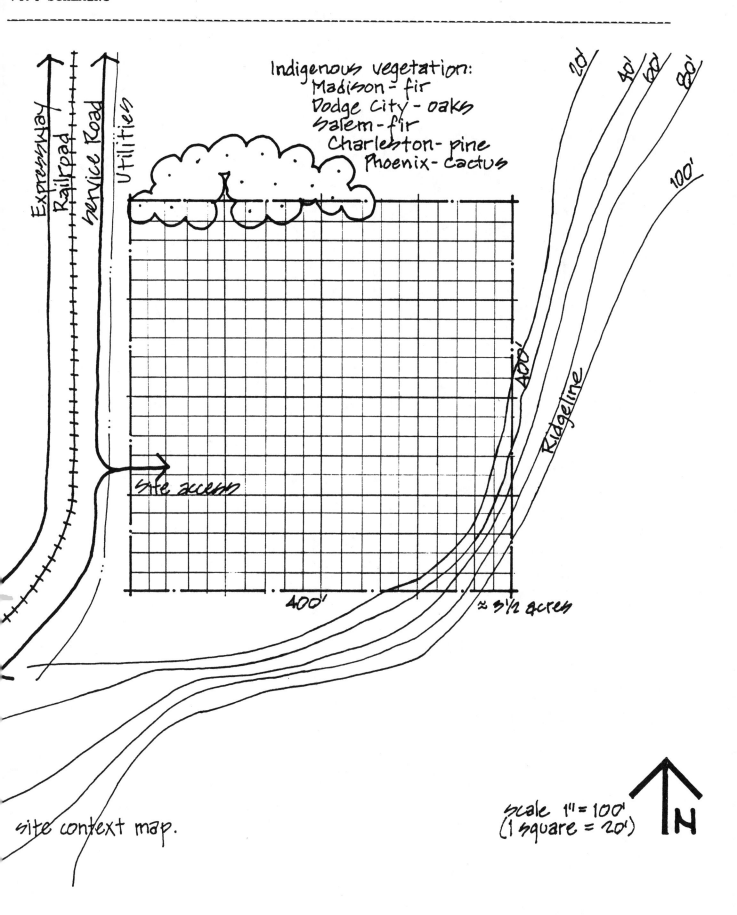

Indigenous vegetation:
 Madison - fir
 Dodge City - oaks
 Salem - fir
 Charleston - pine
 Phoenix - cactus

Expressway
Railroad
Service Road
Utilities

20' 40' 60' 80' 100'

400'

Ridgeline

Site access

400'

≈ 3½ acres

site context map.

scale 1" = 100'
(1 square = 20')

N

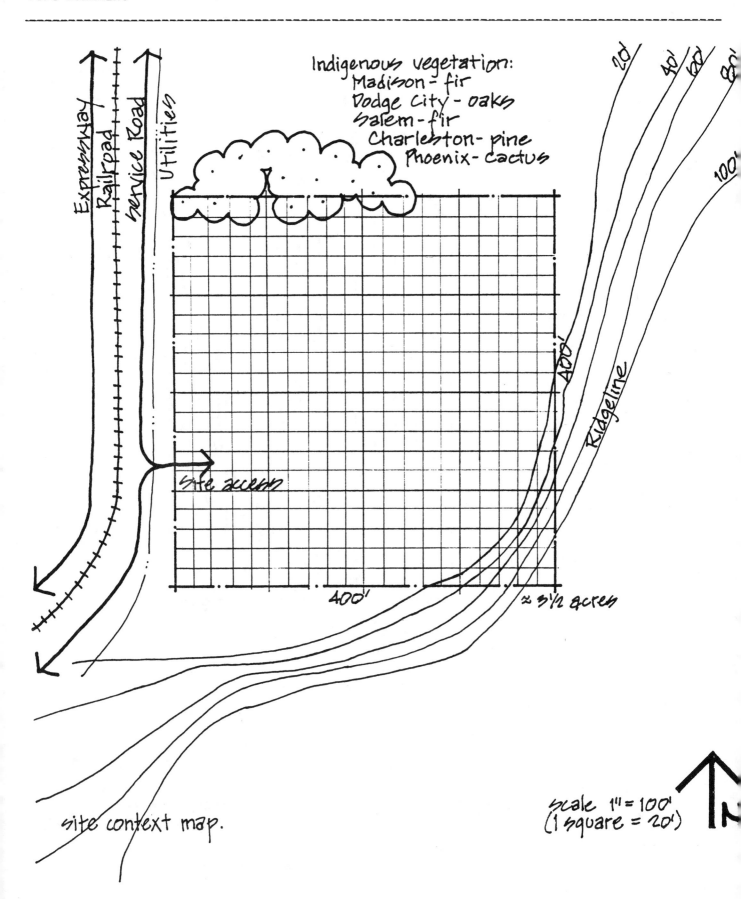

Indigenous vegetation:
- Madison - fir
- Dodge City - oaks
- Salem - fir
- Charleston - pine
- Phoenix - cactus

20' 40' 60' 80'
100'

Expressway
Railroad
Service Road
Utilities

400'

Ridgeline

site access

400'

≈ 3½ acres

site context map.

scale 1" = 100'
(1 square = 20')

N

14.5 DEVELOPING _____

GOALS

A. Make spaces acoustically comfortable with your choice
of interior materials and finishes.
B. Achieve necessary sound isolation characteristics with
your choice of wall construction and finish materials.

CRITERIA

A. Criterion A: Spaces are live, neutral or dead as ap-
propriate for their function.
B. Criterion B: All barrier walls meet the recommended
STC rating.

DESIGN STRATEGIES

A. Component Scale
Spaces with predominantly hard surface materials
(including wood, masonry, plaster, glass and concrete)
are reverberant; a sound stays "alive" longer within
these spaces. Spaces with predominantly soft surface
materials (rugs, drapes, acoustic tile, exposed insulation boards or batts) do not reinforce sounds, so they
"die" away very soon, as they would outdoors.

"Live" spaces encourage sound creation, such as whistling
in concrete stairwells, or singing in the shower. They
also blend sounds together, and can become acoustically
exhausting with prolonged exposure to a high level of
mixed and largely unwanted sounds.

"Dead" spaces discourage the creation of sound, and tend
to keep sounds distinct from one another. A space may be
chosen to be alive or dead simply to provide for
acoustical variety in a collection of spaces. Contrasts
between spaces can be enhanced by liveness or deadness; a
quiet space that is "dead" seems especially quiet if
entered from a noisy, "live" one.

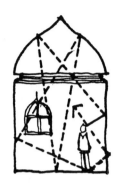

Dead Space   Live Space

Adjacent and very different NC spaces will require
acoustic barriers, as will adjacent spaces with low
NC's.

Privacy is a factor in providing acoustic barriers - you
may want to provide a barrier to overhearing speech from
an adjacent space, even if it has the same NC.

Noise transmission can be reduced by increasing the mass
of the barrier up to the point beyond which diminishing
returns may be expected.

Noise transmission can be reduced by structural discon-
tinuity within the barrier.

A very stiff partition will not resist transmission as
well as might be expected from its mass alone.

Materials chosen for high sound absorption will generally
have little value as sound barriers, due to their porous
and lightweight character.

Acoustical barriers can be rated by their Sound Transmission Class (STC), with more effective barriers having
higher STC's. As a guide:

STC 35 Normal speech can be heard and understood through
barrier
STC 40   Normal speech is heard as a murmur through
barrier
STC 50   Very loud sounds can be heard only faintly
through barrier

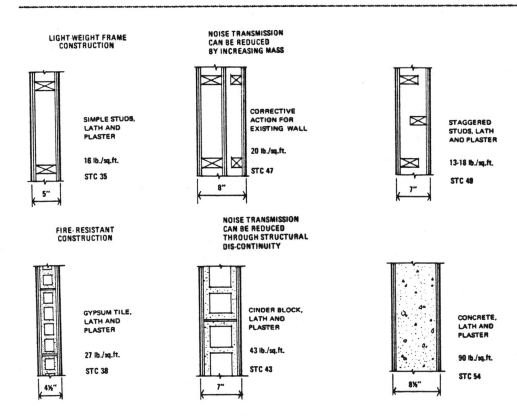

Figure 14-2 STC Ratings for Typical Wall Constructions
from Architectural Interior Systems by J. E. Flynn and A.
W. Segil (c) 1970 by Litton Educational Publishing, Inc.
Reprinted by permission of Van Nostrand Reinhold Co.
For other partition alternatives, check tables A.1 and
A.2 in the back of MEEB.
DESIGN

Develop your design for the factory, one 200 sq.ft.
office and the display conference room to achieve the
acoustical goals above. Document your design:
A. A sectional perspective of each room indicating finish
materials.
B. A section at 3/4"=1'-0" of each acoustical barrier and
key these to the cluster plan.
C. Indicate on the following chart your design intentions
for each of these three spaces:

Space	Live or Dead	Reason
Factory		
Office		
Display/ Conference		

--14-

for daylighting contours, enlarged plans: if 1/8"= 1'-0", 1 square = 10'.

14.0 ACOUSTICS

14.5 DEVELOPING

----

for daylighting contours, enlarged plans:		if 1/8"= 1'-0",	1 square = 10'.		

INSIDEOUT

construction sections: if 3/4" = 1'-0", 1 square = 1'-0".

construction sections: if 3/4" = 1'-O", 1 square = 1'-O".

EVALUATION

A. Criterion A: Spaces are live, neutral or dead as appropriate for their function.
Evaluation Tool for Criterion A - "Room Absorbency"
An early step in nearly all acoustics calculations is to determine the total absorbency in each space in which acoustics are important.
Absorbency is measured in SABINS, and is equal to the coefficient of absorption times the surface area (essentially, the % of sound which will be absorbed times the area absorbing it):

absorbency = (α) x (sf)

Total room absorbency is simply the sum of the absorbencies of all surfaces. For buildings which are naturally ventilated, the openable window areas present a potentially substantial increase in absorbency since an open window's co-efficient of absorption is 1.0, far greater than most building materials. (Of course, $\alpha = 1.0$ also means TL=0. Thus no acoustic barrier value.) Therefore, for ventilated buildings, two absorbency calculations will be performed. Although such absorbency calculations can be done at a variety of frequencies, we will continue to concentrate on the mid-frequency range of 500 Hz.
Use the coefficients of absorption MEEB Table 27.1, p. 1206 - 1208 to calculate the total absorbencies of your rooms on the tables below.

Trial 1:

Factory	Area (sf)	Coefficient of Absorption at 500Hz		Absorbency: Area x α (Sabins)	
		"closed"	"open"	"closed"	"open"
floor					
walls					
openings:					
closed		NA		NA	
open (such as windows)		NA	1.0	NA	
ceiling					
people, # occupants					
			TOTAL =	_____	_____ Sabins

Trial 2:

Factory	Area (sf)	Coefficient of Absorption at 500 hz		Absorbency: Area x α (Sabins)	
		"closed"	"open"	"closed"	"open"
floor					
walls					
openings:					
closed			NA		NA
open (such as windows)		NA	1.0	NA	
ceiling					
people, # occupants					
			TOTAL =	_____	_____ Sabins

Trial 1:

Display/Conf	Area (sf)	Coefficient of Absorption at 500Hz		Absorbency Area x α (Sabines)	
		"closed"	"open"	"closed"	"open"
floor					
walls					
openings:					
closed			NA		NA
open (such as windows)		NA	1.0	NA	
ceiling					
people, # occupants					
			TOTAL =	_____	_____ Sabins

Trial 2:

Display/Conf	Area (sf)	Coefficient of Absorption at 500 Hz		Absorbency Area x α (Sabins)	
		"closed"	"open"	"closed"	"open"
floor					
walls					
openings:					
closed		NA		NA	
open (such as windows)		NA	1.0	NA	
ceiling					
people, # occupants					
			TOTAL =	_____	_____ Sabins

Trial 1:

Office	Area (sf)	Coefficient of Absorption at 500Hz		Absorbency Area x α (Sabins)	
		"closed"	"open"	"closed"	"open"
floor					
walls					
openings:					
closed		NA		NA	
open (such as windows)		NA	1.0	NA	
ceiling					
people, # of occupants					
			TOTAL =	_____	_____ Sabins

Trial 2:

Office	Area (sf)	Coefficient of Absorption at 500Hz		Absorbency Area x α (Sabins)	
		"closed"	"open"	"closed"	"open"
floor					
walls					
openings:					
closed		NA		NA	
open (such as windows)	NA		1.0	NA	
ceiling					
people, # occupants					
		TOTAL =	_____	_____	Sabins

Evaluation Tool for Criterion A – "Room Liveness Graph"
1. Complete this table for your rooms:

	Volume ft^3	Total room absorption in sabins			
		1st Trial		2nd Trial	
		"closed"	"open"	"closed"	"open"
Factory					
Display/conference					
Office					

2. Use this information with the graph below (Flynn and Segil, 1970, p. 54) to evaluate your rooms.

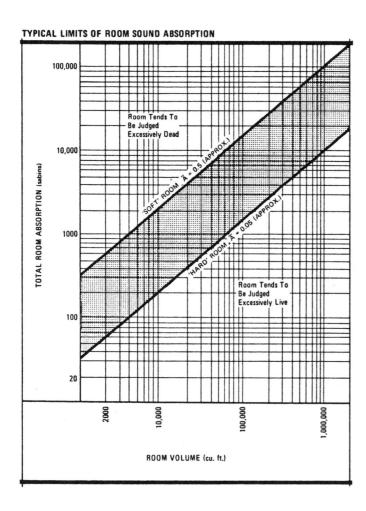

Figure 14-3 Room Liveness Graph from <u>Architectural</u> <u>Interior Systems</u> by J. E. Flynn and A. W. Segil (c) 1970 by Litton Educational Publishing, Inc. Reprinted by permission of Van Nostrand Reinhold Co.

Does your first trial design meet Criterion A?

If not, what will you change in your second trial?

B. Criterion B: All barrier walls have STC ratings greater than or equal to the ratings in the MEEB Table 27.12, p. 1260-61.

Evaluation Tool for Criterion B - "STC Rating of Barrier Walls"

1. List all your barriers as keyed to the cluster plan.

---

2. Enter the recommended STC for each barrier from MEEB Table 27.12.
3. Find the actual STC and transmission loss (TL) for each barrier using MEEB Appendix A (use index A, then table A.1, then table A.2). See also Table 27.3, p. 1220, 27.5, p. 1230, and 27.6, p. 1231 and the discussion of walls with windows p. 1229 - 1231.

Some TL values at 500 Hz for glass are:
1/8" single plate glass	26
1/8" with rubber gasket	30
Laminated glass	
(viscoelastic between glass)	33
1/4" & 1/8" double glazing;	
with 2" air space	35
with 4" air space	42

(From Egan, 1972, p. 75)

		1st Trial		2nd Trial	
Barrier	Recommended STC	Actual STC	TL @ 500 Hz	Actual STC	TL @ 500 Hz

Does your first trial design meet Criterion B?

If not, what will you change in your second trial?

14.6 FINALIZING

GOALS

A. The offices and factory are acoustically comfortable places to work.

CRITERIA

A. Criterion A: Noise levels in the factory are within allowable dB levels.
B. Criterion B: The office can be naturally ventilated and stay within the PNC goal.

DESIGN

Present full documentation of the design being evaluated in this section if it has changed from the design as documented in 14.4 Scheming and 14.5 Developing.

EVALUATION

A. Criterion A: Noise levels in the factory are within allowable dB levels.

Evaluation Tool for Criterion A - "Reverberation Time T_R"
The time during which sounds reverberate within a space is another useful characteristic for acoustic analysis. In "live" spaces, reverberation lasts longer than in "dead" ones. Although none of our spaces is critical for stage/audience situations for which T_R is usually calculated, T_R is also used as a guide in some noise control calculations.

$$T_R = \frac{.049 \times \text{Volume (cf)}}{\text{Absorbency (SABINS)}}$$

1. Determine recommended T_R for your spaces using MEEB, Fig. 26.26, p. 1188.
2. Calculate the T_R for your spaces using the formula above.

Space	Recommended T_R	Actual T_R			
		1st Trial		2nd Trial	
		closed	open	closed	open
Factory					
Office					
Display/ Conference					

Do your spaces meet Criterion A in the first trial?

If not, what will you change for your second trial?

Evaluation Tool for Criterion A - "Room Noise Level-Sound Source in Room"

The factory contains 16 machines, each producing sound as well as manufactured goods and heat. The inherent noisiness of a given machine is called its Sound Power level, PWL, and is typically referenced to 10^{-12} watt (see Egan, 1972, p. 50, 51 for a more thorough discussion).

Once this PWL is known, the resulting noise level, the SPL (Sound Pressure Level), made by this machine in any given space can be determined. The graph below shows how the PWL of a noise source is changed by placing that source in a space of given volume and T_R. (In highly reverberant, smaller rooms the resulting noise level can actually be greater than PWL.)

1. Determine allowable dB level at 500 Hz, using MEEB Fig. 26.23, p. 1184 or Table 27.11, p. 1255.

factory NC = _____

allowable dB at 500 Hz = _____

2. Given: PWL of one machine 500 Hz = 70dB. Use graph below and the T_R from above, to find the resulting SPL of one machine.
closed _____
open _____

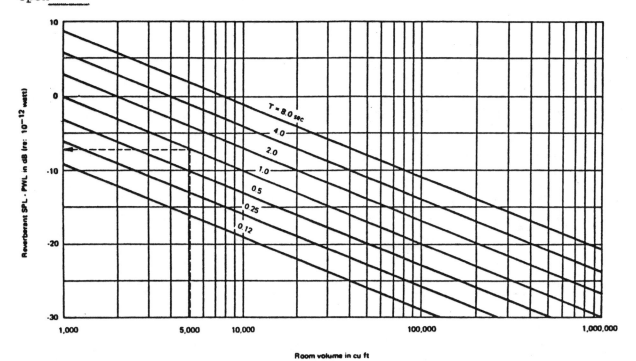

Figure 14-4   Sound   Absorption   reference,   from Concepts in Architectural Acoustics, p. 51 by Egan. Copyright 1972.   Used by permission of McGraw-Hill Book Company.

3. Once the SPL in the factory from one machine is known, the other 15 machines' noise can be added. For the closed building condition, use McGuinness, Fig. 26.12 (p. 1171) to add dB sources. Each machine adds progressively less dB to the previous subtotal.

14.0 ACOUSTICS

14.6 FINALIZING

	1st trial	2nd trial	
Machine #1	_____	_____	dB
add #2	_____	_____	dB added
	_____	_____	SUBTOTAL
add #3	_____	_____	dB added
	_____	_____	SUBTOTAL
add #4	_____	_____	dB added
	_____	_____	SUBTOTAL
add #5	_____	_____	dB added
	_____	_____	SUBTOTAL
add #6	_____	_____	dB added
	_____	_____	SUBTOTAL
add #7	_____	_____	dB added
	_____	_____	SUBTOTAL
add #8	_____	_____	dB added
	_____	_____	SUBTOTAL
add #9	_____	_____	dB added
	_____	_____	SUBTOTAL
add #10	_____	_____	dB added
	_____	_____	SUBTOTAL
add #11	_____	_____	dB added
	_____	_____	SUBTOTAL
add #12	_____	_____	dB added
	_____	_____	SUBTOTAL
add #13	_____	_____	dB added
	_____	_____	SUBTOTAL

---

add #14	_____	_____	dB added
	_____	_____	SUBTOTAL
add #15	_____	_____	dB added
	_____	_____	SUBTOTAL
add #16	_____	_____	dB added
TOTAL	_____	_____	dB

Note: A much less tedious, though more mathematically complicated, approach is illustrated in Egan, 1972, p. 15.

4. What is the difference between the dB level of one machine and the total created by 16 machines?

1st trial_____   2nd trial_____

5. To estimate the SPL of the open building, as a safe estimate, add the above difference to the SPL of one machine in the open building:

_____1st Trial   _____2nd Trial

Is your first trial design within allowable dB levels determined in step 1 for both open and closed buildings?

If not, what will you change in your second trial design?

B. Criterion B: The office can be naturally ventilated and stay within the NC goal.

Evaluation Tool for Criterion B - "Final Site Noise Contours"
In the scheming phase it was assumed that 60 dB came through the exterior walls from the factory interior. Now that you know the actual SPL in the factory as well as the characteristics of the wall, you can finalize the amount of sound that escapes.
1. SPL in open factory: _____ dB at 500 Hz

2. Use MEEB Fig. 27.23, p. 1229, to determine how the open windows affect the barrier characteristics of the exterior factory wall:

a. Which wall are you evaluating?

b. From 14.5 Developing Criterion B, p. 14-41, TL of this wall:
_____ dB (500 Hz)

c. % of wall in ventilating openings

	1st trial	2nd trial
	_____ %	_____ %

d. Resulting TLc _____ dB _____ dB

e. Resulting noise level at factory exterior wall

   SPL - TLc = _____ dB _____ dB

Evaluation Tool for Criterion B - "Room Noise Sound Level-Sound Source Outside"
1. Establish the noise levels outside the office window.

a. The factory is _____ feet away from the office.

b. Add the constant sources using MEEB Fig. 26.12, p. 1171:
factory noise at office _____ dB (assume a 3dB drop 1 foot away from the factory as before in 14.2 Question A.)

   freeway at office _____ dB

   total _____ dB constant

----

c. Add intermittent sources

train at office _____ dB

total _____ dB intermittent & constant

2. Use MEEB Fig. 27.23, p. 1229 to find the resulting "predicted noise level" <u>inside</u> the office.

a. Which office wall are you evaluating?

b. PNC from 14.2 Analysis: _____

c. Allowable dB at 500 Hz from MEEB Fig. 26.23, p. 1184 _____ dB

d. TL of this wall from 14.5 Developing Criterion B
_____ dB (500 Hz)

	1st trial	2nd trial
e. % wall in ventilating openings	_____ %	_____ %
f. Resulting TLc	_____ dB	_____ dB
g. Predicted constant noise indoors; constant exterior dB–TLc =	_____ dB	_____ dB
h. Predicted intermittent noise indoors; intermittent exterior–TLc =	_____ dB	_____ dB

Is the indoor noise level of your first trial design within allowable dB for Criterion B?

If not, what will you change for your second trial?

In this section, we repeat the exercises in climate, heating, cooling, thermal finalization, daylighting and lighting finalization. Now, however, the blanks are filled in and illustrations are prepared, showing the evolution of a design for the office building and factory. The climate is the rather mild, if often cloudy, climate of Salem, Oregon.

A few notes about the solution which follows:

a. We didn't do the more general parts of the exercises, which involve finding examples from existing architecture, and don't pertain specifically to Salem's climate.

b. We didn't do the parts of certain exercises which were inappropriate to Salem's climate.

c. We elected to do elevations and roof plans, in place of paraline drawings, to explain our designs. However we did paralines for Section 5.3 Heating to give a clearer understanding of the buildings.

d. If you find our explanations insufficiently referenced to our information sources, remember to document your own work all that much more carefully. We tried to be complete; if you are more thorough, this will be an even more useful reference in the future.

Finally, we have tried to be clear about how we did the Salem example. However because we wrote the book we may have made some assumptions that aren't apparent to you. Please let us know about these and offer a positive suggestion of how the example can be improved.

3.1 INTRODUCTION

A beginning step in many design projects is analyzing the context of the problem; identifying what you have to work with, what is demanded in the solution, and where the potential assets and liabilities lie in the interaction of these. This exercise is structured to explore the climate as context, using a range of analytical tools to answer questions.

You need to determine when your climate is comfortable for people, potential thermal benefits and problems of the sun and wind on the site, what potential benefits and problems that daylighting offers on the site, and the implications of building program and configuration for choosing a design strategy.

The exercise is structured as a series of goals which are reached by analyzing available data and information about the site and program. Summary questions help you establish design strategies for your climate that you will use throughout the exercises.

One evaluation tool involves building a model of the site and observing sun shadow patterns for different sun positions (summer morning, winter noon, etc.). These observations require direct sun. Given the unpredictability of the weather, we recommend that you build your model now and have it ready to test when the sunny weather occurs.

Remember that a sample of how a climate is analyzed with these tools is found in "Climate; the Salem Oregon Example".

A good way to begin is to read the narrative summaries of the five cities to get a comparative sense of your climate.

3.0 CLIMATE

3.1 INTRODUCTION

Contents

3.2 THERMAL COMFORT

GOAL

Determine how people stay thermally comfortable in your climate.

A. Analysis tool - "Climate Plotted on Bioclimatic Chart"
Use the NOAA climatological data for your city (found at the end of this exercise section 3.7), to <u>plot</u> the average maximum and minimum temperatures for each month against the relative humidity (R.H.) that occurs approximately the same time (generally minimum temperature occurs approximately 3-4 a.m., maximum temperature 3-4 p.m.). <u>Connect</u> the 2 points for each month to get a picture of how the temperature/RH condition changes over a day.

For summer temperatures above the average maximum, find the summer dry bulb (DB) from MEEB 6th edition (<u>Mechanical and Electrical Equipment for Buildings</u>), Appendix C table C.2a. Then <u>extend</u> your hottest month line until it crosses this temperature on the Bioclimatic Chart. This will represent the worst case design condition.

(A full description of the bioclimatic chart and how to use it will be found in <u>Design with Climate</u>, Victor Olgyay, pp. 19-23.)

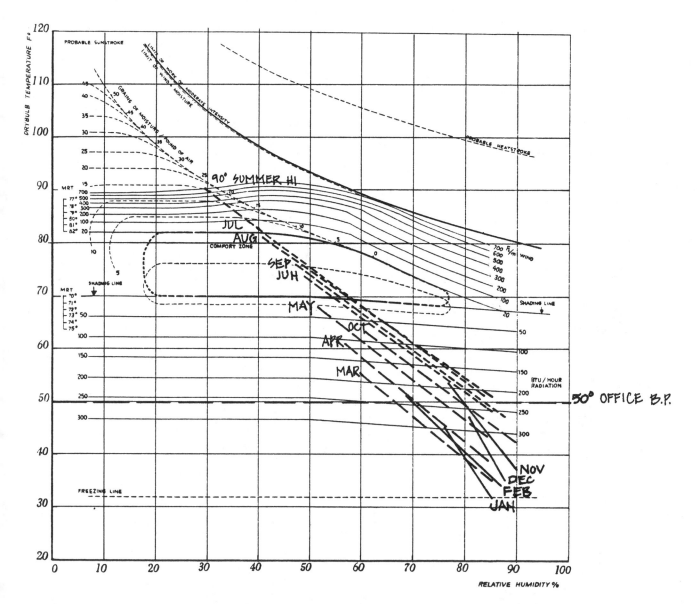

Figure 3-1 Bioclimatic Chart. From Victor Olgyay, *Design With Climate: Bioclimatic Approach to Architectural Regionalism* (copyright (c) 1963 by Princeton University Press) Fig. 45, p. 22. Reprinted by permission of Princeton University Press.

Remember: these are averages, not extremes!

Office building balance point, from section 3.6, approx. 50°

Cool, wet Oregon winter...

QUESTIONS

Use what you have plotted to answer the following:
A. In your climate, most of the time people will be (too hot, too cold, comfortable, too dry, etc.)?

Damp and too cold; need sun

B. In what months does the difference between high and low temperatures seem significant? (For example, on opposite sides of the comfort zone?) Such diurnal changes can be used to advantage by storing heat or coolth in periods when they are not needed for use in periods when they are.

Yes, in the dry summer weather (June through September)

C. At what times of the year will you want to use the wind for cooling?

Most likely in July and August

D. At what times of the year will you want to use the wind for cooling <u>only if</u> you also add moisture?

Extreme days in July and August

E. At what times of the year will you want to block wind?

September through June

F. At what time of year will you need to shade yourself by blocking the sun?

June through September

G. At what time of year will you need to use the sun to stay warm?

October through May

H. What will be the role of the building in keeping people comfortable
 - in summer?

Provide shade and moderate air motion; occasional increased air flow on hotter days

 - in fall?

Admit sun on September mornings, and all day October and November; block wind

 - in winter?

Admit all the sun you can get, and block wind

 - in spring?

Admit sun all day in April, May; and on June mornings; block wind

3.0 CLIMATE

3.3 SITE THERMAL POTENTIAL

_____

3.3 SITE THERMAL POTENTIAL

GOAL

Determine how you can use the natural energy provided by
sun and wind to keep people comfortable.

A. Analysis Tool - "Wind Roses and Site Wind Flows"
Wind "roses" (found in 3.8) show two important
characteristics for each city in each month: a) what
directions are most prevalent each month, and b) what %
of the time is calm. Another important characteristic
which must be added is the average speed of the wind,
which may be obtained from the yearly climate summaries
in 3.7. (For those months in which data is not given for
Madison, use the month immediately following or
preceeding.)
1. From your answers in 3.2
Months to admit wind: *July and August*
Months to block wind: *September through June*
2. On one of the following site plans, show the
approximate wind flows in the months where wind is to be
admitted. Show directionality, approximate flow around
obstacles, and indicate monthly % calm and average wind
speeds for these "admit" months.
3. On the other site plan, show the approximate wind
flows in the months where wind is to be blocked. Show
directionality and flow around obstacles.
4. In months in which wind is to be admitted, how
reliable a resource is it? (Consider % calm, consistency
of direction, and average speed; higher speeds mean more
thorough and rapid "flushing" of heat from a building.)

Relatively reliable; % calm
about 10% (less than the other,
non-overheated months), with a
distinct shift in direction
from the other months, and an
average speed of a little over
6 mph.

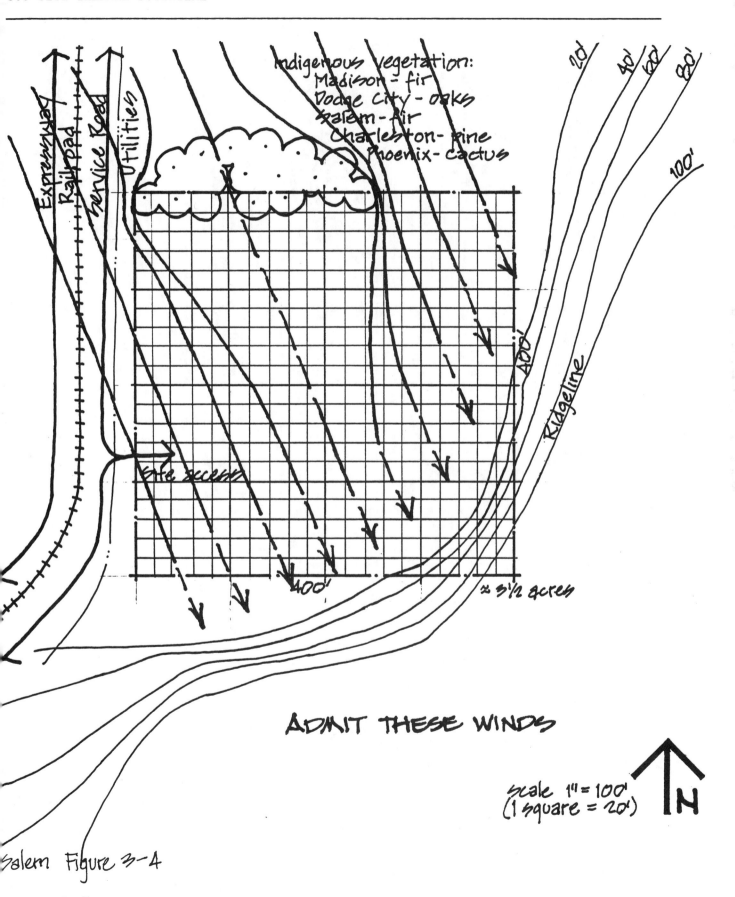

Indigenous vegetation:
Madison - fir
Dodge City - oaks
Salem - fir
Charleston - pine
Phoenix - cactus

Expressway
Railroad
Service Road
Utilities

20' 40' 60' 80'

100'

400'

Ridgeline

Site access

400'

≈ 5½ acres

ADMIT THESE WINDS

scale 1" = 100'
(1 square = 20')

N

Salem Figure 3-4

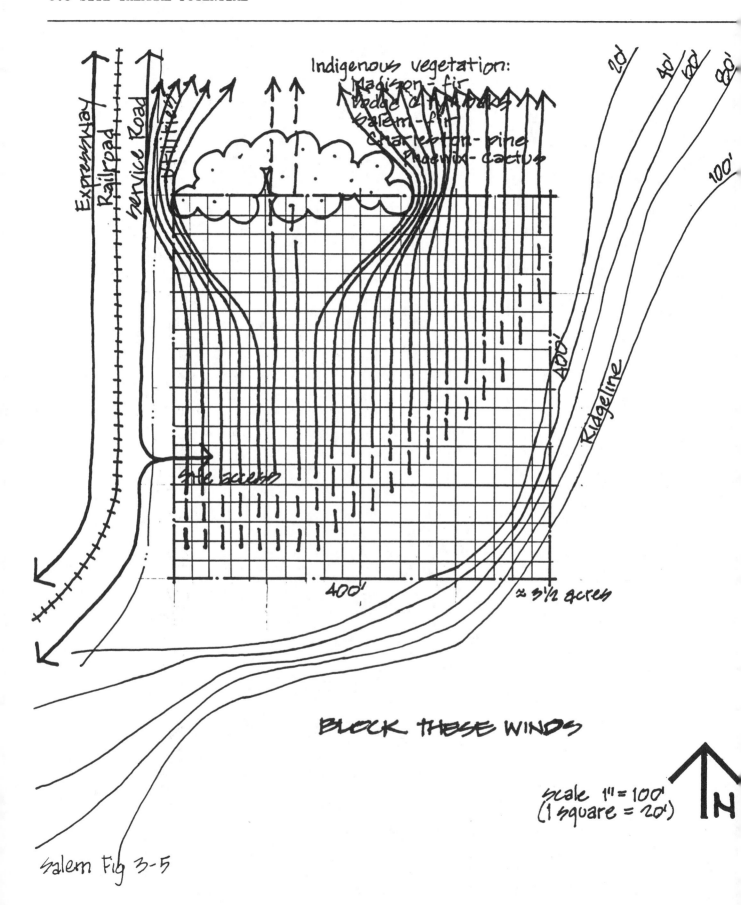

Indigenous vegetation:
Madison - fir
Lodge d - ???????
Salem - fir
Charleston - pine
Phoenix - cactus

20' 40' 60' 80'

100'

400'

Ridgeline

fire access

400'

≈ 3½ acres

BLOCK THESE WINDS

scale 1" = 100'
(1 square = 20')

N

Salem Fig 3-5

B. Analysis Tool - "Sun Peg Shadow Plot"

Using the sun peg will give you information on how the sun and your site conditions will interact daily and seasonally.

1. Make a site contour model, scale 1"=40' from the site context map in the building/site program, section 2.2. Mark the surface into a 10' grid and include all vertical obstructions such as vegetation and land forms. Leave approximately 9" of horizontal space across the south end of your model base for mounting the chart.

Use the "peg shadow plot chart" for your latitude in section 3.9.
Mount the chart perfectly horizontally on your model base, south of the site. The north arrow at the center of the chart must face due north on the model.
Mount a peg exactly the height shown on the shadow plot where indicated, so that the peg remains perfectly vertical.

2. Find a sunny day and take the model and mounted chart out under full sun (indoor lamps do not work; they will give you divergent shadows). By tilting your model in the sun, you can make the end of the peg's shadow fall on any intersection on the shadow plot. Each intersection represents a given time on a given day. When the peg shadow falls on this intersection, the shadows and sun penetration in your model are the actual condition for that time of day and year.

3. Explore how the sun and shading on the site changes over the course of a day and seasonally over the year. On your model record the shadows cast at 9 a.m., noon, and 3 p.m. for the month you most need to block the sun and the month you most need to admit the sun. Transfer this information onto the two site plans which follow, one for each month you looked at.

4. Indicate (on the admit sun site plan) in which areas you could locate a 30' high building without blocking someone else's access to the sun during these hours. A more thorough discussion of solar rights (and the resulting "solar envelope" within which buildings can exist without shading their neighbors) can be found in Ralph Knowles Energy and Form.

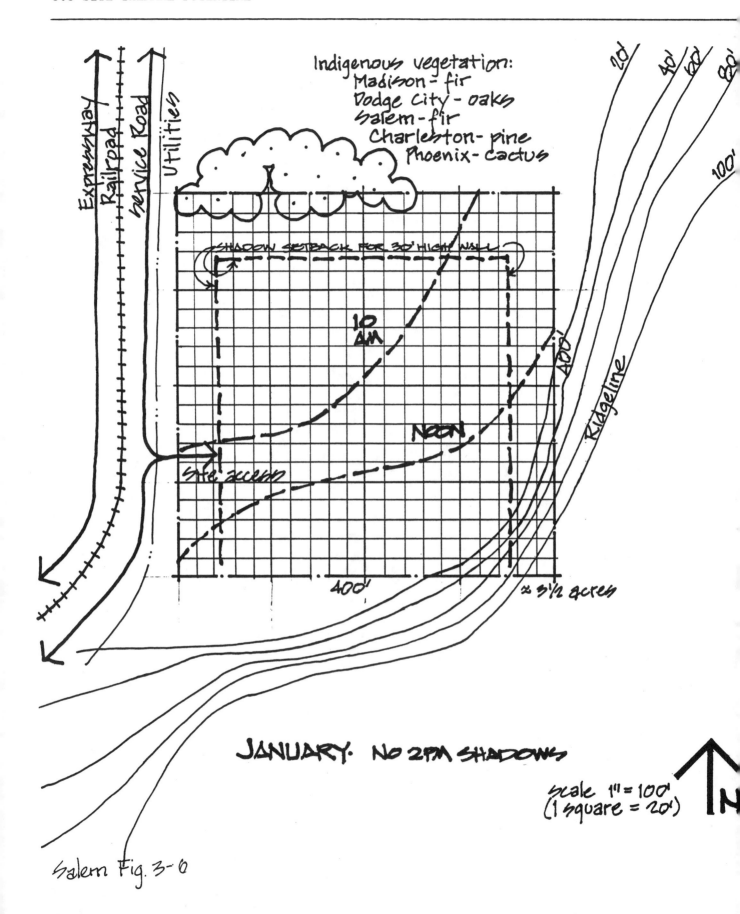

Indigenous vegetation:
Madison = fir
Dodge City - oaks
Salem - fir
Charleston - pine
Phoenix - cactus

SHADOW SETBACK FOR 30' HIGH WALL

10 AM

NOON

Site access

400'

≈ 5½ acres

Ridgeline

JANUARY: NO 2PM SHADOWS

scale 1" = 100'
(1 square = 20')

N

Salem Fig. 3-0

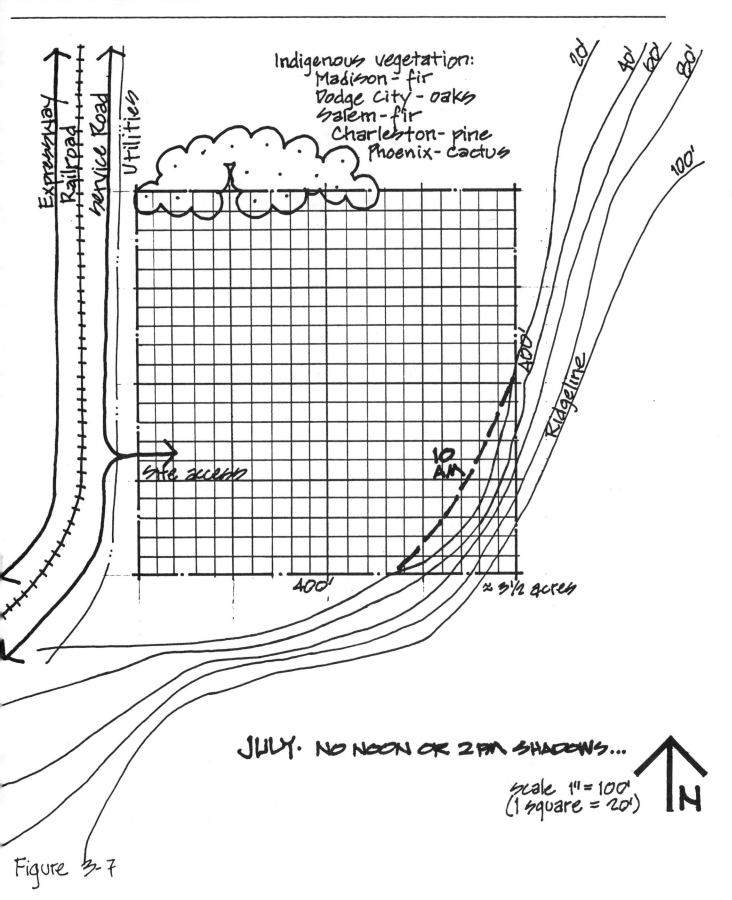

Indigenous vegetation:
Madison = fir
Dodge City - oaks
Salem - fir
Charleston - pine
Phoenix - cactus

JULY· NO NOON OR 2PM SHADOWS...

scale 1" = 100'
(1 square = 20')

Figure 3-7

C. Analysis Tool - "Climatic Response Matrix"
From the Bioclimatic Chart, you have found how to extend
the "comfort zone", by admitting or blocking sun or wind
at appropriate temperature/R.H. combinations. These conditions can be represented by a matrix:

	ADMT SUN	BLOCK SUN
ADMT WIND	A	B
BLOCK WIND	C	D

Diagram 2

As a result of sun position, wind direction, topography
and vegetation, these conditions may occur on your site,
resulting in "microclimates" that may be favorable building sites.
1. On the first of the following site plans, for your
hottest month, at noon, label each 100' x 100' square on
the site with the letter that describes conditions
occuring naturally.
2. On the other site plan repeat this procedure for noon
of your coldest month.

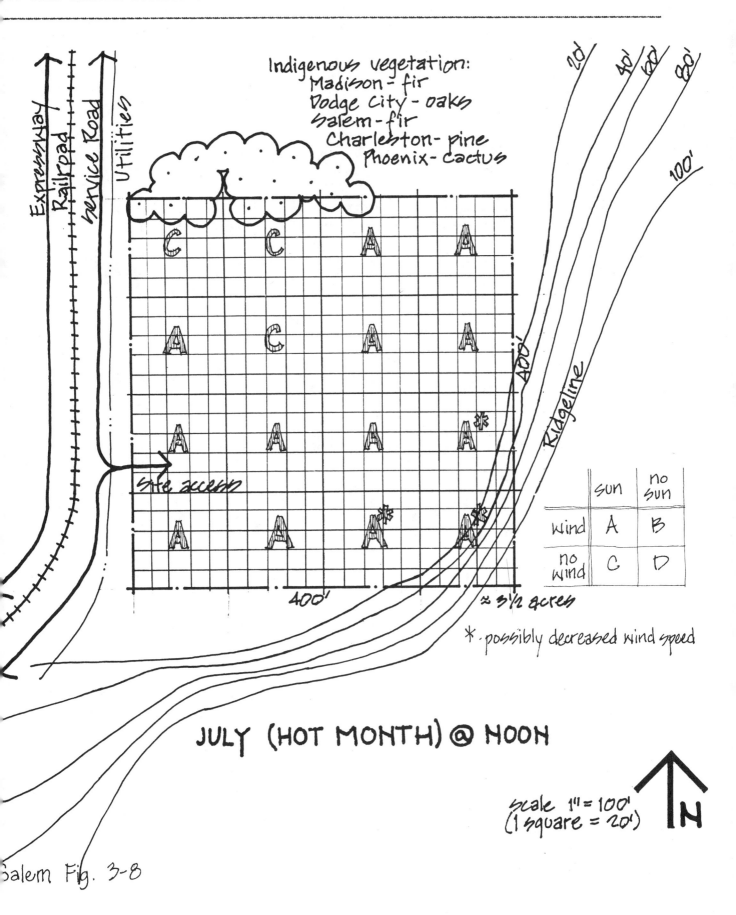

Indigenous vegetation:
Madison - fir
Dodge City - oaks
Salem - fir
Charleston - pine
Phoenix - cactus

Utilities

Service Road

Railroad

Expressway

20' 40' 60' 80'

100'

400'

Ridgeline

C C A A

A C A A

A A A A*

Site access

A A A* A*

400'

≈ 3½ acres

	sun	no sun
wind	A	B
no wind	C	D

* · possibly decreased wind speed

JULY (HOT MONTH) @ NOON

scale 1" = 100'
(1 square = 20')

N

Salem Fig. 3-8

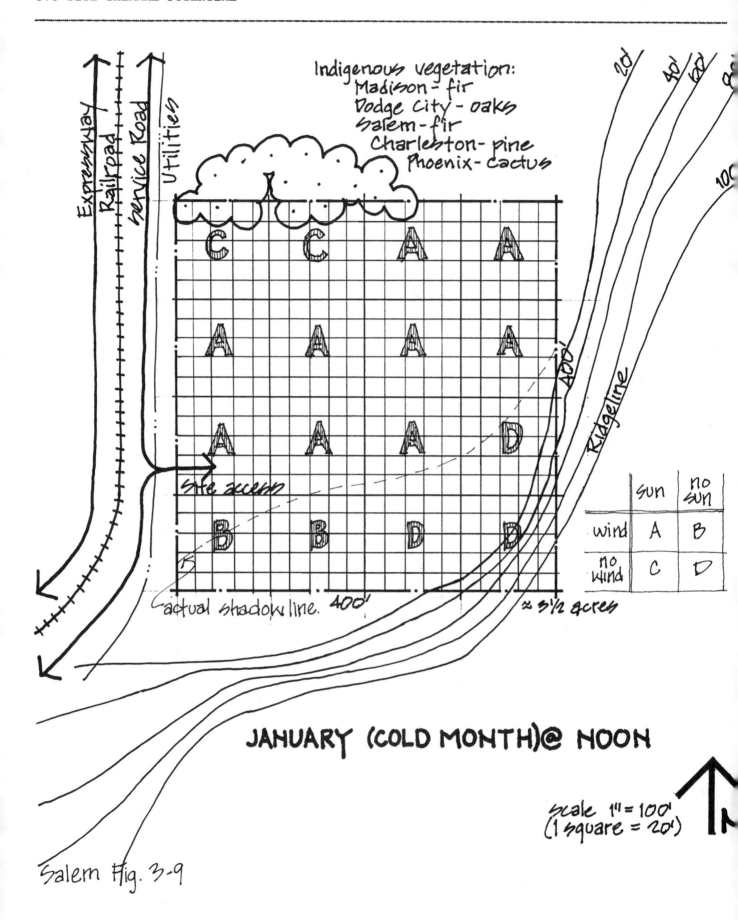

Indigenous vegetation:
  Madison - fir
  Dodge City - oaks
  Salem - fir
  Charleston - pine
  Phoenix - cactus

Expressway
Railroad
Service Road
Utilities

Site access

actual shadow line. 400'

≈ 3½ acres

Ridgeline

	sun	no sun
wind	A	B
no wind	C	D

JANUARY (COLD MONTH)@ NOON

scale 1" = 100'
(1 square = 20')

Salem Fig. 3-9

QUESTIONS

A. Which block/admit condition (A, B, C or D from the matrix) is the most favorable for heating?

C

B. Which is most favorable for cooling?

B

C. Considering the results of the last three analysis tools, which site location is optimum for heating?

Northwest quarter, due south of the fir trees

D. Which site location is optimum for cooling?

Northeast quarter, past wind shadow of trees in July and August

3.4 VISUAL COMFORT

GOAL

Determine how much light people need to do their work.

A. Analysis Tool - "Recommended Footcandle Levels"
Refer to the energy conservation lighting level recommendations, such as those in MEEB (p. 736) to complete this table:

Task	Recommended Footcandles	
Circulation, as in hallways	*10*	fc
Conference Areas	*30*	fc
Display areas	*50*	fc
Office Work	*50*	fc
Toilet rooms	*20*	fc
Work Stations and their immediate surroundings	*30*	fc

Assumed no more critical than "work areas"

on task only

See "casual work", MEEB table 18.5

3.5 ILLUMINATION POTENTIAL

GOAL

Determine if you will need lots of big windows, or a few little ones, or somewhere between to provide adequate daylighting in your climate.

A. Analysis Tool - "Sky Condition"
Use the NOAA climate data in 3.7 to graph your climate in terms of sky conditions:

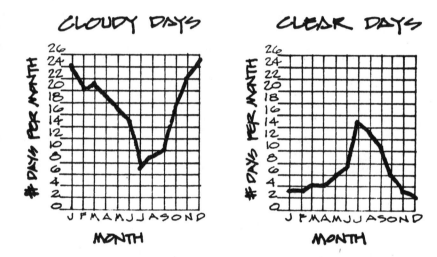

Mostly overcast in winter, when the least daylight is available.

Diagram 3

B. Analysis Tool - "Available Illumination"
The previous analysis tool indicates whether clear or overcast skies are prevalent in your climate throughout the seasons.
1. From tables of available daylight at various latitudes (MEEB Table 19.2a for overcast skies, Table 19.2c for clear skies), graph the following information on diagram 4:

3.5 ILLUMINATION POTENTIAL

Available exterior footcandles over a typical day (in each case, specify sky condition--overcast or clear?) for:
December 21
March or September 21
June 21

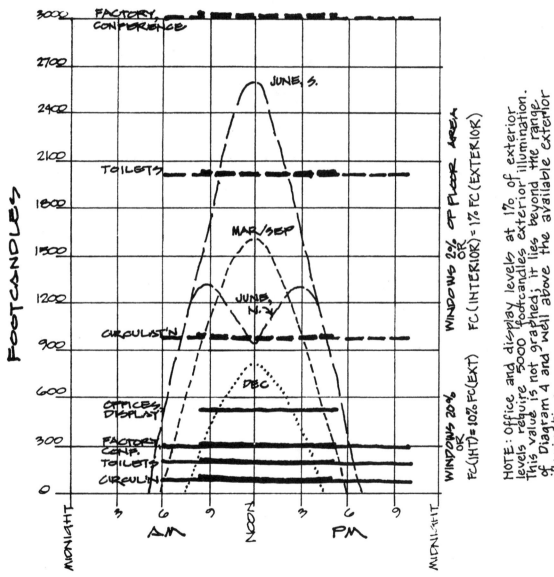

Diagram 4

Note: For seasons where clear sky predominates, graph both north and south windows.

Only June is dominantly clear sky.

C. Analysis Tool - "Required Exterior Illumination"
Roughly, "lots of big windows" (which amount to a total window area of about 1/5 of the total floor area) will produce <u>indoor</u> daylight levels of about 10% of outdoor levels. "A few little windows" (total window area about 1/50 of total floor area) will produce indoor daylight levels of about 1% of outdoor levels.

1. Extend the table you began in the previous analysis tool:

Task	Recommended Ft.C. (Interior)	Min. Required Exterior Illumination	
		If interior=10% exterior level	If interior=1% exterior level
Circulation	10	100	1000
Conference	30	300	3000
Display	50	500	5000
Office	50	500	5000
Toilets	20	200	2000
Work Stations	30	300	3000

2. Graph these minimum required exterior illumination levels, according to the daily hours of operation of your factory and office building, on diagram 4 in Analysis Tool B.

QUESTIONS

A. What percentage of exterior illumination would you attempt to provide <u>indoors</u> to meet the minimum task light requirements on your darkest days? *At least 10%*

What time of year is this? *December*

What is the sky condition? *Overcast*

B. What percentage of exterior illumination would you attempt to provide indoors to meet the minimum task light requirements on your brightest days? *About 2% for south windows; 5% for north windows*

What time of year is this? *June*

What is the sky condition? *Clear*

3.6 BUILDING LOAD

GOAL

Determine whether the buildings have an internally dominated load (IDL) or a skin dominated load (SDL)?

"Load" refers to an imbalance of heat between inside and outside that makes a building skin necessary. If a building has a "heating load" that means you need to provide heat. Conversely a building with a "cooling load" will need to be cooled.

IDL - means that the major thermal problem over the year in the building is caused by what's going on <u>inside</u> (internal), <u>not</u> by what's going on outside the building skin. What's going on inside the building is that people, lights, and machines are giving off large quantities of heat compared to how fast it is dissipated. Therefore the major problem for an internally dominated load building over the year will be removing heat, or cooling.

SDL - means that the major thermal problem over the year will be caused by forces acting <u>on the skin</u> of building. There may still be internal heat generated; however, the skin effects will be relatively more important.

Depending upon the weather, the forces acting on skin make it lose heat (cold environment) or gain heat (hot environment). The building's thermal problem will be either heating or cooling or both depending on the season. The distinction between an SDL and IDL building depends on how the building interacts with the climate and therefore on many factors: how hot and cold the climate is, how much heat is generated out of the building through the walls and with the air flow, and the overall building configuration.

A. Analysis Tool - "Internal Heat Gain Estimate"
1. Heat from people:

Factory: the number of people is determined by the number of work stations, _____people, total.   _16_

Office/conference building: In this case, the number of people is determined by total floor area. When the spaces defined in the program are added together, along with circulation, the total floor area will probably range from 1,000 sq.ft. to 2,000 sq.ft. Only the actual areas occupied by the functions of office and conference need to counted when determining occupancy.

From Program, Section 2:

Display and Conference space: *10,000 cu.ft. including mezzanine.*
Assume average height is 20' (10' below mezzanine, 30' or more in remainder).

$$\text{Floor area then} = \frac{10,000 \text{ cu.ft.}}{20' \text{ high}} = 500 \text{ sq.ft.}$$

$$\text{Occupancy} = \frac{500 \text{ sq.ft.}}{15 \text{ sq.ft. per person}} = 33 \text{ people, maximum}$$

Office space, total: *4,000 cu.ft.*
Assume average height is 10'

$$\text{Floor area then} = \frac{4,000 \text{ cu.ft.}}{10' \text{ high}} = 400 \text{ sq.ft.}$$

$$\text{Occupancy} = \frac{400 \text{ sq.ft.}}{100 \text{ sq.ft. per person}} = 4 \text{ people}$$

Total occupancy = 33 + 4 = 37 people, maximum.

For total floor area including "non-occupied areas", assume midway between the 1,000 to 2,000 sq.ft. estimate, or 1,500 sq.ft.

Factory floor area: volume 72,000 cu.ft. Assume average height is 30'

$$\text{Floor area then} \frac{72,000 \text{ cu.ft.}}{30' \text{ high}} = 2,400 \text{ sq.ft.}$$

Use	sq.ft. floor area per occupant
1. Aircraft Hangars (no repair)	500
2. Auction Rooms	7
3. Assembly Areas, Concentrated Use (without fixed seats)	7
Auditoriums	
Bowling Alleys (Assembly areas)	
Churches and Chapels	
Dance Floors	
Lodge Rooms	
Reviewing Stands	
Stadiums	
4. Assembly Areas, Less-concentrated Use	15
Conference Rooms	
Dining Rooms	
Drinking Establishments	
Exhibit Rooms	
Gymnasiums	
Lounges	
5. Children's Homes and Homes for the Aged	80
6. Classrooms	20
7. Dormitories	50
8. Dwellings	300
9. Garage, Parking	200
10. Hospitals and Sanitariums - Nursing Homes	80
11. Hotels and Apartments	200
12. Kitchen - Commercial	200
13. Library Reading Room	50
14. Locker Rooms	50
15. Mechanical Equipment Room	300
16. Nurseries for Children (Day-care)	50
17. Offices	100
18. School Shops and Vocational Rooms	50
19. Stores - Retail Sales Rooms	
Basement	20
Ground Floor	30
Upper Floors	50
20. Warehouses	300
21. All others	100

Figure 3-2 Floor Area Per Occupant.

Use the above table from the Uniform Building Code to figure

office area occupancy: ____people   *4*

conference area occupancy: ____people   *33 (maximum)*

Total number of people: factory + office + conference = people.   *53*

Estimate the heat generated by these people from MEEB Table 4.30 (p. 175). Use sensible heat gains, assuming moderately active office work (____Btuh per person) and   *250 Btuh per person*
light bench work in the factory (____Btuh per person).   *275 Btuh per person*
2. Heat from electric lighting. Compare recommended footcandles, section 3.4, to MEEB Figure 20.2 (p. 867).
With reasonably efficient sources such as fluorescents, the work and surrounding circulation area footcandle levels can be obtained by installing ____watts/sq.ft. of   *office: 1.5 watts/sq.ft.*
electric lighting. To convert to heat (col. 5) 1 watt =   *(goal: 30 footcandles, general)*
3.41 Btuh.)
3. Business machinery in the office building may be   *factory: 3.3 watts/sq.ft.*
assumed at 1 watt/sq.ft. (half that for which electric   *(program specifies 500 watts*
circuits are designed). See MEEB Table 17.1, p. 645)   *per 150 sq.ft. station)*
Again for col. 5, 1 watt = 3.41 Btuh.
4. Factory machinery heat generation for col. 5 may be figured by the standard conversion factor of 1 hp = 746 watts = 2544 Btuh.

---

5. Summarize your internal heat gains as follows:

Col.1 Building	Col. 2 Heat Source	Col. 3 Area(sq.ft.) x	Col. 4 Unit per sq.ft. x	Col. 5 Btuh per unit =	Col. 6 Btuh total
Office:	. People	.(number of people= 37)		. 250 Btuh	9,250
	. Lights	. 1,500 total	1.5 watts/sq.ft.	3.41 Btu/watt	7,670
	. Machines	. 400 office	1.0 watts/sq.ft.	3.41 Btu/watt	1,360
				18,280/1,500 = 12.2 .Total Btuh/sq.ft. office	
Factory:	. People	.(number of people= 16)		. 275 Btuh	4,400
	. Lights	. 16 stations	500 watt/station	3.41 Btu/watt	27,280
	. Machines	. 16	. 2.5 hp	. 2,544 Btu/hp	101,760
				133,440/2,400 = 55.6 .Total Btuh/sq.ft. factory	

B. Analysis Tool - "Balance Point Charts"
The external temperature at which internal loads are enough to adequately heat a building is called the "balance point" for that building. So, if the balance point of a given building is 40°: the building must be cooled when the external temperature is above 40°, and the building must be heated when the external temperature is below 40°.

The balance point depends on the climate, the internal gains, and the skin/volume characteristics. Once the internal heat gain has been estimated, the skin area must be examined to see how it interacts with the climate outside and the heat gains inside. The volume is also important, because it so often directly relates to the amount of internal heat produced. A long, thin cylinder has relatively more skin, and less volume, than a sphere; the cylinder's skin is likely to dominate its thermal behavior, the sphere's volume is likely to dominate its thermal behavior.

A building with relatively little skin area and a lot of
internal gain will be an "internal-dominated load" (IDL)
building. The hotter the climate, the more severely the
internal loads will dominate, and the more difficult it
will be to get rid of excess internal heat in summer.

A building with relatively great skin area and not much
internal gain will be a "skin-dominated load" (SDL)
building. The colder the climate, the more severely the
skin loads will dominate and the more difficult it will
be to keep the interior warm enough in winter.

1. Skin/volume characteristics. Shown here are several
ratios of skin area to total enclosed volume for buildings similar to the office and factory:

Total volume 20,000-27,000 cuft (similar to office
building)
Low skin/volume: (30' high, 30' wide, 30' long): .167
High skin/volume: (20' high, 10' wide, 100' long): .27

Total volume 72,000-75,000 cuft (similar to factory)
Low skin/volume: (30' high, 50' long, 50' wide): .113
High skin/volume: (30' high, 20' wide, 120' long): .15

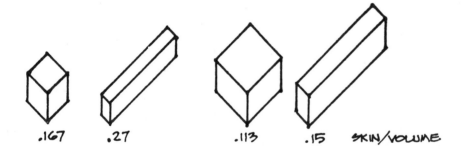

.167   .27   .113   .15   SKIN/VOLUME

Diagram 5

---

2. Use these skin/volume ratios for each building and the internal heat gains from the previous analysis tool to plot the buildings on the charts below. The charts will give you the estimated balance point temperature. Plot both high and low skin/volume ratios and the occupied hourly gains. Then, plot the balance point(s) of your buildings on the Chart of Average Temperatures in the next analysis tool, Diagram 7, and on the Bioclimatic Chart, back in section 3.2.

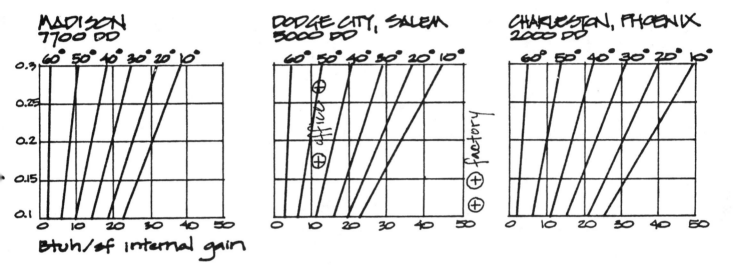

Diagram 6

Figure 3-3 Balance Point Charts

Technical note: To generate these charts, the following assumptions were made:
Skin heat loss: roof and wall U_o were half those allowed for "type B" buildings, by ASHRAE 90-75 (see Heating/Cooling Finalization Exercise), in keeping with recent more strict standards. Balance points vary with climate because ASHRAE 90-75 recommendations for Uo vary with climate. Further, roof area assumed to be 1/3, walls 2/3, of total skin area.
Infiltration (lung heat loss): assumed at 1 ACH (air change per hour)
The floor area was kept constant as the skin/volume ratios changed.
The internal temperature was held constant at 65°F.

C. Analysis Tool – "Graph Average Temperatures"

On the graph below, plot the average daily maximum and minimum temperatures, by month, from the NOAA data you used earlier.

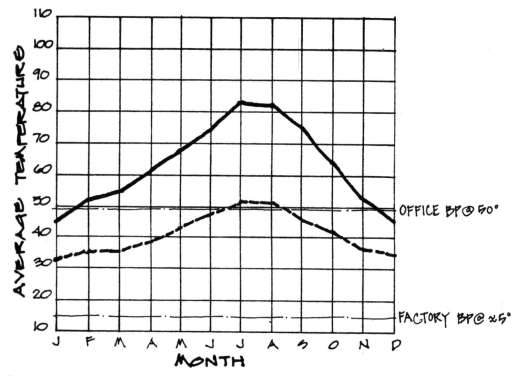

Diagram 7

QUESTIONS

A. Will the skin/volume ratio affect the Balance Point of the office significantly?

No; could move from low of 45° to high of about 50°

Of the factory?

No; balance point is well below the lowest average monthly temperature

B. On which months will the balance point for each building be <u>greater than</u> the temperature outside (heat required)?
office:

factory:

Assume balance temperature of 50°; this would require heat November through April.

Balance point is always low; heat is never required.

C. In which months will the balance point for each building be <u>less than</u> the temperature outside (cooling required)?

office:

    May through October

factory:

    All year

D. Which is the dominant problem, heating or cooling, for each building?

office:

    Heating, slightly

factory:

    Cooling, absolutely

E. Daylighting is particularly important for buildings with cooling problems. Does your daylight analysis indicate enough exterior daylight to adequately daylight tasks inside?

    Yes, with windows equal to about 10% of floor area in the factory

F. Would you design to achieve a high or low skin/volume for the factory?
Why?

    High – Although it appears less significant, it still helps expose the building to cooling breezes.

For the office?
Why?

    Medium – Heating slightly dominates; will compromise between high and low

G. What % of internal gains is due to lighting in the office, factory? What % is due to people and machines in the office, factory?

    Office lighting:
    7670/18280 = 42%
    Factory lighting:
    27280/133440 = 20%
    Office people & machines 58%
    factory people & machines 80%

H. Will considerations of heating, cooling or lighting have the greatest influence on the form of your office, factory?

    Office: <u>Heating</u>; then Daylighting, <u>Cooling</u>.

    Factory: <u>Cooling</u>; then Daylighting, Heating.

SALEM, OREGON

Narrative Climatological Summary

Salem is located in the middle Willamette Valley some 60 airline miles east of the Pacific Ocean. The valley here is approximately 50 miles wide with the City about equidistant from the valley walls formed by the Coast Range on the west and the Cascade Range on the east.

The usual movement of very moist maritime air masses from the Pacific Ocean inland over the Coast Range produces near its crest some of the heaviest yearly rainfall in the United States. An annual total of nearly 170 inches has been recorded, and one station established a period-of-record annual average of approximately 130 inches. From the ridge crest of the Coast Range, approximately 3,000 feet above sea level, there is a gradual decrease of rainfall downslope to the valley floor where annual totals average between 35 and 45 inches. As these marine conditioned air masses continue to move farther inland they are forced to ascend the west slopes of the Cascades to approximately 5,000 feet above sea level and again rainfall amounts substantially increase with elevation.

Most of this precipitation in both the valley and its bordering mountain ranges occurs during the winter. At Salem 70 percent of the annual total occurs during the five months of November through March while only 6 percent occurs during the three summer months, with practically all of it falling in the form of rain. In the immediate area, on the average, there are only three or four days a year with measurable amounts of snow. Its depth on the ground rarely exceeds 2 or 3 inches, and it usually melts in a day or two. The few thunderstorms that occur each year are not generally severe and seldom do they, or the hail that occasionally accompanies them, cause any serious damage. A tornado in the immediate metropolitan area has never been recorded.

The seasonal difference in temperatures is much less marked than that of precipitation. There is only a range of about 28° between the mean temperature for January, the coldest month, and July, the warmest. Maximums as high as 100° or more seldom occur, and in only four years during the station's 84 years of record have minimums of 0° or lower been observed. A 30-year average for the dates of the last occurrence in spring and first in fall of temperatures of 32° or lower are respectively April 14 and October 28, though these have been recorded as late as May 28 and as early as September 12. This provides an average growing season of six and a half months.

The mild temperatures, long growing season and plentiful supply of moisture are ideally suited for a wide variety of crops. In dollar value of agricultural returns, this is the most productive land in Oregon. Large orchards of sweet cherries are grown and processed here for maraschino cherries. Hops, filberts, walnuts, cane, and strawberries each contribute many millions of dollars to the annual farm income. A wide variety of vegetables is raised for both the fresh market and to support a large number of processing plants located in Salem. This climate is also suitable for the production of a number of specialty crops including mint, several seed crops, and nursery stock, particularly roses and ornamental shrubs.

noaa   NATIONAL OCEANIC AND   / ENVIRONMENTAL DATA AND / NATIONAL CLIMATIC CENTER
        ATMOSPHERIC ADMINISTRATION   / INFORMATION SERVICE   / ASHEVILLE, N.C.

Normals, Means, And Extremes

Salem, Oregon (McNary Field); Pacific Standard Time.
Latitude: 44°55'N; Longitude: 123°00'W; Elevation (ground) 196 feet.

Month	Temperatures °F Normal Daily maximum	Normal Daily minimum	Normal Monthly	Extremes Record highest	Year	Record lowest	Year	Normal Degree days Base 65°F Heating	Cooling	Precipitation Water equivalent Normal	Maximum monthly	Year	Minimum monthly	Year	Maximum in 24 hrs	Year	Snow, Ice pellets Maximum monthly	Year	Maximum in 24 hrs	Year
(a)				42		42					42		42		42		42		42	
J	45.3	32.2	38.8	64	1971	-10	1950	812	0	6.90	15.40	1953	0.57	1949	3.07	1972	32.8	1950	10.8	1943
F	51.4	34.4	42.9	72	1968	-4	1950	619	0	4.79	12.31	1949	0.78	1964	3.16	1949	8.4	1962	5.4	1962
M	54.9	35.4	45.2	80	1947	12	1971	614	0	4.33	8.42	1938	0.87	1965	3.03	1943	10.9	1951	8.5	1960
A	61.0	38.5	49.8	88	1957	23	1968	456	0	2.29	5.18	1955	0.39	1939	2.22	1971	0.1	1972	0.1	1972
M	68.1	43.3	55.7	95	1956	25	1954	295	7	2.09	4.58	1942	0.18	1947	1.84	1963	T	1978	T	1978
J	74.0	48.4	61.2	102	1942	32	1976	133	19	1.39	3.60	1947	0.01	1951	1.60	1950	0.0		0.0	
J	82.4	50.7	66.6	108	1941	37	1962	43	92	0.35	1.80	1974	0.00	1967	0.87	1961	0.0		0.0	
A	81.3	50.9	66.1	106	1978	38	1969	53	87	0.75	4.17	1968	T	1970	1.25	1943	0.0		0.0	
S	76.5	47.3	61.9	103	1944	26	1972	120	27	1.46	3.98	1971	0.00	1975	1.86	1951	0.0		0.0	
O	64.1	42.5	53.2	93	1970	23	1971	366	0	3.98	11.17	1947	0.37	1978	2.84	1955	T	1974	T	1974
N	53.0	37.4	45.2	72	1958	9	1955	594	0	6.08	15.23	1973	0.84	1939	2.82	1950	6.1	1977	6.1	1977
D	47.1	34.7	40.9	64	1958	-12	1972	747	0	6.85	12.40	1964	1.26	1976	2.72	1964	14.6	1972	9.4	1972
YR	63.3	41.3	52.3	108 JUL 1941		-12 DEC 1972		4852	232	41.08	15.40	1953	0.00	1975	3.16	1949	32.8	1950	10.8	1943

(Handwritten annotations: "← Jan: coldest month"; "← Jul: hottest month")

(Handwritten annotation above RH Hour 16 column: "% relative humidity @ max. temp.")

Month	Relative humidity pct. Hour 04	Hour 10	Hour 16	Hour 22	Wind Mean speed m.p.h.	Prevailing direction	Fastest mile Speed m.p.h.	Direction	Year	Pct. of possible sunshine	Mean sky cover, tenths, sunrise to sunset	Mean number of days Clear	Partly cloudy	Cloudy	Precipitation .01 inch or more	Snow, Ice pellets 1.0 inch or more	Thunderstorms	Heavy fog, visibility ¼ mile or less	Temperatures °F Max. 90° and above	Max. 32° and below	Min. 32° and below	Min. 0° and below	Average station pressure mb. Elev. 201 feet m.s.l.
	17	17	17	17	31	15	30	30		35	42	42	42	42	42	42	42	17	17	17	17	7
JAN	85	83	76	85	8.5	S	40	18	1966		8.3	3	4	24	19		•	6	0	2	11	0	1011.8
FEB	87	81	69	84	7.8	S	46	18	1958		8.1	3	5	20	17		•	2	0	0	11	0	1009.7
MAR	86	75	60	80	8.1	S	40	19	1971		7.9	6	6	21	17		•	2	0	0	7	0	1009.1
APR	85	69	57	78	7.2	S	44	18	1962		7.4	7	7	19	14	0	1	0	0	0	7	0	1010.9
MAY	85	65	52	75	6.6	S	28	25	1975		6.8	6	8	17	11	0	1	0	0	0	1	0	1010.9
JUN	85	61	49	73	6.5	S	25	18	1974		6.4	7	8	15	8	0	1	2	0	0	•	0	1010.7
JUL	84	57	40	68	6.5	N	26	24	1979		4.0	15	9	7	3	0	0	0	0	0	0	0	1009.9
AUG	85	59	41	71	6.3	NW	24	29	1969		4.7	13	9	9	5	0	1	6	0	0	0	0	1009.0
SEP	87	65	47	78	6.1	S	31	18	1969		5.1	11	9	10	7	0	1	2	0	0	•	0	1008.8
OCT	90	77	61	85	6.2	S	58	18	1962		6.8	6	8	17	13	0		7	0	0	3	0	1010.8
NOV	90	85	77	88	7.3	S	38	18	1951		8.1	3	5	22	18	•		7	0	1	8	0	1011.3
DEC	88	86	81	87	8.2	S	45	23	1953		8.8	2	4	25	20	1		7	0	1	12	•	1012.5
YEAR	86	72	59	79	7.1	S	58	18 OCT 1962			6.9	77	82	206	149	2	5	38	17	3	68	•	1010.5

(Handwritten annotations: "← Jan: coldest month"; "← Jul: hottest month"; below table: "% relative humidity @ min. temp.")

Means and extremes above are from existing and comparable exposures. Annual extremes have been exceeded at other sites in the locality as follows: Maximum monthly precipitation 17.54 in December 1933; maximum precipitation in 24 hours 4.30 in December 1933; maximum snowfall in 24 hours 25.0 in February 1937.

(a) Length of record, years, through the current year unless otherwise noted, based on January data.
(b) 70° and above at Alaskan stations.
* Less than one half.
T Trace.

NORMALS - Based on record for the 1941-1970 period.
DATE OF AN EXTREME - The most recent in cases of multiple occurrence.
PREVAILING WIND DIRECTION - Record through 1963.
WIND DIRECTION - Numerals indicate tens of degrees clockwise from true north. 00 indicates calm.
FASTEST MILE WIND - Speed is fastest observed 1-minute value when the direction is in tens of degrees.

5.1 INTRODUCTION _____

This exercise deals with heating, which is a major end-use of energy for many buildings in most of the United States.

Your climate analysis in Exercise 3.0 helped to identify whether heating, cooling, or daylighting (or some combination) was the dominant concern in each building. This should give you a sense of to what degree you need to keep in mind cooling and daylighting objectives in the design for each building.

Note that the sun peg evaluation tool is used to check your design for a number of criteria in the scheming and design section. This tool requires both a model and sunshine, so plan your work accordingly.

Contents

_____

A. Criterion A: Sun Peg Check, p. 5-34
B. Criterion B: Aperture & Mass; Sizing for Solar
   Savings, p. 5-34
C. Criterion C: Overall Heat Loss - ASHRAE 90-75
   Procedure, p. 5-38

5.2 CONCEPTUALIZING

GOALS

A. Use the sun for heating.
B. Prevent losing the heat you have collected.

DESIGN STRATEGIES

A. Site Scale
Sunny spaces for winter places.
B. Cluster Scale
Make winter places warmer by arranging buildings to form
sun traps.

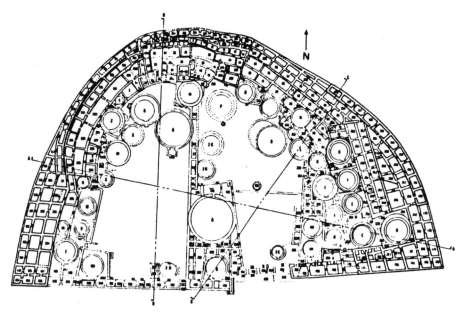

New Pueblo Bonito.   Reprinted from Energy and Form by
Ralph Knowles by permission of the MIT Press, Cambridge,
Massachusetts.
*(This more general part of these exercises is not
included in the Salem, Or example solution.)*

C. Building Scale
Increase the south facing skin area.

St. George's School, Wallasey, England.   From <u>Process</u>
<u>Architecture</u> No. 6 "Solar Architecture".

D. Component Scale
Windows for solar gain are potentially different from
windows for light, view, or ventilation.
The greater the wall mass the less the indoor temperature
fluctuates.

DESIGN

Choose an existing building/site which has a clear conceptual approach, at any scale, to achieving these goals.
Document your choice as follows:
A. Identify the location, program, architect (if any),
and the source of your information.
B. Include photocopies or drawings (whatever is quick and
easy for you) to explain the design.

EVALUATION

Evaluation tool - "Conceptual Diagrams".
Diagram how this design is organized to achieve these
goals.

5.3 SCHEMING

GOALS

A. Expose yourself to the south in proportion to the heat you need.
B. Protect yourself from the wind to the extent you need heat.

CRITERIA

A. Buildings are located on the site so they don't cast shadows off the site.
B. The buildings are clustered and sited to create optimal microclimates for heating in the months you need heat.
C. The necessary quantity of south aperture for heating in your climate is provided.
D. The winter sun can reach all the south aperture you have provided for collection.
E. (If you have completed the cooling exercise) Your buildings still meet the schematic phase criteria for cooling, section 5.3.

DESIGN STRATEGIES

A. Site Scale
Use wind breaks to protect building clusters without shading them in winter.

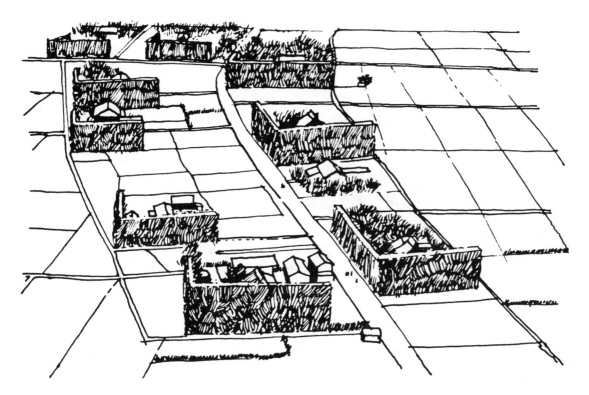

Windbreaks in Shi Mani prefecture, Japan. Reprinted by permission of the publisher from <u>Introduction to Landscape Architecture</u> by Michael Laurie, p. 176, copyright 1975 by Elsevier North Holland, Inc.

B. Cluster Scale
Arrange buildings so that winter wind is blocked from sunny exterior spaces.
C. Building Scale
Put the spaces with higher internal gains on the colder side of the building.

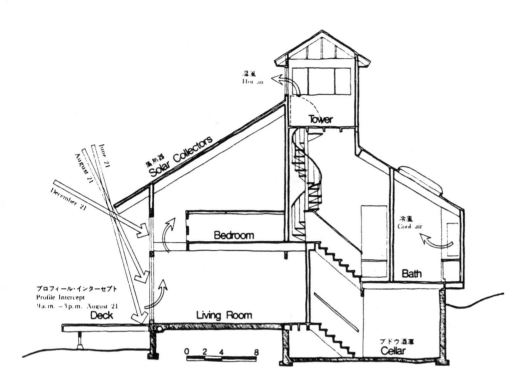

Interactive Resource Inc., Neihuser House (Process 6, 1978, p. 148)

Use spaces that can tolerate greater temperature variations as buffer areas.

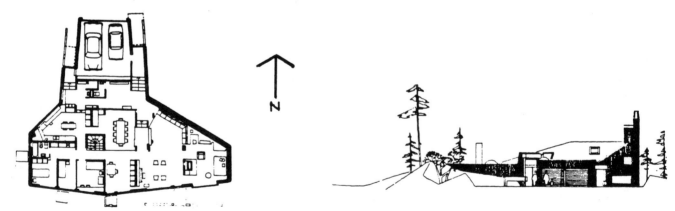

Erskine, Villa Gadelius, 1961 (A.D., vol. 47, p. 784)

The orientation of the solar glazing should be between 20° east and 32° west of south (Balcomb, 1980, p. 28).

Recommended areas of south facing glazing ("solar aperture") for a small, SDL building (Balcomb, 1980, Table D-1) are:

Location	Aperture as % of floor area
Madison	20-40%
Dodge City	12-23
Salem	12-24
Phoenix	6-12
Charleston	7-14

We choose 24% aperture to get a higher solar input (but higher indoor temperature swings...).

Figure 5-1 Solar Aperture Sizing for Heating

Note: a similar rule of thumb does not exist for predominantly IDL buildings, although they may have a heating load under some conditions. The % solar aperture required would be proportionally smaller, as the % heat provided by internal sources was larger. One unsubstantiated way to adjust these numbers to reflect internal gains is to use a 24 hr. avg. Balance Point Temp. that is the assumed Balance Point for the Btuh daily <u>average</u> gain. (Use the total Btuh and balance Point Charts from Climate 3.6, Building Load.)

$$\text{Btuh daily avg. gain} = \frac{(\text{Btuh gain while occupied}) \times (\text{\# hours occupied})}{24 \text{ hours}}$$

$$\text{\% Adjusted aperture} = \text{\% Aperture} \times \frac{\text{Balance Point Temp (24 hr. avg.)}}{65^{\circ}}$$

Office daily avg. gain:
$$\frac{18,280 \times 8 \text{ hours per day}}{24}$$
6,090 Btuh ÷ 1,500 sq.ft. = 4.1 Btuh/sq.ft.

Office % aperture:
$$24\% \times (\frac{55^{\circ} \text{ approx. balance pt.}}{65^{\circ}})$$
= 20% adjusted aperture

DESIGN

Design the office, factory and outside areas according to program requirements in 2.1 building/site description. Document your design as follows using the grids provided:

A. Site plan including parking and access drive(s)
B. Cluster plan including outdoor space
C. Floor plans of office and factory
D. Section(s) of office and factory
E. Roof plan and elevations, or paralines which show all sides and top of the buildings

Note: Your design may change in the course of the evaluation. The final schematic design which you show to meet all criteria should be fully documented here.

Factory daily avg. gain:
$$\frac{133,440 \times 16 \text{ hours per day}}{24}$$
88,960 Btuh ÷ 2,400 sq.ft. = 37.1 Btuh/sq.ft.

Factory % aperture:
$$24\% \times (\frac{5^{\circ} \text{ approx. balance pt.}}{65^{\circ}})$$
= 2% adjusted aperture

See Salem example Figures 5-1 - 5-6 for design documentation.

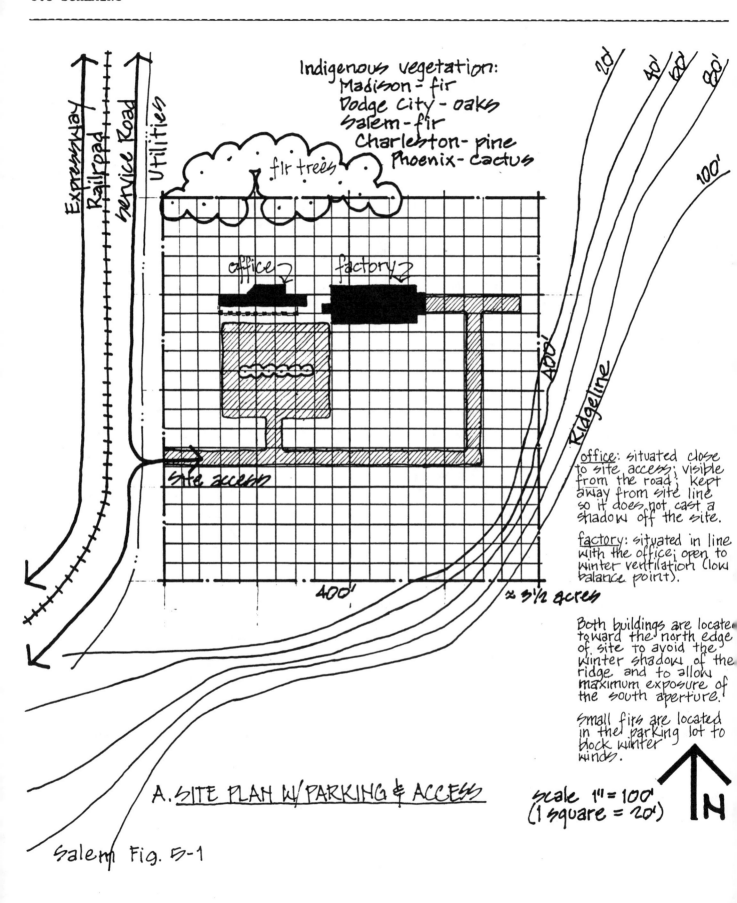

Indigenous vegetation:
Madison - fir
Dodge City - oaks
Salem - fir
Charleston - pine
Phoenix - cactus

office: situated close to site access; visible from the road; kept away from site line so it does not cast a shadow off the site.

factory: situated in line with the office; open to winter ventilation (low balance point).

Both buildings are located toward the north edge of site to avoid the winter shadow of the ridge and to allow maximum exposure of the south aperture.

Small firs are located in the parking lot to block winter winds.

A. SITE PLAN W/ PARKING & ACCESS

scale 1" = 100'
(1 square = 20')

Salem Fig. 5-1

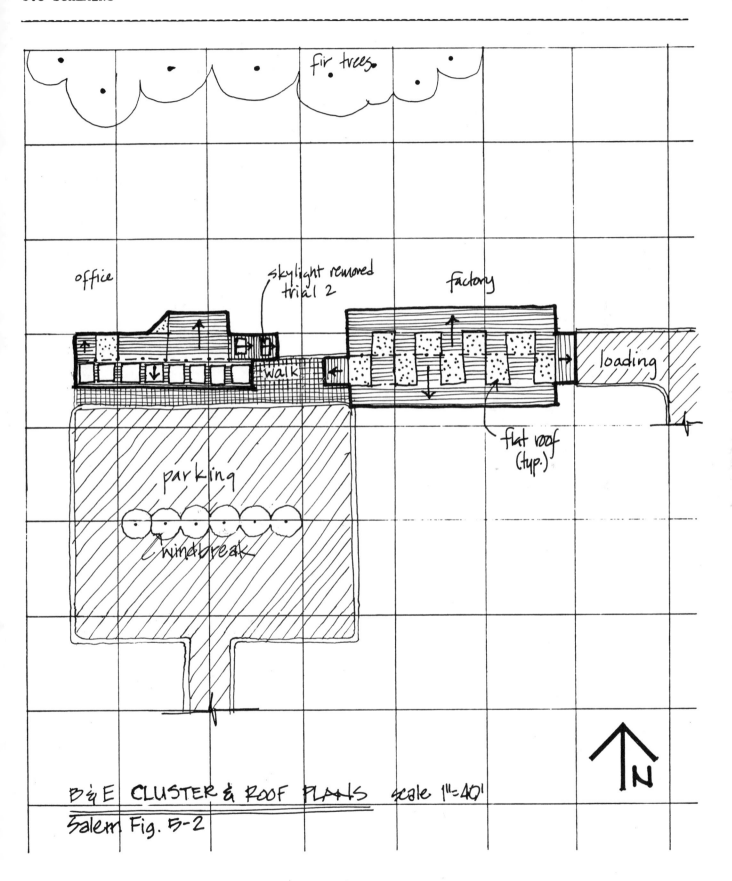

fir trees

office

skylight removed
trial 2

factory

walk

loading

flat roof
(typ.)

parking

windbreak

N

B & E CLUSTER & ROOF PLANS scale 1"=40'

Salem Fig. 5-2

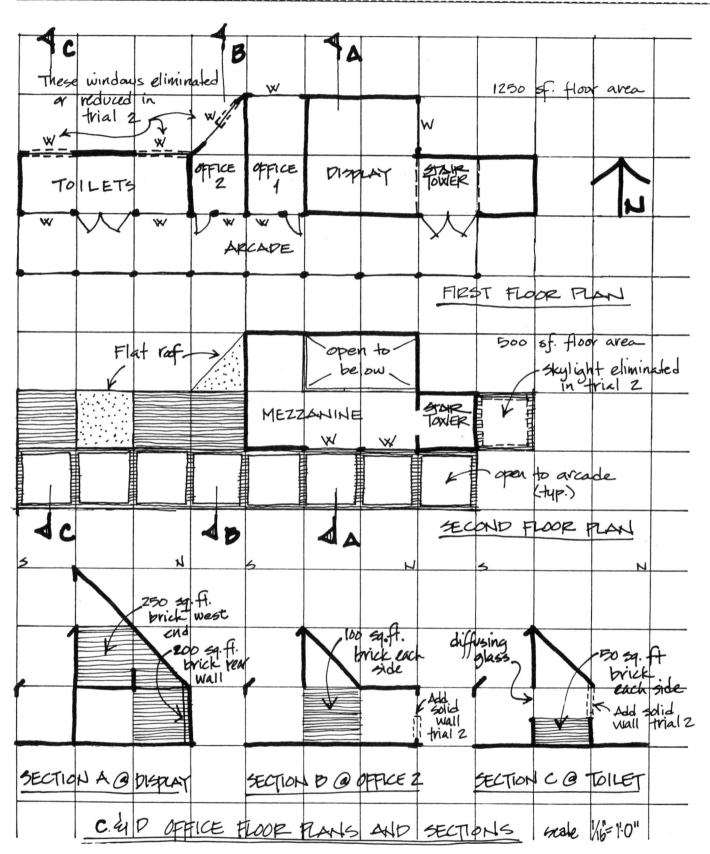

These windows eliminated or reduced in trial 2

1250 sf. floor area

W W W W

TOILETS   OFFICE 2   OFFICE 1   DISPLAY   STAIR TOWER

W W W W

ARCADE

N

FIRST FLOOR PLAN

Flat roof

open to below

500 sf. floor area

skylight eliminated in trial 2

MEZZANINE   STAIR TOWER

W W

open to arcade (typ.)

SECOND FLOOR PLAN

S   N   S   N   S   N

250 sq.ft. brick west end

200 sq.ft. brick rear wall

100 sq.ft. brick each side

diffusing glass

50 sq.ft. brick each side

Add solid wall trial 2

Add solid wall trial 2

SECTION A @ DISPLAY   SECTION B @ OFFICE 2   SECTION C @ TOILET

C & D OFFICE FLOOR PLANS AND SECTIONS   scale 1/16"=1'-0"

Salem Fig. 5-3

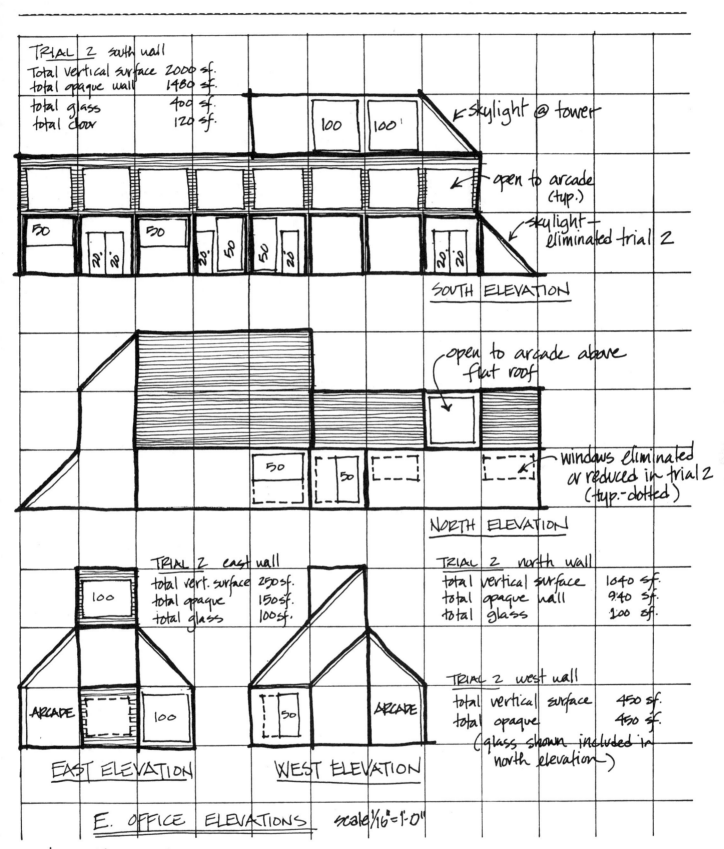

TRIAL 2 south wall
Total vertical surface 2000 sf.
total opaque wall   1480 sf.
total glass   400 sf.
total door   120 sf.

100   100'

← skylight @ tower

← open to arcade (typ.)

50   50

← skylight —
eliminated trial 2

20" 20"   20" 50   50 20"   20" 20"

SOUTH ELEVATION

open to arcade above
flat roof

50   50

← windows eliminated
or reduced in trial 2
(typ.-dotted)

NORTH ELEVATION

100   TRIAL 2 east wall   TRIAL 2 north wall
                     total vert. surface 250 sf.   total vertical surface   1040 sf.
                     total opaque   150 sf.   total opaque wall   940 sf.
                     total glass   100 sf.   total glass   200 sf.

ARCADE   100   50   ARCADE   TRIAL 2 west wall
                                           total vertical surface   450 sf.
                                           total opaque   450 sf.
                                           (glass shown included in
                                           north elevation)

EAST ELEVATION   WEST ELEVATION

E. OFFICE ELEVATIONS   scale 1/16"=1'-0"

Salem Fig. 5-4

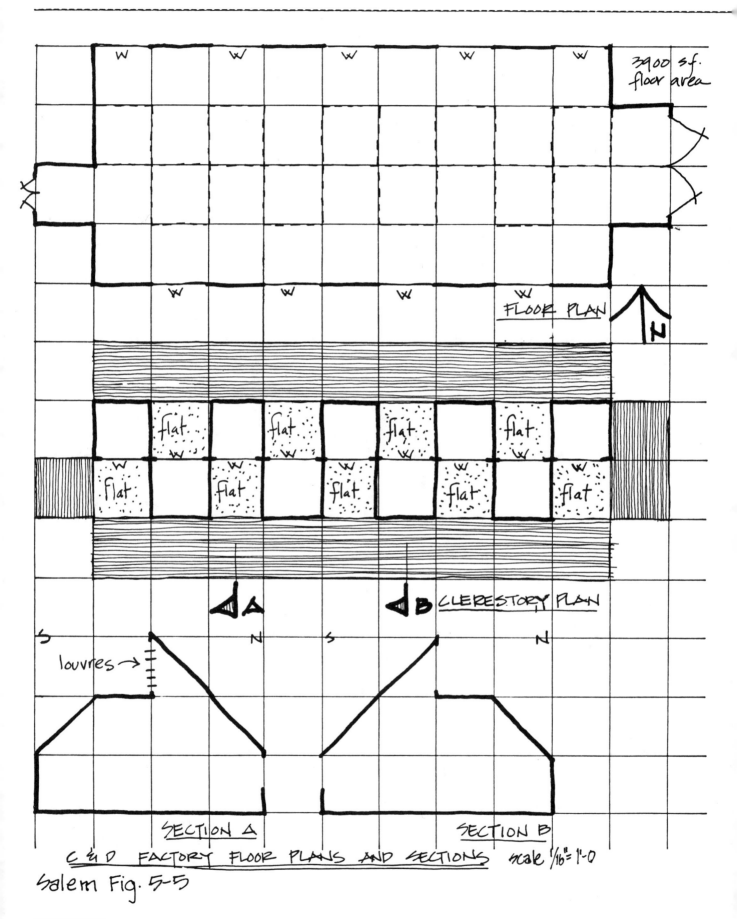

3900 s.f. floor area

FLOOR PLAN

N

flat flat flat flat flat
flat flat flat flat flat

A   B CLERESTORY PLAN

S   N   S   N

louvres →

SECTION A   SECTION B

C & D FACTORY FLOOR PLANS AND SECTIONS scale 1/16"=1'-0

Salem Fig. 5-5

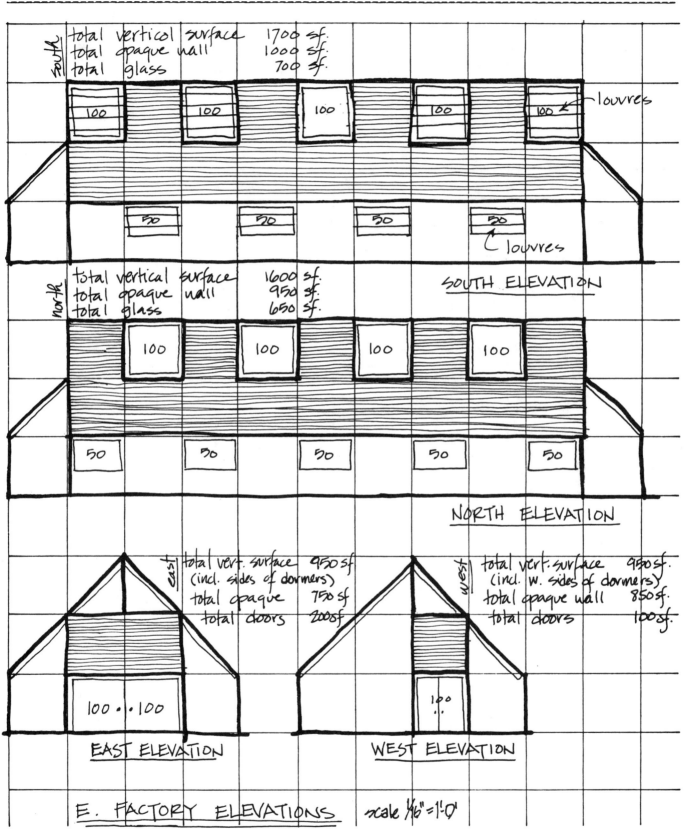

total vertical surface 1700 sf.
total opaque wall 1000 sf.
total glass 700 sf.

south

100   100   100   100   100 ← louvres

50   50   50   50 ← louvres

SOUTH ELEVATION

total vertical surface 1600 sf.
total opaque wall 950 sf.
total glass 650 sf.

north

100   100   100   100

50   50   50   50   50

NORTH ELEVATION

east   total vert. surface 950 sf
(incl. sides of dormers)
total opaque 750 sf
total doors 200sf

100 •.• 100

EAST ELEVATION

west   total vert. surface 950 sf.
(incl. w. sides of dormers)
total opaque wall 850 sf.
total doors 100sf.

100
..

WEST ELEVATION

E. FACTORY ELEVATIONS   scale 1/16"=1'-0"

Salem Fig. 5-6

for plans, sections, & elevations: if 1/16"=1'-0", 1 square = 10'.

for plans, sections, & elevations: if 1/16"=1'-0",   1 square = 10'.

for plans, sections, & elevations: if 1/16"=1'-0", 1 square = 10'.

for plans, sections, & elevations: if 1/16"=1'-0", 1 square = 10'.

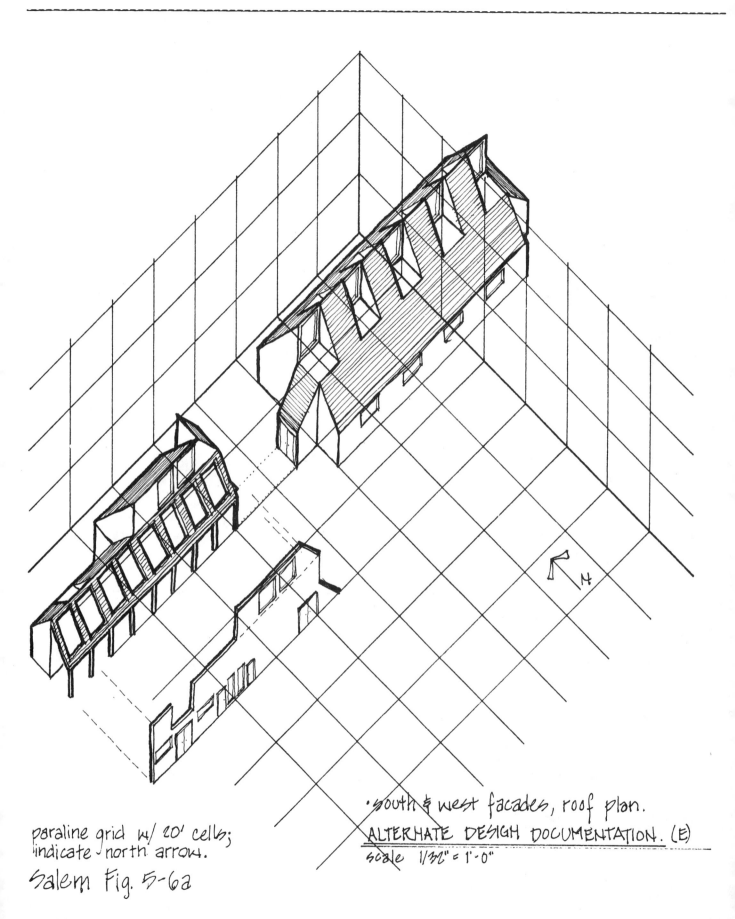

paraline grid w/ 20' cells;
indicate north arrow.

Salem Fig. 5-6a

· south & west facades, roof plan.
ALTERNATE DESIGN DOCUMENTATION. (E)
scale 1/32" = 1'-0"

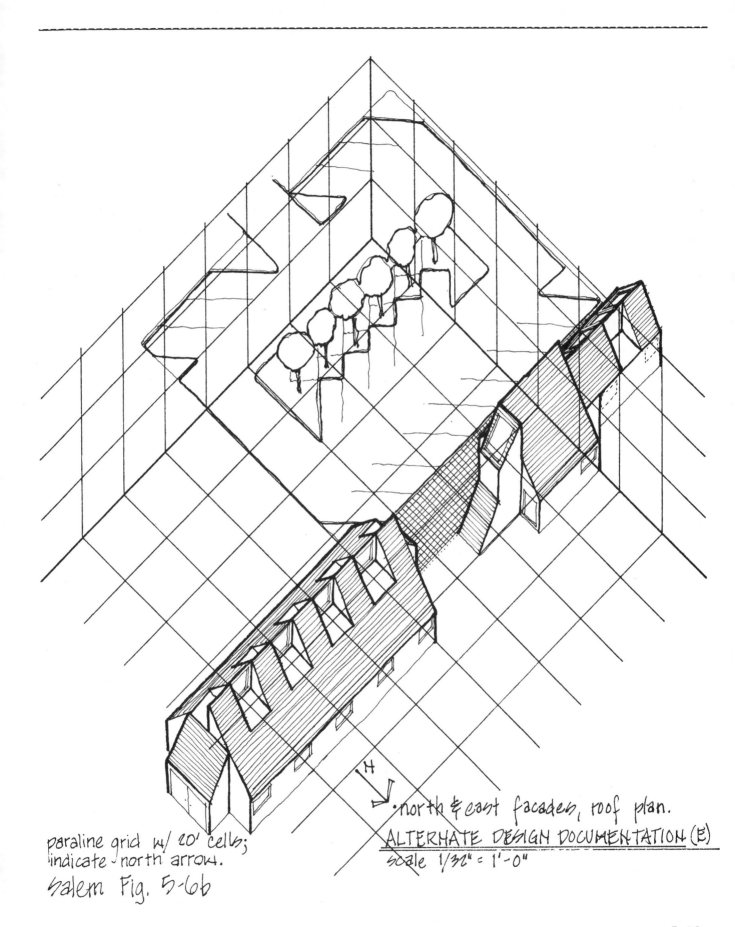

paraline grid w/ 20' cells;
indicate north arrow.
salem Fig. 5-6b

·north & east facades, roof plan.
ALTERNATE DESIGN DOCUMENTATION (E)
scale 1/32" = 1'-0"

EVALUATION

Criterion A: Buildings are located on the site so they don't cast shadows off the site.

Evaluation Tool for Criterion A - "Sun Peg Check"
1. Make a model of your buildings at 1"=40' of clay or cardboard to go on your site model from 3.3 Climate Exercise. The models must be accurate in size and massing, as well as location on the site.
2. Using the shadow plot and sun peg (as described in 3.3 Climate Exercise) look at the shadows cast by your buildings throughout the year.
3. Draw the shadows cast on December 21st at 10 a.m. and 2 p.m. on the site plan.

Does your first trial design meet the criteria?
Yes; we stayed within the portion of the site that we identified as "safe" in the Climate exercise.

We used 1/4" foam core - it works very well, the 1/4" corresponds to the 10' vertical module.

Combined w/winter wind flow diagram.
See Salem example Figure 5-7.

If not, what will you change in your second trial design?

Remember that the design documented for this phase should be the trial which you have shown to meet <u>all</u> criteria.

Criterion B: The buildings are clustered and sited to create optimal microclimates for heating in the months you need heat.
According to the climate analysis, what months will you need to block wind/admit sun to heat:
the outdoor area?
office building?
factory?

September through June
November through April
No months

Evaluation Tool for Criterion B - "Draw Wind Flows"
Draw the wind flows on your site plan <u>and</u> cluster plan during the months you need heating.

Evaluation Tool for Criterion B - "Sun Peg Check"
1. Use the 1/40th model of the site and buildings to check your design for the months you need heating, 10 a.m. to 2 p.m.
2. Draw and clearly label the shadows cast during these times on your cluster plan.

Here we used the prevailing winds from the NOAA climate data; check the wind roses - some climates are more complex than Salem.

See Salem example Figures 5-7 and 5-8.

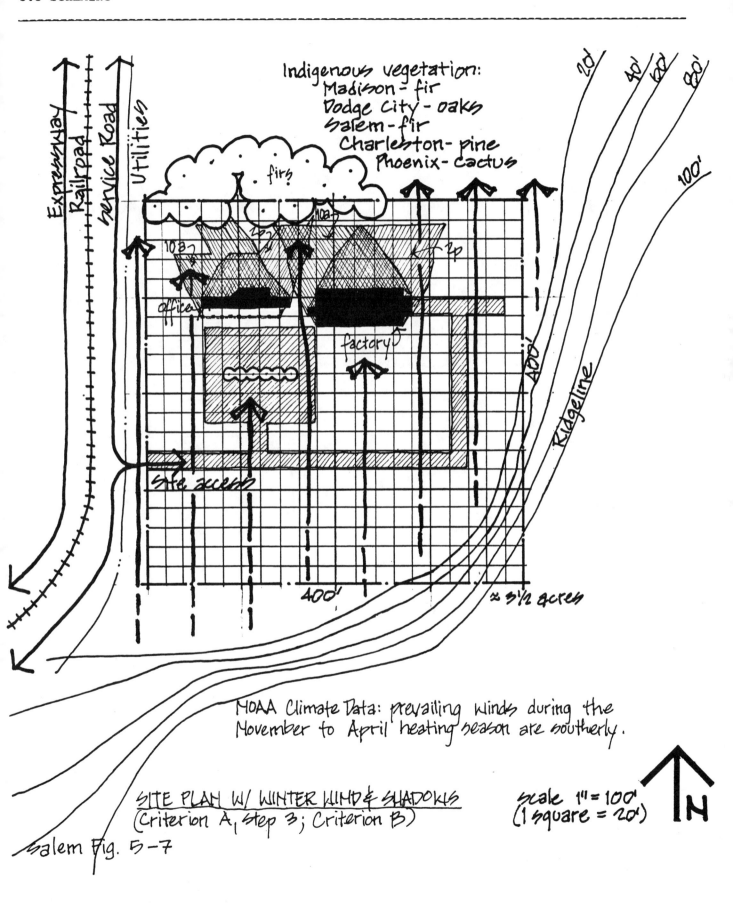

Indigenous vegetation:
Madison - fir
Dodge City - oaks
Salem - fir
Charleston - pine
Phoenix - cactus

firs

officers

factory

site access

400'

≈ 3½ acres

20' 40' 60' 80'
100'
400'
Ridgeline

Expressway Railroad Service Road Utilities

MOAA Climate Data: prevailing winds during the
November to April heating season are southerly.

SITE PLAN W/ WINTER WIND & SHADOWS
(Criterion A, step 3; Criterion B)

scale 1" = 100'
(1 square = 20')

N

Salem Fig. 5-7

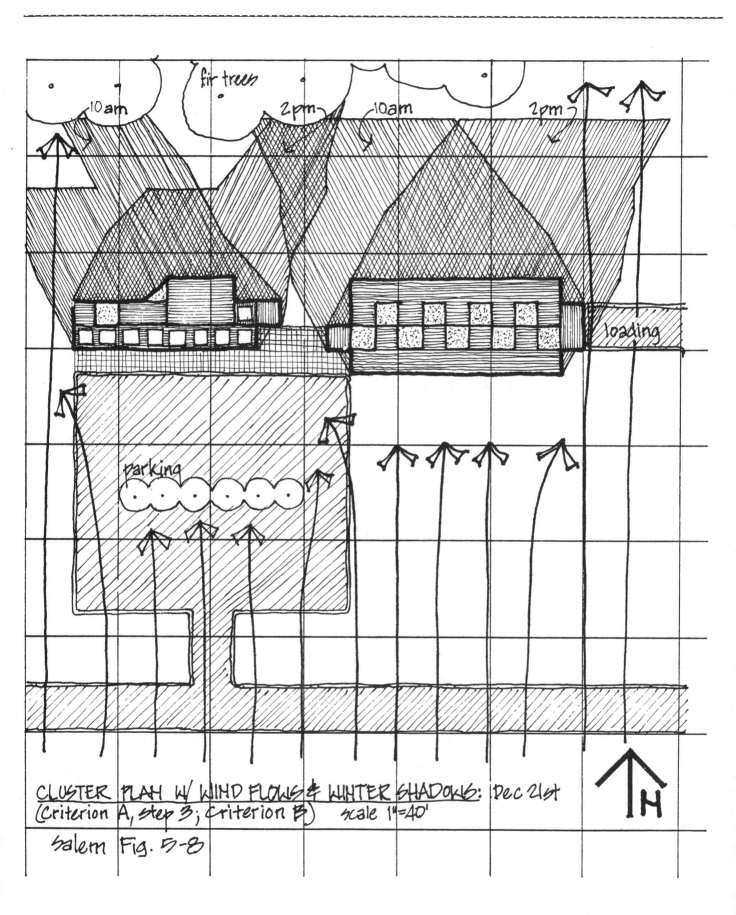

fir trees

10am
2pm
10am
2pm

loading

parking

CLUSTER PLAN w/ WIND FLOWS & WINTER SHADOWS: Dec 21st
(Criterion A, step 3; Criterion B)   scale 1"=40'

Salem Fig. 5-8

_____

Does your first trial design meet the criteria?
Yes.

If not, what will you change in your second trial?
No change needed.

Criterion C: The necessary quantity of south (solar collecting) aperture is provided in your office design for heating in your climate.

Evaluation Tool for Criterion C - "Rule of Thumb - Solar Aperture Sizing"
Providing the recommended aperture to floor area ratios for your climate in Figure 5-1 will allow you to achieve a reasonable % solar heating without overheating on clear days.
We have included this as a schematic criterion to help organize your basic building form and orientation. Assumptions involved in the table and further references are in the developing section of this exercise, 5.4 Criterion B.

If you have adjusted the criterion to reflect internal gains, your new criterion is:
office: _20_ %
factory: _0_ %

	Trial 1		Trial 2	
	Office	Factory	Office	Factory
Total floor area	1750	3900	.	.
Total south aperture	400	100 (unshaded)		.
sq.ft. south aperture / sq.ft. floor area	23%	2.5%	.	.

_____

Does your first trial design meet the criteria?
Yes. For the office building, we are sticking with our 23% rather than the "adjusted%" because we're skeptical about all that internal gain. We don't expect 33 people in the conference space very often, and we expect that daylighting will result in much less usage of electric lighting.

If not what will you change in the next design trial?

Criterion D: The winter sun can reach all the south aperture you have provided for solar collection.
From the climate exercise, what months will you need to use sun to heat
the office?
the factory?

November through April
(Rarely need it)

Evaluation Tool for Criterion D - "Sun Peg Check"
1. Clearly identify the areas of solar aperture on your 1/40th models.
2. Use the sun peg and model to see if the sun reaches these areas between 10 a.m. and 2 p.m. during the months you have identified above.
3. Draw the shadows on your cluster plan for December 21, 10 a.m. and 2 p.m. or include photographs of the study.

Drawing done with Criterion B above.
See Salem example Figure 5-8.

Does your first trial design meet the criteria?
Yes.

If not, what will you change for your second trial?

Criterion E: Cooling
If you have already done the cooling exercise your design must still meet the criteria in the Developing section 5.4 for cooling.

If you have not yet done the cooling exercise but you know that cooling will be a concern in either building in your climate, check these criteria now so that you don't make things impossible in the next exercise.

5.4 DEVELOPING

GOALS

A. Use the sun and mass together to heat buildings as much as necessary.

B. Use materials and construction techniques to control heat loss as necessary in your climate.

CRITERIA

A. The sun reaches the thermal mass during months your buildings need heating.

B. Mass and aperture in the office building are sized so solar heat will reduce fuel consumption by the following percentage over a non-solar conserving building:

Madison	50%
Dodge City	50%
Salem	40%
Phoenix	40%
Charleston	30%

These values are derived from a range of values given in Balcomb, 1980, Table D-1. Values for Madison, Dodge City, and Salem are low-end values assuming night insulation; for Phoenix and Charleston, low-end values without night insulation. We tabulate the range of values in Fig. 5-7 on p. 5-37.

C. The heat loss from the office building must be less than or equal to the following:

Heating Degree Days	Btu/DD SF	
Less than 1,000	9	
1,000 - 3,000	8	
3,000 - 5,000	7	*Salem, Or, has 4,850 DD*
5,000 - 7,000	6	
Greater than 7,000	5	

DESIGN STRATEGIES

A. Building Scale

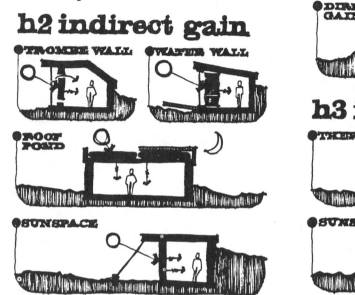

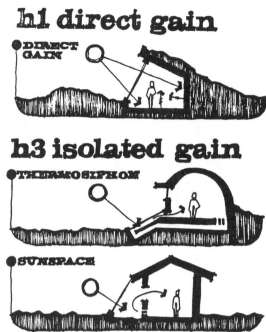

From "Natural Solar Cooling" by David Wright, 3rd
National Passive Conference Proceedings 1979 by
permission of the author.

	Mass Compared to Aperture Area	
	sq.ft. of 4" Masonry	Lbs. of Water
Location	sq.ft. of Aperture	sq.ft. of Aperture
Madison	4	36
Dodge City	3	30
Salem	2.5	24
Phoenix	3	30
Charleston	2.5	24

Figure 5-2 Mass Sizing for Heating (Balcomb, 1980, p. 26)

Note: Because you have already adjusted your % aperture
to reflect internal gains you will not have to adjust
these numbers.

Lesser U values mean slower heat flow through walls, an
energy conservation benefit in cold climates.

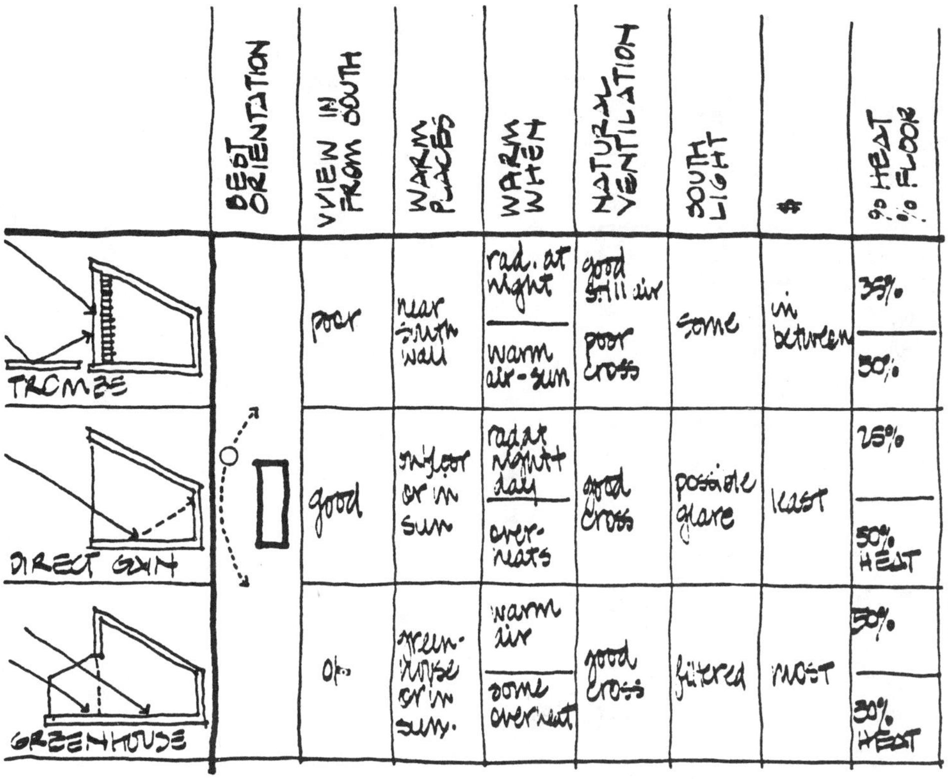

Figure 5-3   Passive Space Heating Systems

Office:
Choose direct gain
+view of ridge
+early AM warm-up
+ventilation and daylight

Factory:
Choose direct gain
+ventilation and daylight

B. Component Scale
Consider the possibility of reducing the need for heating by hours of use; if the work schedule corresponds to the hours of greatest solar heat utilization less auxiliary heat would be needed.

Nearly horizontal reflectors in front of south glass can substantially increase winter solar collection, but may also become sources of glare.

Thermal mass should be kept dark in color, and inside wall or floor insulation whenever solar heating is a design objective.

Winter heat conservation and summer heat rejection are both aided by well-insulated roofs.

Note (warning!):
The next two design tools help you choose wall, window, and roof constructions that are energy conserving. They are based on ASHRAE 90-75, a set of recommendations drawn up in the mid-70's, and the first

serious U.S. building design response to energy conservation. They are for conventionally designed buildings. The criteria you will be asked to meet, however, are based on much tougher and newer ideas about energy conservation. Although a direct comparison is impossible, you can expect the tougher criteria to allow about 1/3 as much heat loss as do the ASHRAE 90-75 standards.

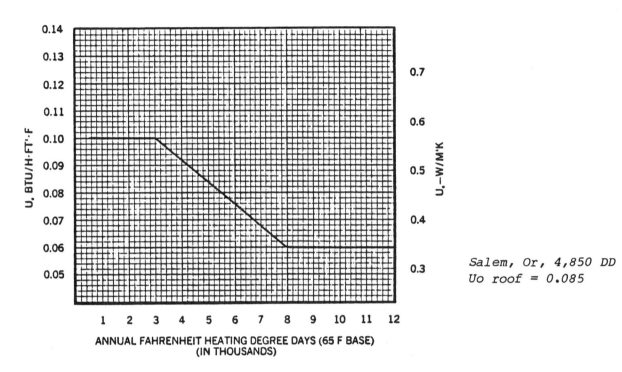

Salem, Or, 4,850 DD
Uo roof = 0.085

Figure 5-4   Uo - Roofs, Commercial Buildings.   From ASHRAE 90-75.

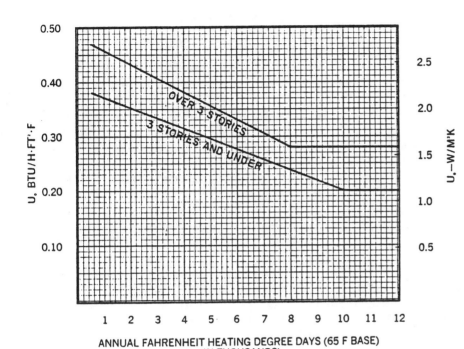

Salem, Or, 4,850 DD
Uo walls = 0.30

Figure 5-5   Uo - Walls, Commercial Buildings.   From
ASHRAE 90-75.

Wall insulation is common to nearly all buildings.   The
maximum U-value recommended (which is averaged for walls
and windows and called U_o,) is listed below.   Since a
double glazed window has a relatively high U-value, about
.60, we have also listed the U-values the wall itself
must have to still reach Uo and have windows in it:

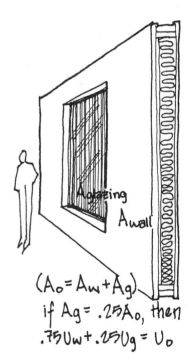

Heating DD	U_o	U Wall, if 25% of area is double glazed window	
2,000	0.35	.26	
4,000	0.31	.21	
6,000	0.28	.17	Salem, U wall about .19
8,000	0.24	.11	

Figure 5-6   Recommended Wall U-values

DESIGN

Develop your design for the office and factory to achieve
the goals.   Document your design as follows:

$(A_o = A_w + A_g)$
if $A_g = .25 A_o$, then
$.75 U_w + .25 U_g = U_o$

A. State your choice of passive heating system or combination of systems (if any) for each building and draw a sectional perspective to show the relation of south aperture, mass and room. Include at least one person for scale.

B. Draw typical sections at 3/4" = 1'-0" indicating materials of the following:
 - non south wall at opening
 - south wall at opening
 - roof
 - thermal mass

C. State what hours per day and months per year that thermal shutters, if you have any, will be open and closed.

Note: Your design may change in the course of the evaluation. The final development design which you show to meet all criteria should be fully documented here.

Passive strategy:
Office Building: direct gain with mass in concrete floor and brick veneer on walls.

Factory: very small direct gain, with mass in concrete floor.

See Salem example Figures 5-9 and 5-10

Office Building: shutters open during hours of operation, 8 am to 5 pm; at least in Dec, Jan and Feb, closed 5 pm to 8 am.

Factory: no thermal shutters.

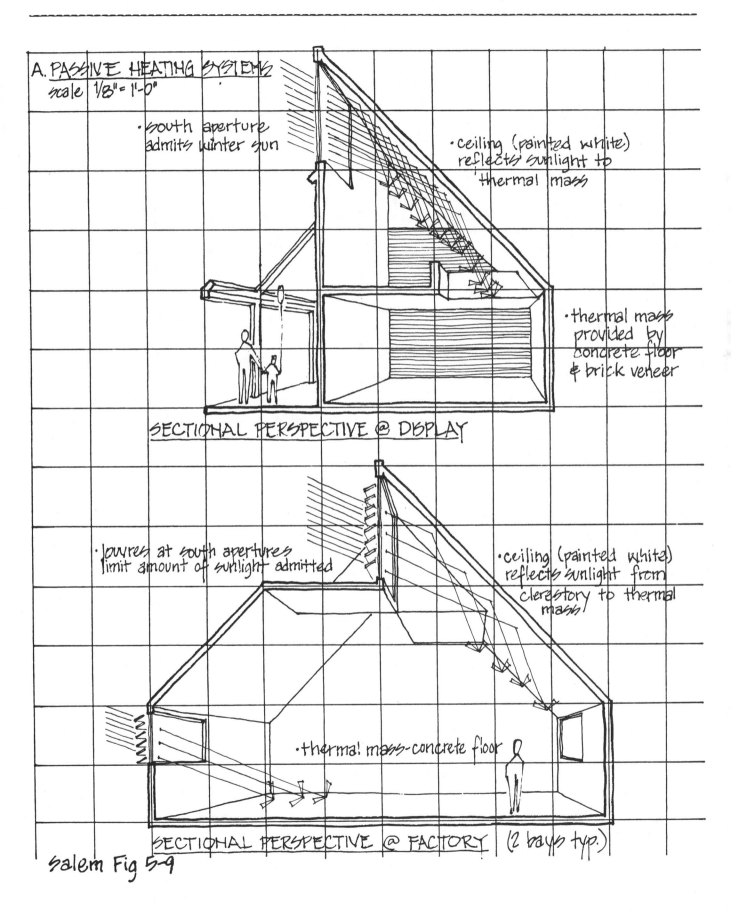

A. PASSIVE HEATING SYSTEMS
scale 1/8" = 1'-0"

• south aperture admits winter sun

• ceiling (painted white) reflects sunlight to thermal mass

• thermal mass provided by concrete floor & brick veneer

SECTIONAL PERSPECTIVE @ DISPLAY

• louvres at south apertures limit amount of sunlight admitted

• ceiling (painted white) reflects sunlight from clerestory to thermal mass

• thermal mass-concrete floor

SECTIONAL PERSPECTIVE @ FACTORY (2 bays typ.)

Salem Fig 5-9

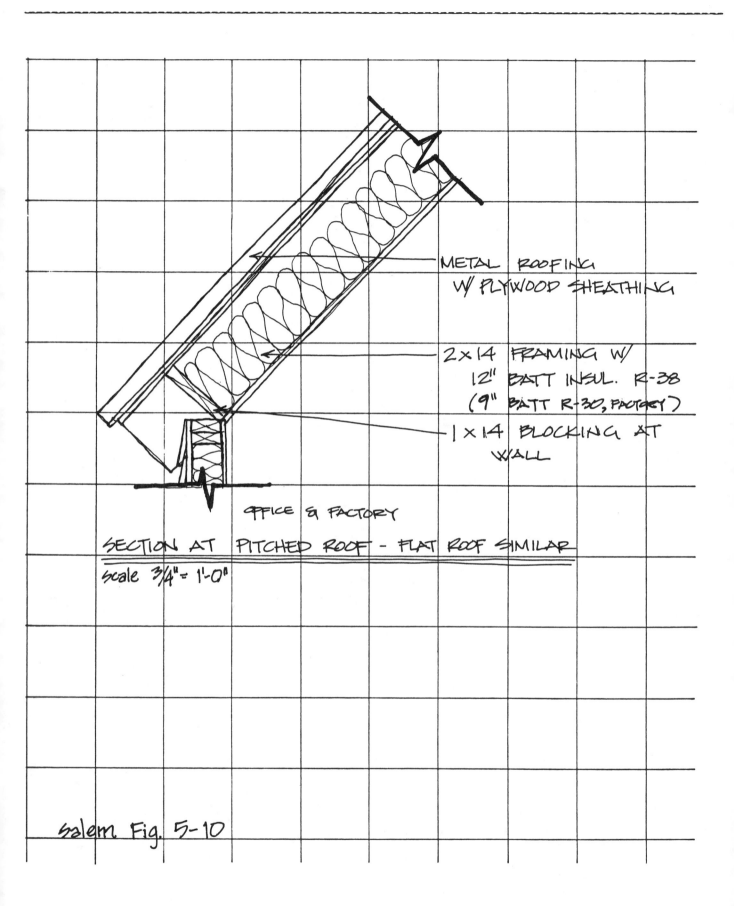

METAL ROOFING
W/ PLYWOOD SHEATHING

2×14 FRAMING W/
12" BATT INSUL. R-38
(9" BATT R-30, FACTORY)

1×14 BLOCKING AT
WALL

OFFICE & FACTORY

SECTION AT PITCHED ROOF - FLAT ROOF SIMILAR
scale 3/4" = 1'-0"

salem Fig. 5-10

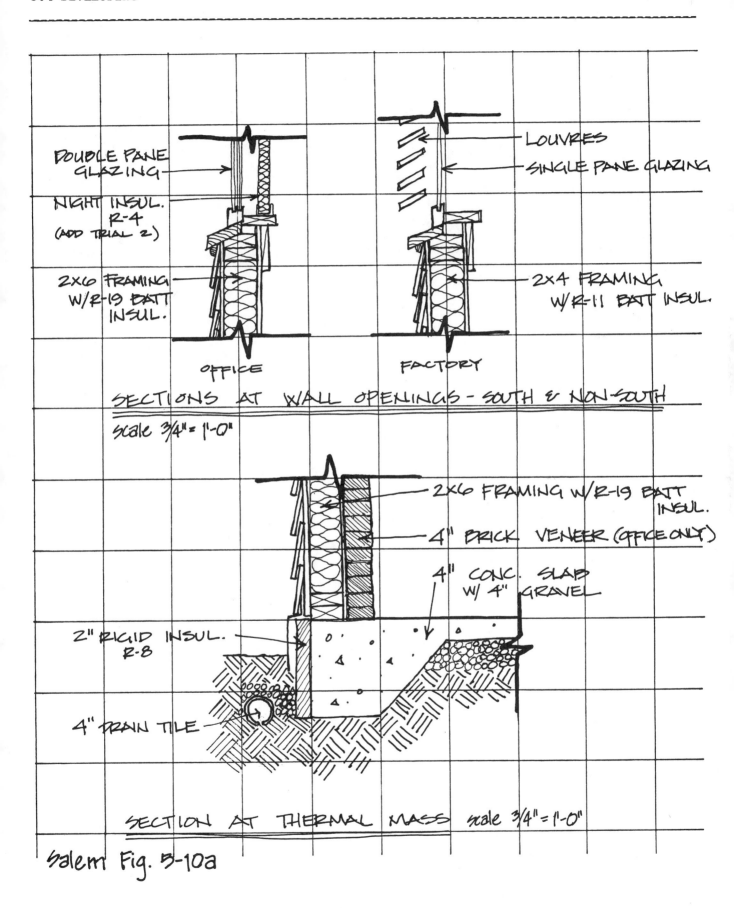

DOUBLE PANE GLAZING

NIGHT INSUL. R-4 (ADD TRIAL 2)

2X6 FRAMING W/R-19 BATT INSUL.

OFFICE

LOUVRES

SINGLE PANE GLAZING

2X4 FRAMING W/R-11 BATT INSUL.

FACTORY

SECTIONS AT WALL OPENINGS - SOUTH & NON-SOUTH

scale 3/4" = 1'-0"

2X6 FRAMING W/R-19 BATT INSUL.

4" BRICK VENEER (OFFICE ONLY)

4" CONC. SLAB W/ 4" GRAVEL

2" RIGID INSUL. R-8

4" DRAIN TILE

SECTION AT THERMAL MASS   scale 3/4" = 1'-0"

Salem Fig. 5-10a

EVALUATION

Criterion A: The sun reaches the thermal mass during months you need heating.

Evaluation Tool for Criterion A - "Sun Peg Check"
1. Build a quick cardboard model at 1/8"=1'-0" of the building(s) which use the sun for heat. The location and size of the openings and thermal mass must be accurate.
2. Use the sun peg and shadow plot to check sun penetration for the months you need heating, at 10 a.m. and 2 p.m.
3. Show the results by drawing the penetration patterns on your plans and section or your perspective.

Does your first trial design meet the criteria?
Yes.

If not, what will you change for your second trial?

We found it necessary to take "notes" on tracing paper while checking the sun angles - it was very hard to record angles on the model.

We are showing the AM penetration for the extreme months in the heating season (April and December). Since our building is fairly symetrical we found the p.m. patterns to be similar to the am and aren't showing them to avoid clutter.

(Be sure that the final design you document is the one that meets all developing criteria.)

See Salem example Figures 5-11 and 5-12.

Criterion B: Mass and aperture in your buildings are sized so that solar heat will reduce heating fuel consumption over a non-solar conserving building by:

Madison   50%
Dodge City   50%
Salem   40%
Phoenix   40%
Charleston   30%

Evaluation Tool for Criterion B - "Aperture and Mass Sizing for Solar Savings"

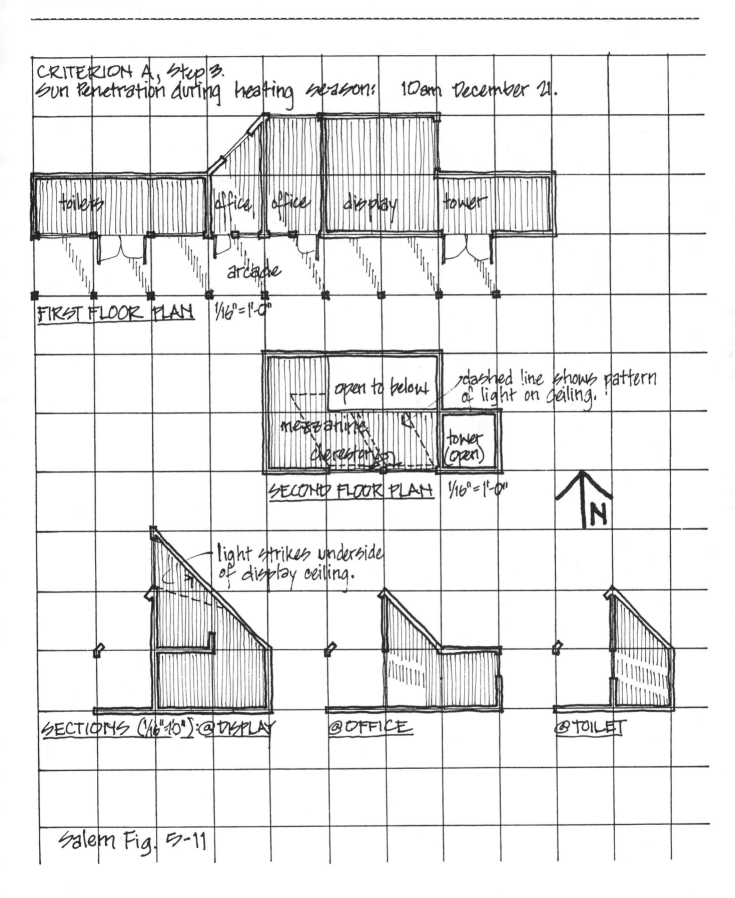

CRITERION A, Step 3.
Sun Penetration during heating season: 10am December 21.

toilets   office office display   tower

arcade

FIRST FLOOR PLAN   1/16"=1'-0"

open to below   dashed line shows pattern of light on ceiling.

mezzanine
clerestory   tower (open)

SECOND FLOOR PLAN   1/16"=1'-0"

N

light strikes underside of display ceiling.

SECTIONS (1/16"=1'0"): @DISPLAY   @OFFICE   @TOILET

Salem Fig. 5-11

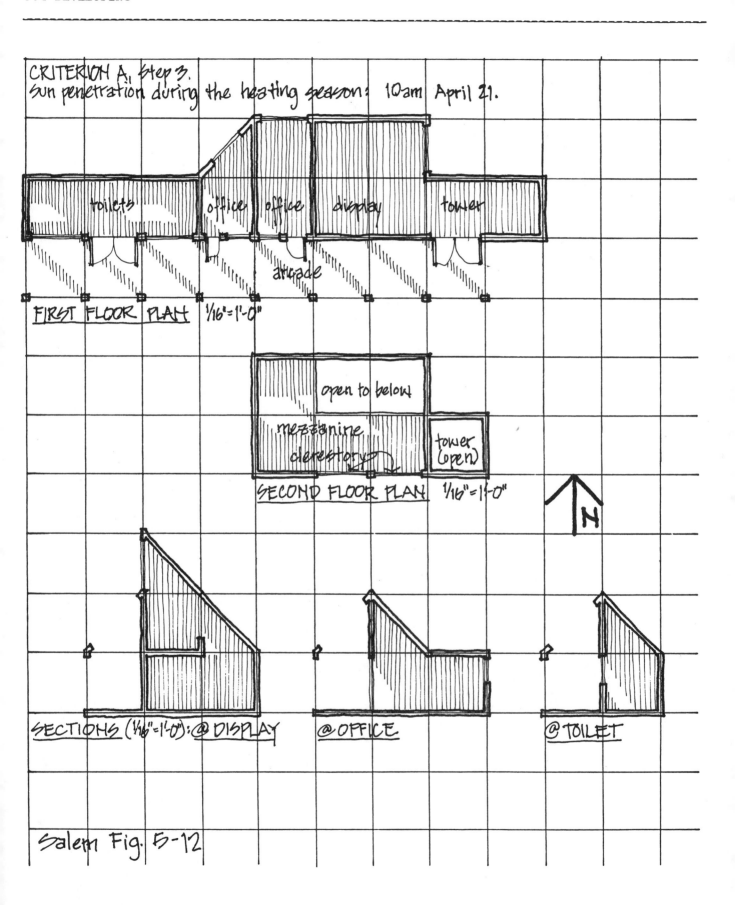

CRITERION A. step 3.
Sun penetration during the heating season: 10am April 21.

toilets   office   office   display   tower

arcade

FIRST FLOOR PLAN   1/16"=1'-0"

open to below

mezzanine
clerestory

tower
(open)

SECOND FLOOR PLAN   1/16"=1'-0"

N

SECTIONS (1/16"=1'-0"): @DISPLAY   @OFFICE   @TOILET

Salem Fig. 5-12

---

1. Complete the following for your building:

	Trial 1		Trial 2	
	Office	Factory	Office	Factory
Floor area (sq.ft.)	• *1750*	*3900* •	•	
South aperture (sq.ft.)	• *400*	*100* •		
Masonry mass (sq.ft. surface)	• *2200*	*3900 conc. floor* •		
Water mass (lbs)	•	•	•	
sq.ft. south aperture / sq.ft. floor area	• *23%*	*2.5%* •	•	
sq.ft. masonry mass / sq.ft. aperture	$\frac{2200}{400} = 5.5$ •	*39* •		
lbs water mass / sq.ft. aperture	•	•		

Office Building:

Offices: wall, brick 400 sf
Display: wall, brick 450 sf
Toilet: wall, brick 100 sf
Conc. floor slab   1250 sf
                          2200 sf

Factory: well over the minimum recommended! Office: 5.5 more than twice the 2.5 recommended. (For brick alone,

$$\frac{950}{400} = 2.4$$

about equal to recommendation) This increased mass should help control thermal swing.

2. Is insulation used for the south aperture? *(Yes, in trial 2)*
3. According to the following rules of thumb (Balcomb, 1980, pp. 20-26), the office building will use the sun to reduce fuel costs by what % (over a conventional design)? *32% without night insulation, 59% with night insulation;*

4. Does this meet Criterion B?
Some night insulation needed to meet 40% criteria.

Location	sq.ft. so. aperture[2] / sq.ft. floor area	sq.ft. masonry / sq.ft. aperture	lbs water / sq.ft. aperture	% Reduction in Heating Load (SSF) w/o night insulation	w/night insulation
Madison	20-40%	4	36	15-17%	51-74%
Dodge City	12-23	3	30	27-42	46-73
Salem	12-24	2.5	24	21-32	37-59
Phoenix	6-12	3	30	37-60	48-75
Charleston	7-14	2.5	24	25-41	34-59

Figure 5-7   Sizing for Solar Savings[1]

Note 1: These rules of thumb were developed assuming that solar glazing was half direct gain and half water wall, well insulated walls and double glazing. The upper limit was chosen to prevent overheating on <u>clear</u> January days and internal gains were assumed to be low. See Balcomb for further details.

Note 2: If you have adjusted these percentages to reflect internal gains and have provided the correct aperture as adjusted, you can use the rest of this table as it is to answer to questions above.

Criterion C: The heat loss from your office building must be less than or equal to the following (Balcomb, 1980, p. 24):

Heating Degree Days	Heat Loss in Btu/DD SF	
Less than 1,000	9	
1,000 – 3,000	8	
3,000 – 5,000	7	*Salem, Or; 7*
5,000 – 7,000	6	
Greater than 7,000	5	

If your buildings have greater internal gains than a small residential building for which these numbers were generated, you can adjust the criteria as follows using the average Balance Point Temperature found (for sizing solar aperture) on p. 5-7.

$$\text{Required Btu/DD sq.ft.} \times \frac{65^{\circ}}{\text{Balance Point Temp (24 hr. avg.)}} = \text{adjusted criterion}$$

$$office: 7 \times \frac{65}{50} = 9.1$$

$$factory: 7 \times \frac{65}{5} = 91$$

Office criterion: _9.1_
Factory criterion: _91_

Note: When your design meets this criteria your building will achieve the Solar Savings Fraction specified in Criterion B above.

Evaluation Tool for Criteria C - "Overall Heat Loss - ASHRAE 90-75 Procedure" (See ASHRAE Standard <u>90-75</u> for a more complete derivation and explanation.)

5.0 HEATING

5.4 DEVELOPING

Note: This heat loss calculation is done for the whole building rather than room by room and will be done for a 24 hour period in order to use the DD data and calculate internal temperatures in the Thermal Finalization Exercise, 7.2. If you used earth berms in your design to reduce heat loss, you will not be able to calculate their effect using the procedure as written or use them to meet Criterion C numerically. However, we have included more information at the end of this tool so that you can estimate the potential contribution.

1. Document the U Values
a. Find the U Values for your walls and roof:

WINDOW U-values are listed in MEEB, Table 4.16, p. 128-129. For windows utilizing <u>night insulation</u>, as is typical for passive solar buildings in colder climates, these U-values may be used (Balcomb, 1980, p. 142):

	No Night Insulation	Night Insulation: R 4	R 9
Direct Gain	0.55	0.30	0.24
Trombe Wall	0.22	0.15	0.12
Water Wall	0.33	0.20	0.17

Note: R4 insulating shades are rather common; R9 are more expensive and available in less variety.

WALL U-values are listed in MEEB, Tables 4.7-4.9, p. 119-121 (with added insulation in Table 4.13, p. 125)

Note: Berms are not calculated by their U-value but by their effect on the ΔT. The procedures used in this workbook use Degree Days rather than ΔT so you will not be able to calculate the effect of berming on heat loss in your buildings.

SKYLIGHT U-values are found in MEEB, Table 4.16, p. 128-129. (Also see U-values, above, for windows using night insulation.)

ROOF U-values are listed in MEEB, Tables 4.10-4.12, p. 122-124, with added insulation in tables 4.13-4.14.

b. Document these by labelling the 3/4" sections in the design documentation. If you use values not found in MEEB, please reference the source.
See Salem example Figures 5-11 and 5-12

U-value Selection
Office window:
Trial 1, use double glazed with 1/2" air space; U = .49 (MEEB p. 128).
Trial 2, add R-4 night insulation; U = .30
Factory window:
Trial 1, use single glazed, U = 1.10.
Office wall: Trial 1, use frame wall in MEEB p. 119, with R-11 insulation as shown: U = .08 ("Construction No. 2")
Trial 2, go to insulation of R-19; new U = .06

Factory wall: Use same (U=.08) construction, to allow use of single glazed windows under ASHRAE 90-75.
Office Skylight: As in windows, U = .49 vertical, U = .59 horiz; so U at $45°$ = .54 (Trial 2: U = .30 with night insulation)
Office and Factory Roof:
Trial 1:

Item	*R*	*MEEB pg.*
outer air film	*0.17*	*110*
metal roofing	*negligible*	
3/4" plywood	*0.93*	*111*
9" batt insulation	*30.00*	*131*
vapor barrier	*negligible*	
1/2" gypsum board	*0.45*	*111*
inner air film (sloping surface)	*0.62*	*110*

Total R = $\overline{32.17}$

$$U = \frac{1}{R} = .032$$

Trial 2: increase R to 38, 12" batt (MEEB p. 131)
Total R = 40.17;

$$U = \frac{1}{40.17} = .02$$

---

2. Calculate Heat Loss Through Building Components

a. Heat Loss Through Walls

1. Determine A$_o$	Trial 1		Trial 2	
	Office	Factory	Office	Factory
Percent wall area in windows (including clerestories)	23%	25 %	16%	%
Percent wall area in opaque (typical) construction (including doors) (If more than one type, list separately)	77%	75 %	84%	%
Vertical Surface Area Total (A$_o$ wall)	3740 s.f.	5200 s.f.	3740 s.f.	s.f.

Example: Office, trial 1:
$$\frac{(.08)(77) + (.49)(23)}{100} = .17$$

Determine U$_o$

Trial 2:
$$\frac{(.06)(84) + (.30)(16)}{100} = .10$$

Wall U$_o$ = $\dfrac{(U\text{ wall})(\%\ A\text{ wall})+(U\text{ window})(\%\ A\text{ window})}{100\%}$

	Trial 1		Trial 2	
	Office	Factory	Office	Factory
Wall Uo	0.17	0.34	0.09	

Heat Loss

Factory, trial 1:
$$\frac{(.08)(75) + (1.10)(25)}{100} = .34$$

Wall heat loss (Btuh/$^\circ$F) = Wall U$_o$ x A$_o$ Wall

	Trial 1		Trial 2	
	Office	Factory	Office	Factory
Wall Heat Loss	636	1768	374	

Office: .17 x 3740 = 636   *.10 x 3740 = 374*
Factory: .34 x 5200 = 1768

---

b. Heat Loss Through Roofs

Determine A$_o$	Trial 1		Trial 2	
	Office	Factory	Office	Factory
Percent roof area in sky-lights	12 %	0 %	6 %	%
Percent roof area in opaque (typical) construction(s)	88 %	100 %	94 %	%
(If more than one type, list separately.)				
Horizontal/Sloped Surface Area Total (A$_o$ roof)	1700 s.f	3780 s.f	1700 s.f	s.f

Determine U$_o$

Office, trial 1:
$$\frac{(.03)(88) + (.54)(12)}{100} = .09$$

Roof U$_o$ = $\dfrac{\text{(U roof)(\% A roof) + (U skylight)(\% A skylight)}}{100\%}$

Trial 2:
$$\frac{(.02)(94) + (.30)(6)}{100} = .036$$

	Trial 1		Trial 2	
	Office	Factory	Office	Factory
Roof Uo	.09	.03	.036	

Heat Loss

Roof heat loss (Btuh/$^\circ$F) = roof U$_o$ x roof A$_o$

	Trial 1		Trial 2	
	Office	Factory	Office	Factory
Roof Heat Loss	153	113	63	

Office: .09 x 1700 = 153
Factory: .03 x 3780 = 113
Office: .036 x 1700 = 63, Trial 2

---

c. Heat Loss Through Floors

A simplifying technique is to assume that only the perimeter of the floor slab loses heat, and that the ground temperature below the floor slab stabilizes to the extent that all slab losses can be ignored except for the edge. (ASHRAE Fundamentals, 1972, P. 354)

Assuming a 2" slab edge insulation at R = 4 per inch, (MEEB, 6th ed., Table 4.6, p. 113), the approximate U-value at the slab edge =

$$\frac{1}{2 \times 4} = .125$$

The resulting slab edge heat loss per lineal foot for a 6" slab would be

heat loss per foot = UxA = .125 x $\frac{6"}{12"}$ = .06 Btuh/$^\circ$F

Floor heat loss (Btuh/$^\circ$F) = heat loss per foot x lineal feet floor slab

Trial 1 and 2:
Office:
214 lineal feet x .06 = 13
Factory:
300 lineal feet x .06 = 18

	Trial 1		Trial 2	
	Office	Factory	Office	Factory
Floor Heat Loss	13	18	13	

d. Heat Loss by Infiltration

At this development stage, we can assume a rule-of-thumb of 3/4 air change per hour (ACH) for infiltration heat loss calculation. (Balcomb, 1980, p. 37)

Trial 1 and 2:
Office volume: 21,000 cu.ft.
Factory volume: 71,500 cu.ft.

Infiltration heat loss (Btuh/$^\circ$F) = 3/4 (ACH) x volume x .018*
(in Btu/hr $^\circ$F)

*A constant, relating specific heat and density of air (MEEB, 1980, p. 135-136).

	Trial 1		Trial 2	
	Office	Factory	Office	Factory
Infiltration Heat Loss	284	965	189	

Trial 1:
Office: 21,000 x .018 x .75 = 284   *21,000 x .018 x .5 = 189*
Factory: 71,500 x .018 x .75 = 965

---

3. Approximate Total Heat Loss

Add the heat losses due to individual components of the design:

| | Trial 1 | | Trial 2 | |
	Office	Factory	Office	Factory
Wall Heat Loss	636	1768	374	
Roof Heat Loss	153	113	63	
Floor Heat Loss	13	18	13	
Infiltration Heat Loss	284	965	189	
Total loss in Btuh/^{o}F	1086	2864	639	
Total Loss in Btu/DDSF $= \dfrac{\text{(Btuh)}(^o F)(24)}{\text{floor area}}$	$\dfrac{1086 \times 24}{1750\ s.f.}$ = 14.9	$\dfrac{2864 \times 24}{3900\ s.f.}$ = 17.6	$\dfrac{639 \times 24}{1750\ s.f.}$ = 8.8, less than 9.1,ok.	

Note: If you choose to use earth berms to meet the criterion you must show that they significantly reduce the heat loss of your designs. Earth rather than air on the outside wall surface effects the ΔT factor of the heat loss calculation. The room to air ΔT is already assumed in the Degree Day data, however you can also look at heat loss on an hourly basis:

heat loss (Btuh) = UAΔT

To calculate the ΔT, the earth temperature can be estimated as the ground water temperature (see Figure 12. Water and Waste Exercise), and you can use 65°F or whatever you have chosen as the inside temperature.

Does your first trial design meet the criteria?
No, the office building is significantly more heat consuming than allowable: 14.9 Btu/DD SF, compared to the "allowable" 9.1 Btu/DD SF; big changes are needed! The Factory, however, seems quite "safe" at 17.6, compared to a (ridiculously high?) allowable of 91!

If not, what will you change in the second trial?
First, insulate the windows at night (which we will assume cuts infiltration to .5 ACH also); then increase both the wall and roof insulation. Then, reduce skylight and window areas that don't face south, since we already have well over 10% of floor area in windows for daylighting.

The design you document (for the DESIGN section in DEVELOPING) should be the final trial design which meets all criteria.

See Salem example, Figures 5-1 to 5-6. We include notes on changes made during heating exercise.

6.1 INTRODUCTION

In this set of design phases you will explore ways of naturally cooling your buildings. Even in the severe winter climates, such as Madison, summer overheating of buildings can occur. As you design for cooling, it is obviously important to keep your building in good condition for winter solar heating performance. In some climates, for some functions, this may create a very different-looking building in winter when compared to summer. For example you may have minimal openings on all walls but south in winter, but generous north and south openings for summer ventilation.

Contents

6.2 CONCEPTUALIZING

GOALS

A. Use the wind for cooling.
B. Use heat sinks for cooling. (A heat sink is any cold area that is available for the dumping of excess heat. This includes the atmosphere, such as the night sky, water bodies, ground, or massive building materials.)
C. Minimize heat gains.

DESIGN STRATEGIES

A. Site Scale
Cool places for summer spaces.

From Landscape Planning for Energy Conservation, ed. by G. Robinette, 1977 by permission of Environmental Design Press.

B. Cluster Scale
Use building arrangement to provide shading of people and buildings.
Preserve each building's access to cooling breezes during overheated months.

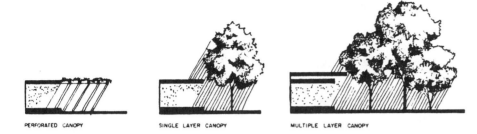

PERFORATED CANOPY   SINGLE LAYER CANOPY   MULTIPLE LAYER CANOPY

From Robinette, 1977 by permission of Environmental Design Press.

C. Building Scale
Orient buildings to the direction of cooling breezes and/or to heat sinks.
D. Component Scale
Openings for ventilation can be separate from those for view and light.

DESIGN

Choose an existing building/site which has a clear conceptual approach, at any scale, to achieving these goals.
Document your choice:
A. Identify the location, program, architect (if any), and the source of your information.
B. Include photocopies or drawings to explain the design.

EVALUATION

Evaluation Tool - "Building Response Diagrams"
Diagram how this building achieves the goals. If the climate of this documented example is different from yours, speculate on how the design responses might change for your climate.

(This more general part of these sections is not included in the Salem, Or example solution.)

6.3 SCHEMING

GOALS

A. Use ventilation and/or mass for cooling, as suitable for your climate.
B. Expose yourself to the cool source in proportion to the cooling you need.
C. When overheated protect yourself from the sun by shading.

CRITERIA

A. The cooling strategy chosen will work in your climate.
B. Enough aperture is provided to ventilate the building as needed.
C. Ventilation flows remove heat from the buildings without leaving hot, dead areas.
D. Glazing is shaded from the sun during overheated months.
E. The building's heat performance has not been damaged by cooling design decisions.

DESIGN STRATEGIES

A. Site Scale
Use trees for shading, particularly on the east and west building faces, and for outdoor spaces.
Schedule building use periods to avoid the hottest times of the day.

Design with the wind and heat, considering that wind tends to keep moving in the same direction, wind flows from high pressure to negative pressure areas, and hot air rises while cool air falls.
B. Cluster Scale
Provide courts which are open to heat sinks.

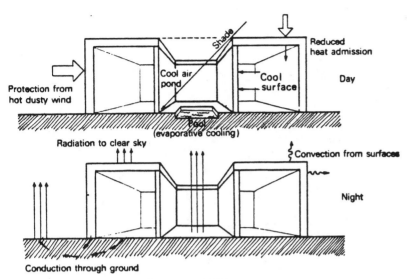

(Koenigsberger, 1974, p. 205)

C. Building Scale
Move activities to cooler places during the hottest periods of the day (or season).
Adopt a "closed", "open" or mixed strategy regarding ventilation:
"Open" buildings are those using natural ventilation.
"Closed" buildings are those using high mass without simultaneous ventilation, those using evaporative cooling, and those using mechanical refrigeration or "air-conditioning".
Mixed buildings can be "closed" during very hot hours, then "opened" for night ventilation to remove stored heat.

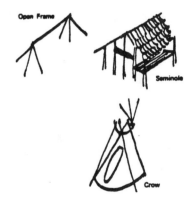

The open building depends upon internal air-flow for interior cooling, and this works only if the inside temperature is higher than the outside temperature. For some climates, natural ventilation can be used even during the hottest months, producing indoor temperatures near the upper limit of comfort with moving air, about 87° in drier climates, less in humid climates (refer to the Bioclimatic Chart, Climate exercise). In other climates, it is preferable to close up the building in the early morning to prevent hot air from entering. In this case, the indoors will remain cooler than outside, although conditions indoors might not be within the comfort zone unless you have auxiliary refrigeration equipment. This is a closed building. The thermal mass you may have provided when solar heating the building can be a useful heat sink for the closed building.

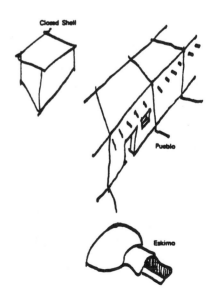

Two basic ways to ventilate a building are:

<u>Cross Ventilation</u> (<u>wind</u>) - depending on the force of the wind to expell hot air from the leeward side of the building, to be replaced by cooler air forced in on the windward side.
<u>Stack (gravity)</u> - depending on the principle of hot air rising to expell hot air from openings high up in a building, to be replaced by cooler air drawn in at much lower openings. This is an important strategy if you have calm wind conditions in overheated months.

Building codes generally allow commercial buildings to be built without mechanical ventilation systems if <u>at least 5%</u> of the floor area is provided in fully openable window area. The more you depend upon natural ventilation for cooling, the larger these window openings should become; approximate sizes are graphed in the Evaluation section.

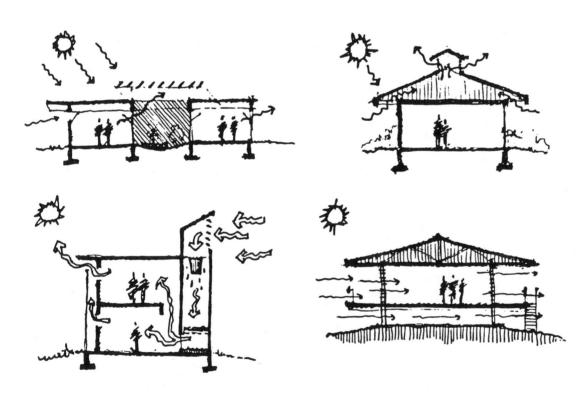

(Holmes, 1979, p. 452-3)

Keep clear paths through buildings for unobstructed ventilation.
While excluding direct sun, utilize daylighting instead of electric lighting.

___

D. Component Scale
Glass areas in the hottest summer exposures (east and west walls, and roofs) should be minimized, or fully shaded from direct sun.
Clerestories protected from direct summer sun are potential sources of daylight, as well as openings to allow heated air to escape. They are especially effective ventilators if they open on the leeward (downwind) side.

DESIGN

Design the office, factory and site to meet the program requirements in 2.0 Site and Building Description. Document your design using the grids provided.
A. State whether your buildings will be open, closed or mixed:
Office: *open*
Factory: *open*
B. Site plan including parking and access drives, and vegetation.
C. Cluster plan including outdoor spaces and vegetation.
D. Floor plans of office and factory.
E. Sections of office and factory.
F. Roof plan and elevations (or paralines) which clearly show all sides and tops of the buildings.
G. Key each glazed aperture to the plan and state at which scale (site, cluster, building or component) you will shade each:

Glazed openings: office   *Keyed to plan and eleva-*
A: Component   *tions, Salem example*
B: Component   *Figures 6-6 - 6-9*
C: Component
D: Building (louvres over arcade beyond the windows)
E: Component
F: Component

Factory:
G: Component

                                                     See Salem example Figures 6-4 -
                                                     6-9

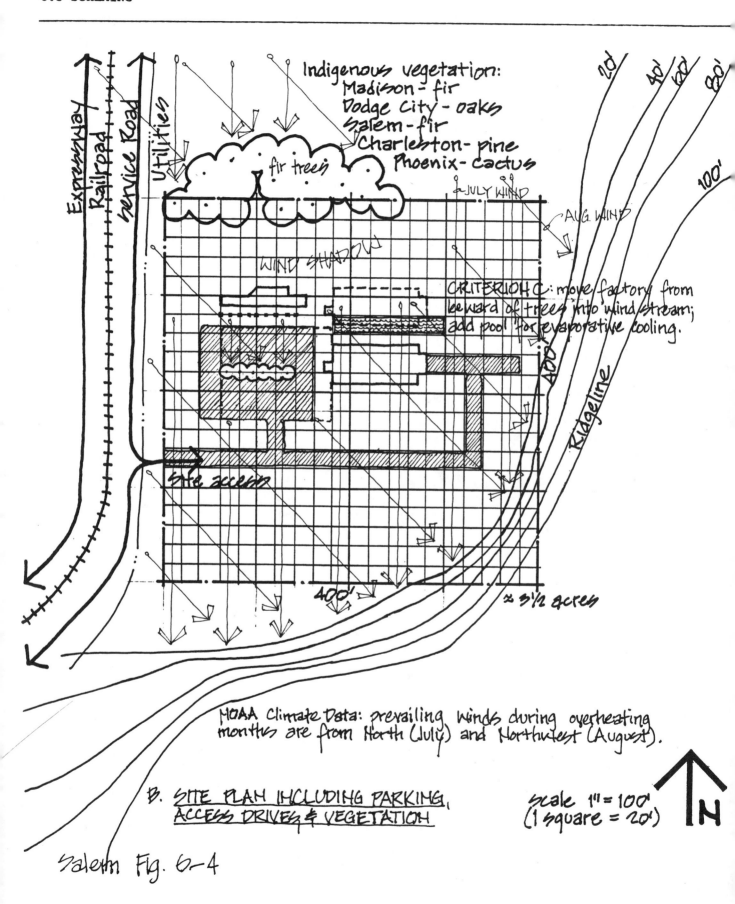

Indigenous vegetation:
Madison – fir
Dodge City – oaks
Salem – fir
Charleston – pine
Phoenix – cactus

fir trees

JULY WIND

AUG. WIND

WIND SHADOW

CRITERION C: move factory from leeward of trees into wind stream; add pool for evaporative cooling.

Expressway
Railroad
Service Road
Utilities

Ridgeline

Site access

20' 40' 60' 80'
100'
400'

400'

≈ 3½ acres

NOAA Climate Data: prevailing winds during overheating months are from North (July) and Northwest (August).

B. SITE PLAN INCLUDING PARKING, ACCESS DRIVES & VEGETATION

scale 1" = 100'
(1 square = 20')

N

Salem Fig. 6-4

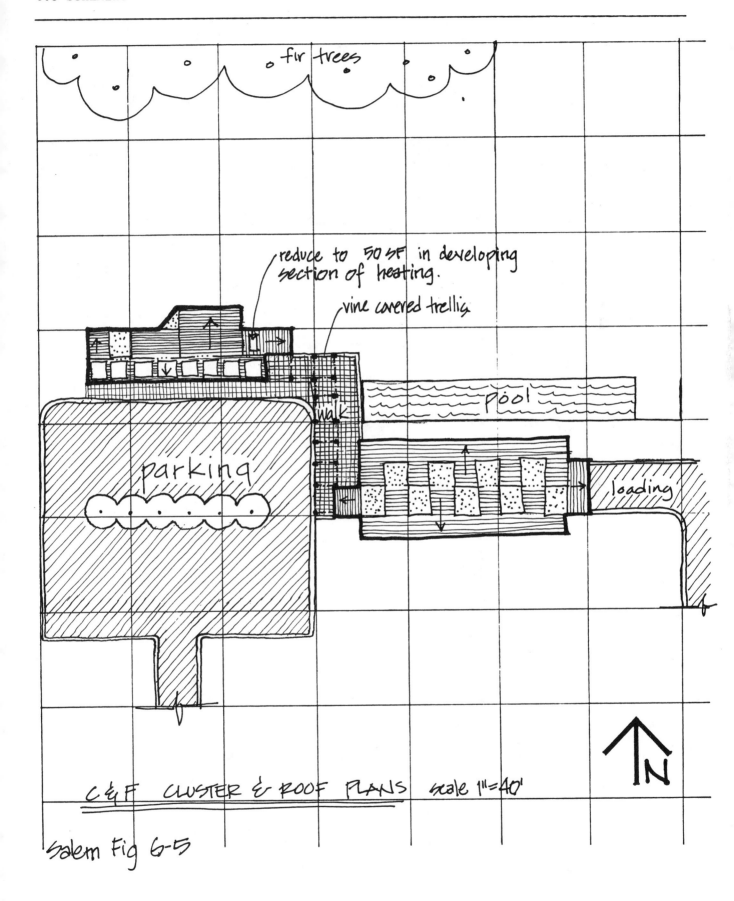

fir trees

reduce to 50 SF in developing
section of heating.

vine covered trellis

pool

walk

parking

loading

C&F CLUSTER & ROOF PLANS scale 1"=40'

Salem Fig 6-5

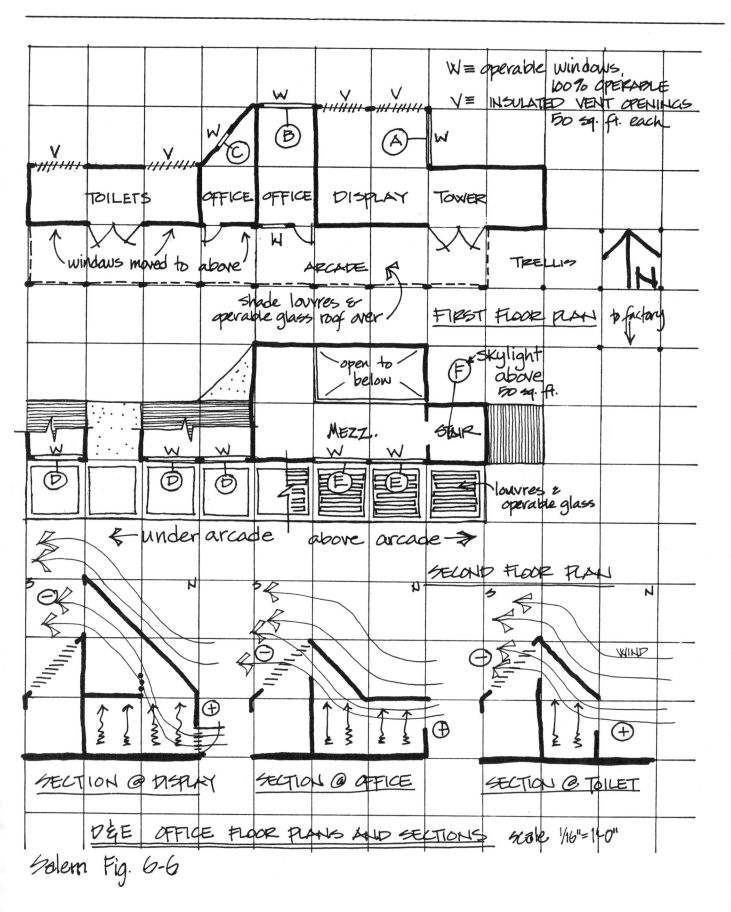

W ≡ operable windows, 100% OPERABLE

V ≡ INSULATED VENT OPENINGS 50 sq. ft. each

TOILETS · OFFICE · OFFICE · DISPLAY · TOWER

windows moved to above

ARCADE

TRELLIS

shade louvres & operable glass roof over

FIRST FLOOR PLAN

to factory

N

open to below

skylight above 50 sq. ft.

MEZZ.

STAIR

louvres & operable glass

under arcade ← · → above arcade

SECOND FLOOR PLAN

SECTION @ DISPLAY

SECTION @ OFFICE

SECTION @ TOILET

WIND

D&E OFFICE FLOOR PLANS AND SECTIONS   scale 1/16"=1'-0"

Salem Fig. 6-6

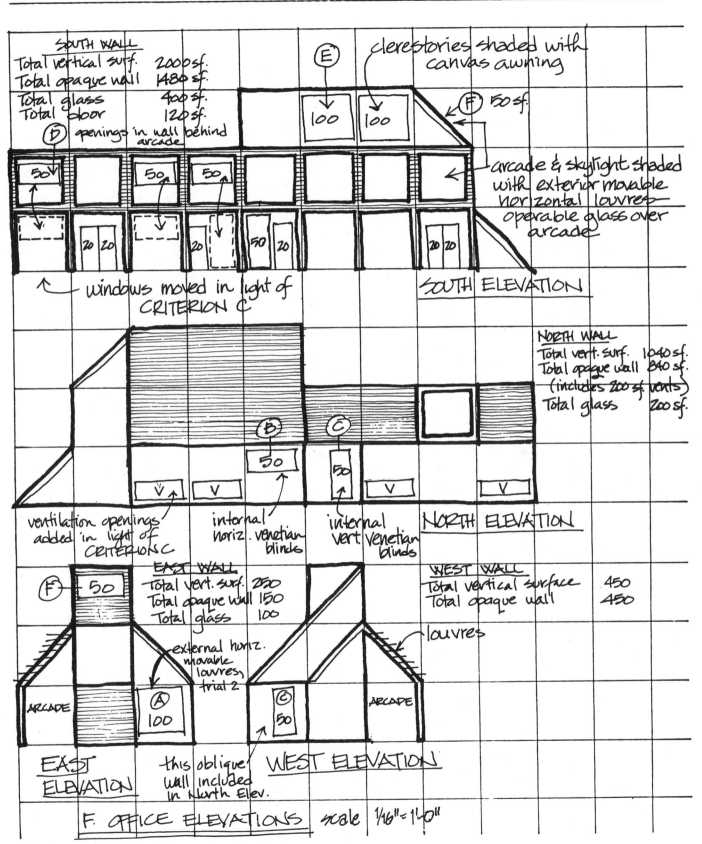

SOUTH WALL
Total vertical surf.   2000sf.
Total opaque wall   1480sf.
Total glass   400sf.
Total door   120sf.
(D) openings in wall behind arcade

(E)

clerestories shaded with canvas awning

(F) 50sf.

arcade & skylight shaded with exterior movable horizontal louvres
operable glass over arcade

windows moved in light of CRITERION C

SOUTH ELEVATION

NORTH WALL
Total vert. surf.   1040sf.
Total opaque wall 840sf.
(includes 200sf. vents)
Total glass   200sf.

(B) (C)

ventilation openings added in light of CRITERION C

internal horiz. venetian blinds

internal vert venetian blinds

NORTH ELEVATION

EAST WALL
Total vert. surf. 250
Total opaque wall 150
Total glass   100

(F) 50

external horiz. movable louvres, trial 2

WEST WALL
Total vertical surface   450
Total opaque wall   450

louvres

ARCADE

(A) 100

(C) 50

ARCADE

EAST ELEVATION

this oblique wall included in North Elev.

WEST ELEVATION

F. OFFICE ELEVATIONS   scale 1/16" = 1'-0"

Salem Fig 6-7

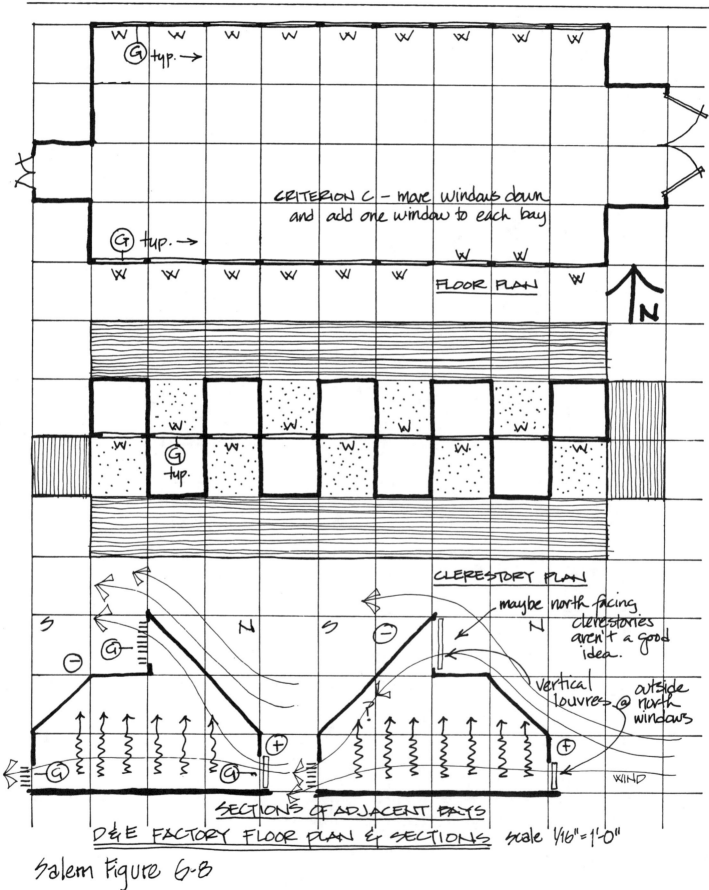

CRITERION C – move windows down
and add one window to each bay

FLOOR PLAN

N

CLERESTORY PLAN

maybe north facing
clerestories
aren't a good
idea.

S   N   S   N

vertical
louvres @ outside
north
windows

WIND

SECTIONS OF ADJACENT BAYS

D & E FACTORY FLOOR PLAN & SECTIONS scale 1/16"=1'-0"

Salem Figure 6-B

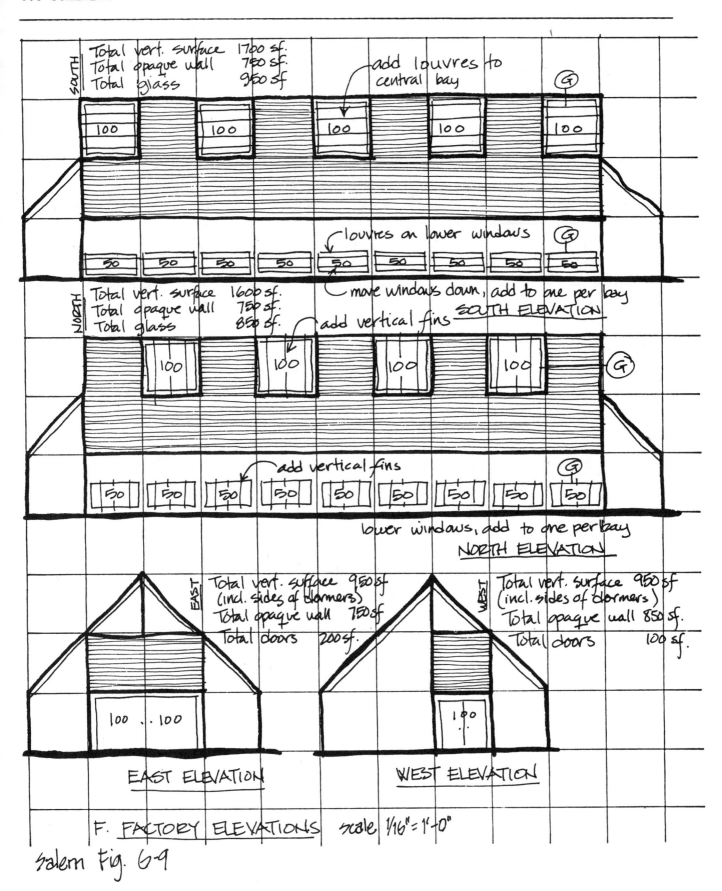

SOUTH:
Total vert. surface 1700 sf.
Total opaque wall 750 sf.
Total glass 950 sf

add louvres to central bay

louvres on lower windows

move windows down, add to one per bay

SOUTH ELEVATION

NORTH:
Total vert. surface 1600 sf.
Total opaque wall 750 sf.
Total glass 850 sf.

add vertical fins

add vertical fins

lower windows, add to one per bay

NORTH ELEVATION

EAST:
Total vert. surface 950 sf
(incl. sides of dormers)
Total opaque wall 750 sf
Total doors 200 sf

EAST ELEVATION

WEST:
Total vert. surface 950 sf
(incl. sides of dormers)
Total opaque wall 850 sf.
Total doors 100 sf.

WEST ELEVATION

F. FACTORY ELEVATIONS scale 1/16" = 1'-0"

Salem Fig. 69

for plans, sections, & elevations :   if   1/16"=1'-0",   1 square = 10'.

for plans, sections, & elevations : if 1/16"=1'-0", 1 square = 10'.

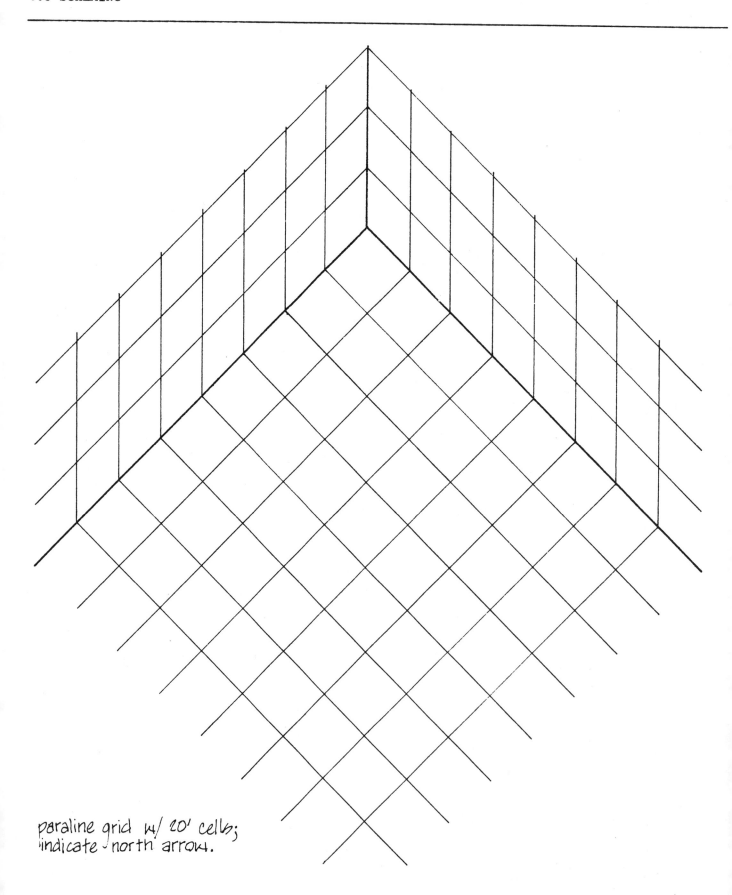

paraline grid w/ 20' cells;
indicate north arrow.

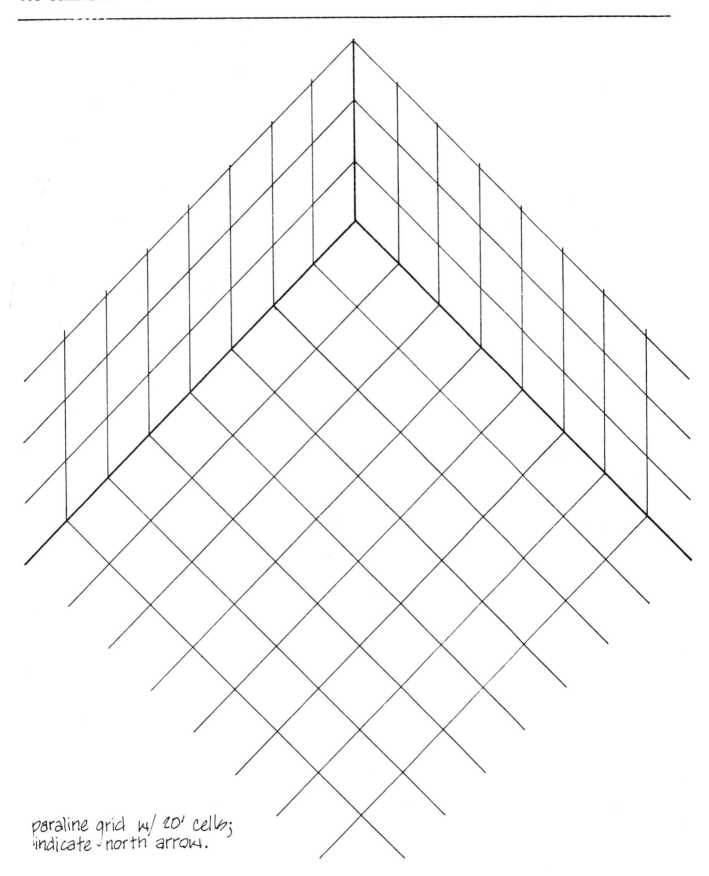

paraline grid w/ 20' cells;
indicate north arrow.

EVALUATION

A. Criterion A: The cooling strategy chosen works in your climate.

Evaluation Tool for Criterion A - "Bioclimatic Chart of Cooling Strategies"
In the Climate exercise section 3.2 you plotted average monthly temperatures on a chart similar to Fig. 6.1. This chart tells you whether buildings in your climate should be open or closed. In the comfort zone, either of these strategies is possible.

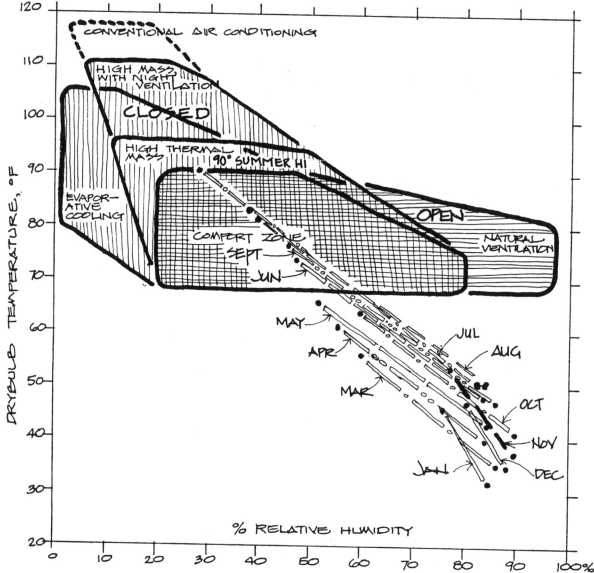

Figure 6-1   Bioclimatic Chart for Cooling Strategies

A. Use overlays to transfer your climate as plotted in the Climate exercise, 3.2 to Fig. 6-1.
B. Is the cooling strategy appropriate for your climate an open or closed building?

Is this the strategy you have chosen?
Open, yes.
If not, what will you do to meet Criterion A?
On extreme days, close it up and count on thermal mass provided for passive heating; and/or add a pond to the north side of the "worst" building (factory).
For a detailed explanation of Fig. 6-1, see Milne and Givoni, "Architectural Design Based on Climate" in Watson, Energy Conservation Through Building Design, 1979.

Open, under average conditions; closed, on extreme days, or open with moisture added to wind.

B. Criterion B: Enough aperture is provided for ventilation.

Evaluation Tool for Criterion B - "Aperture Sizing for Ventilation"
1. Complete the table below. From the Climate exercise, section 3.6, find the Btuh internal gain for each building. For open buildings, use the Btuh gain while occupied. For closed buildings, you will need to ventilate stored heat in addition to heat being produced while venting, so use two times the Btuh while occupied.

Building	Open or Closed	Cross-ventilation or stack ventilation?	Internal gains Btuh	sq.ft. floor area	Revised Btuh/sq.ft.
Office	open	cross ventilation	18,280	1,750	10.4
Factory	open	cross ventilation	133,440	3,900	34.2

Btuh/sq.ft. gain	sq.ft. Aperture as % of floor area*			
	6 mph Salem	7 mph Phoenix	8 mph Madison Charleston	13 mph Dodge City
10	2%	2%	1%	1%
20	3	2	2	1
30	5	4	3	2
40	6	5	4	3
50	7	6	5	3
60	9	8	7	4
70	10	9	8	5
80	12	10	9	5
90	13%	11%	10%	6%

Office will need a bit more than 2% (but code requires 5%).

Factory will need about 8%.

*needed on each of two sides of a building

Figure 6-2 Aperture Sizing For Cross-Ventilation

Btuh/sq.ft. gain	sq.ft. Aperture as% of floor area*
10	11%
20	22
30	34
40	45
50	56
60	67
70	78
80	90
90	100%

If stack ventilation is to be relied upon, office will need about 12%.

Factory will need about 62%.

*needed in three places: lower openings, stack cross-section, and openings at top of stack.

Figure 6-3 Sizing for Stack Ventilation

Technical note: These rules of thumb were generated with these assumptions: $\Delta T=3^{\circ}$, a conservative choice. Wind effectiveness factor = 0.4 (see discussion in Thermal Finalization 7.2, p.). Stack height = 30'.
Heat gain used here is attributable <u>only</u> to estimated internal gains. It seems likely that the additional solar gains will be controlled by shading devices such that they will essentially equal the decrease in internal gains allowed by daylighting.

Cross Ventilation: Trial 1
Office 5% of 1750 = 88 req.
Factory 8% of 3900 = 312 req.

(Stack: Trial 1)
(Office 12% of 1750 = 210 req.)
(Factory 62% of 3900 = 2418 req.)

---

2. Use Figs. 6-2 and 6-3 to complete this table.

	Trial 1		Trial 2	
	Office	Factory	Office	Factory
Sq.ft. aperture necessary as either inlet or outlet	88 (210)	312 (2418)		
Sq.ft. aperture provided for inlets	200	650		
Sq.ft. aperture provided for outlets	400	700		

Does your first trial design provide the necessary aperture? *Yes, for cross ventilation. (Also for stack, in office building)*
If not, what will you change for trial 2?

C. Criterion C: Ventilation flows remove heat from the buildings without leaving hot, dead areas.

Evaluation Tool for Criterion C - "Heat Flow Diagrams"

STREAM OF WIND (open buildings)
On your site plan and your building plans and sections, draw the flow pattern of air around and through the buildings for the months you need to cool your buildings. In rendering those flows, remember: Wind tends to keep going in the same direction. Wind flows from positive to negative pressure areas. Hot air rises, cool air falls.

MASS COOLING (closed buildings, including night ventilation)
In your building plan and section show the locations of thermal mass. If the mass will be cooled primarily by radiation, draw the path of radiation transfer, from mass to sink, and label the sink. If the mass is cooled by conduction/convection (as in night ventilation), draw the path of air currents over the face of the mass, both in plan and section.

We are using open building technique.
Overheating months windflow:
JULY N
AUGUST NW
Prevailing wind from NOAA Climate Data; see Salem example, Figures 6-4, 6-6 and 6-8.

Does your first trial design meet Criterion C?
Office: No; hot dead spots occur in toilet and display areas.
Factory: Too close to trees; inside minor hot spots at walls.

If not, what will you change for the second trial?
Office: Add insulated cover ventilating openings on North wall of toilet rooms (100 s.f.) and display space (100 s.f.)
Factory: Move further from trees, and add 50 s.f. windows each side, to lower walls that are currently without openings; move these windows down to floor slab, to enhance stack effect.

───

D. Criterion D: Glazing is shaded from the sun in the overheated months. (From the balance point analysis in Climate section 3.6, the months your buildings are overheated are when the balance point is lower than the outdoor temperature.)

Factory overheated months are: *All year.*
Office overheated months are: *May through October.*   *(See p. S3-27, question C.)*

Evaluation Tool for Criterion D - "Sun Peg and Site Model"
If you have opted to shade at the site or cluster scale, use the sun chart and peg with your 1/40th scale model to evaluate this decision. Building and component scale shading will be evaluated at a more detailed level in 6.4 DEVELOPING.
1. Check the shading of all glazed apertures for 9 a.m., noon, and 3 p.m. of the overheated months above.
2. Document the shading with photos of your model or elevations or paralines with shadows drawn.

(All our shading is done at building/component scales; therefore no need to check site and cluster scale shading.)

Does your first trial design meet Criterion D?

If not what will you change for your second trial?

E. Criterion E: Heating Check
If you have already done the heating exercise your design must still meet the criteria for 5.3 SCHEMING on p. 5-4.
If you have not yet done the heating exercise, but you know that heating will be a concern in either building in your climate, check the criteria for 5.3 SCHEMING now so that you don't make things impossible in the next exercise.

6.4 DEVELOPING

GOALS

A. Minimize peak daily heat gains during overheated months.

CRITERIA

A. Glazed apertures are fully shaded during overheated months.
B. Daily heat peaks are minimized during overheated months by avoiding simultaneous peaks from more than one source.
C. Heat gain peaks do not occur during the hours of the day when it is hottest outside.

DESIGN STRATEGIES

A. Site/Cluster Scales
Where evaporative cooling is a strategy, incoming air can be brought across the surface of a water body, lowering its temperature while raising its relative humidity.

See site plan Salem example Figure 6-4.

B. Building Scale

  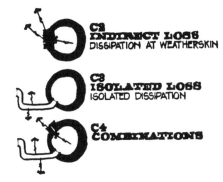

From "Natural Solar Cooling" by David Wright, 3rd National Passive Conference Proceedings, 1979. Used by permission of the author.

Organize your most severely overheated spaces so that they can be cooled without overheating other spaces.

In closed buildings, single-story, high-ceiling spaces with thermal mass allow heat build-up higher in the space, while cooler air remains at occupancy levels nearer the ground.

In open buildings, shape ceilings so that heat from the highest part(s) of the space can readily escape to the outdoors. Consider prevailing wind direction so that heat from one space is not carried to another.

C. Component Scale

The openings for incoming ventilating air should be adjustable so that when such air is colder than comfortable in contact with people, it can be admitted without directly chilling the occupants. Yet when moving air across the body is a requirement for comfort, it can be brought in directly across the occupants.

Choose the shading devices that allow daylighting while preventing direct sun from overheating your space.

The months with highest temperatures (such as July and August) have the same sun paths as much cooler months (such as May and April). To respond to the problem of identical sun paths/different temperatures, use variable shading devices. Deciduous vegetation is one seasonal response (see MEEB p. 54-55). Note that vines may be trimmed back to allow 100% solar gain; deciduous trees are likely to shade 50% of solar gain even in winter. Moveable shading devices (perhaps incorporating moveable thermal insulating devices) can allow daily responses to sunny or overcast weather.

(The chart below is diagrammatic. You should refer to the LOF sun charts for your specific latitude in section 6.5)

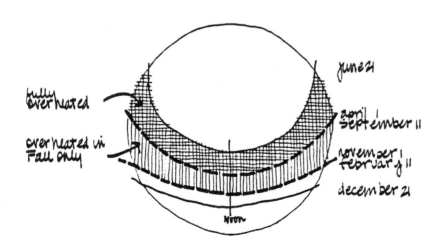

DESIGN

Develop your design for the office and factory to achieve the goals. Document the design using grids provided:

A. Describe the natural cooling system you have chosen for each building: cross or stack ventilation, high mass, or evaporative cooling.

Cross ventilation; some evaporative for factory; thermal mass helps too.

B. Draw sectional perspectives of the rooms which are passively cooled. These should illustrate:
1. size and location of the cooling system components (aperture, mass, etc.) and their relation to the space
2. how the system changes diurnally (if it does)
3. how the heat is removed from the room

- factory, display, office and toilet.

See Salem example Figures 6-10 and 6-11.

C. Draw all the types of shading devices, including vegetation, that you are using in both section and elevation at 1/8th scale. Key them to paralines or elevations and label the devices as fixed or movable.

See Salem example Figure 6-12.

D. Show your office and factory occupancy schedule on the graph below.

Factory   6 am--------------------10 pm

Factory 16 hours occupancy

Office   8 am----------4:30 pm

Office 8 hours occupancy

             Midnight   Noon   Midnight

E. If you have completed the HEATING section 5.4, include your wall sections design documentation. (Revise anything you have changed in the design).
If you have not completed HEATING section 5.4 you must make preliminary material and construction decisions. Specify what basic materials (wood, brick, concrete block, insulation, etc.) your walls and roofs will be. (See MEEB Table 4.22c p. 149 for the level of detail required in this exercise.)

Practically no change. See Salem example Figures 6-13 and 6-14

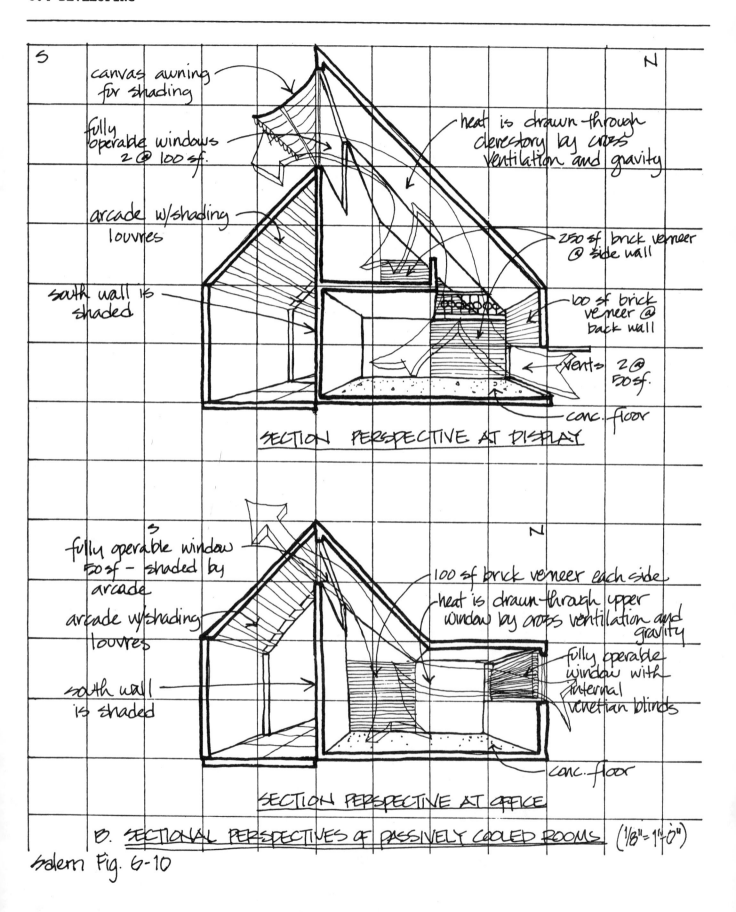

S   N

canvas awning for shading

fully operable windows 2 @ 100 sf.

arcade w/shading louvres

south wall is shaded

heat is drawn through clerestory by cross ventilation and gravity

250 sf brick veneer @ side wall

100 sf brick veneer @ back wall

vents 2 @ 50 sf.

conc. floor

SECTION PERSPECTIVE AT DISPLAY

S   N

fully operable window 50 sf - shaded by arcade

arcade w/shading louvres

south wall is shaded

100 sf brick veneer each side

heat is drawn through upper window by cross ventilation and gravity

fully operable window with internal venetian blinds

conc. floor

SECTION PERSPECTIVE AT OFFICE

B. SECTIONAL PERSPECTIVES OF PASSIVELY COOLED ROOMS (1/8" = 1'-0")

Salem Fig. 6-10

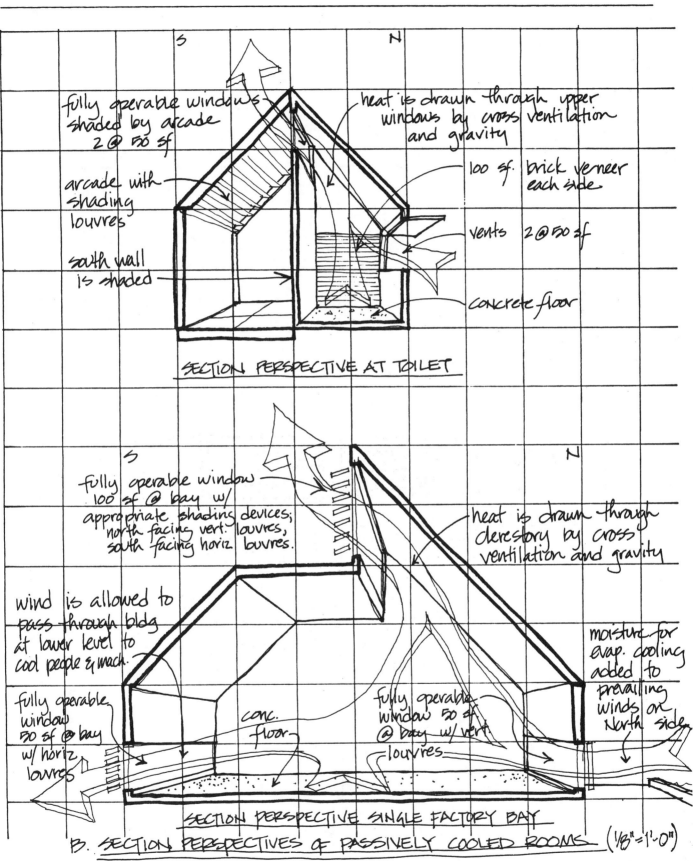

SECTION PERSPECTIVE AT TOILET

fully operable windows
shaded by arcade
2 @ 50 sf

arcade with
shading
louvres

south wall
is shaded

heat is drawn through upper
windows by cross ventilation
and gravity

100 sf. brick veneer
each side

vents 2 @ 50 sf

concrete floor

fully operable window
100 sf @ bay w/
appropriate shading devices;
north facing vert. louvres,
south facing horiz. louvres.

heat is drawn through
clerestory by cross
ventilation and gravity

wind is allowed to
pass through bldg
at lower level to
cool people & mach.

moisture for
evap. cooling
added to
prevailing
winds on
North side

fully operable
window
50 sf @ bay
w/ horiz
louvres

conc.
floor

fully operable
window 50 sf
@ bay w/ vert
louvres

SECTION PERSPECTIVE SINGLE FACTORY BAY

B. SECTION PERSPECTIVES OF PASSIVELY COOLED ROOMS (1/8"=1'-0")

Salem Figure 6-11

for daylighting contours, enlarged plans: if 1/8"= 1'-0", 1 square = 10'.

DESIGN DOCUMENTATION Part C: Shading Devices	1/8" = 1'-0"	elevations:	sections:	
VERTICAL LOUVERS (1) factory ground floor- see fig. 6-6 (Salem), North Elevation.				movable
(2) factory clerestory- Salem 6-6, North elevation.				movable
HORIZONTAL LOUVERS (3) factory ground floor- Salem Example 6-6, South Elevation.				movable
(4) factory clerestory- Salem Example 6-6 South Elevation.				movable
CANVAS AWNING (5) display clerestory- see Salem Example 6-4, South Elevation.				movable

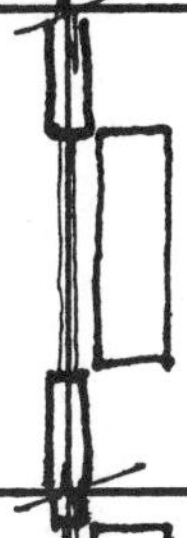

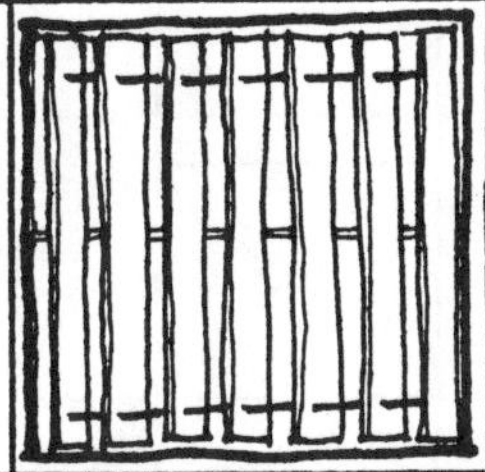

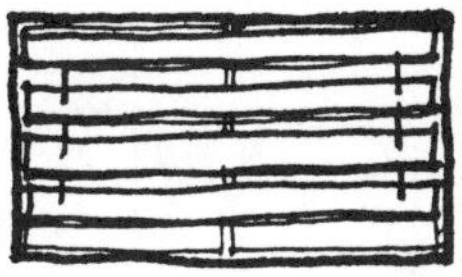

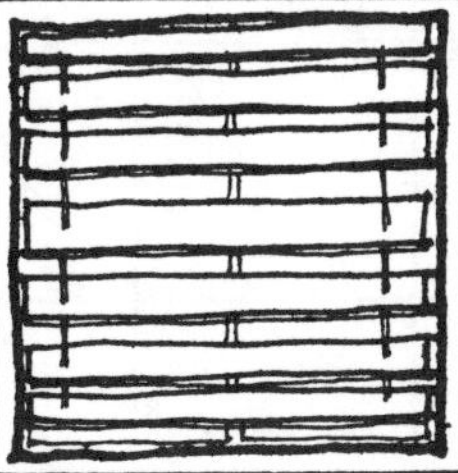

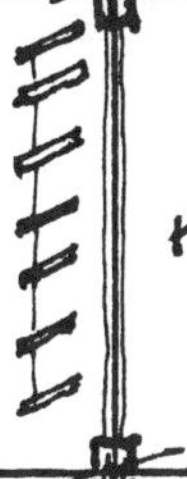

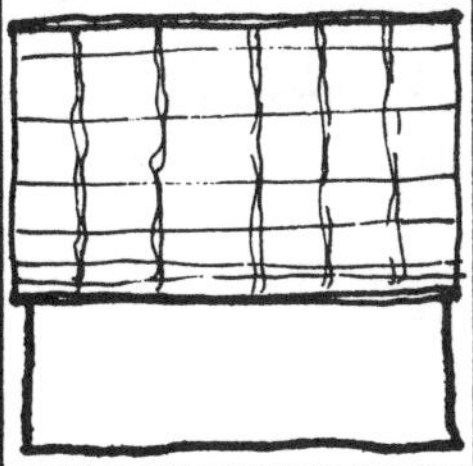

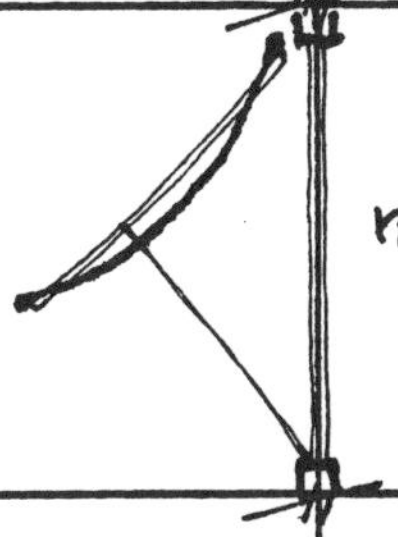

Salem Figure 6-12

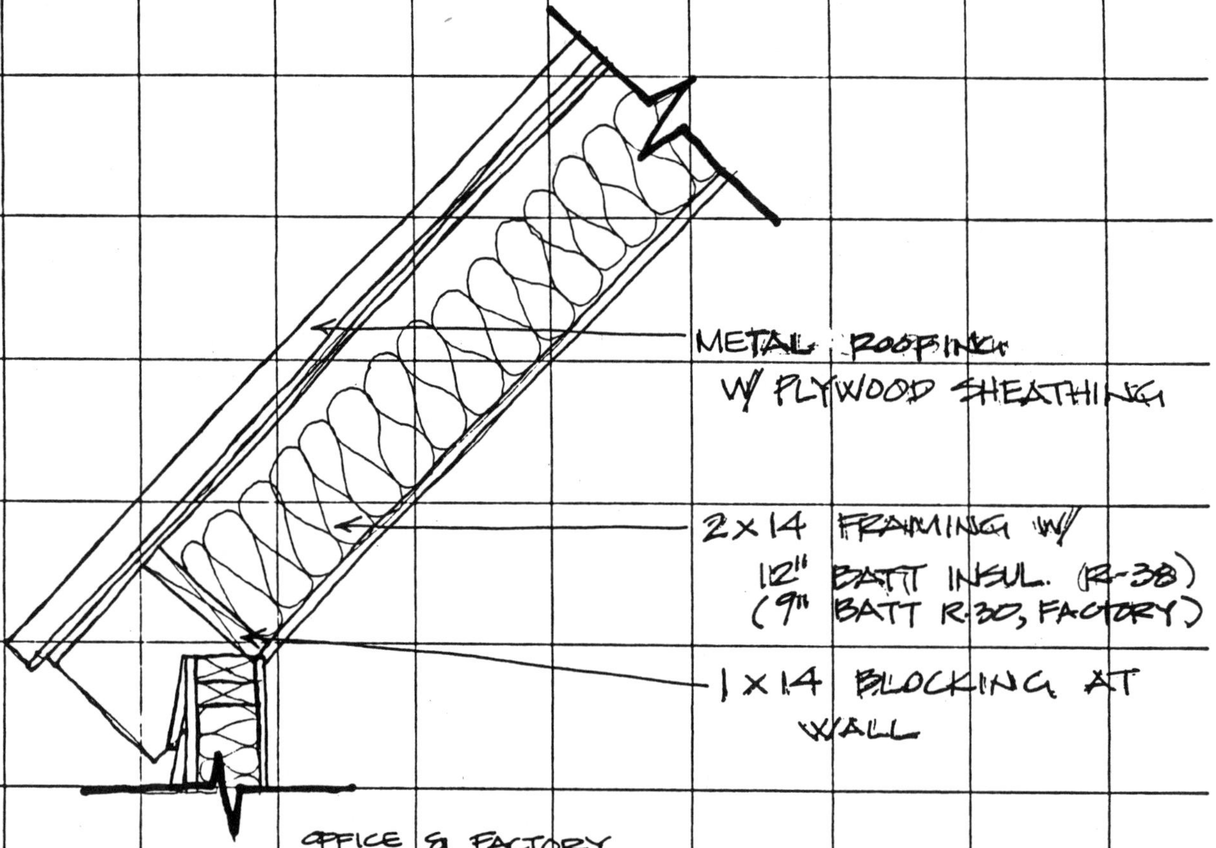

METAL ROOFING
W/ PLYWOOD SHEATHING

2 X 14 FRAMING W/
12" BATT INSUL. (R-38)
(9" BATT R-30, FACTORY)

1 X 14 BLOCKING AT
WALL

OFFICE & FACTORY

SECTION AT PITCHED ROOF - FLAT ROOF SIMILAR

scale 3/4" = 1'-0"

salem Figure 6-13

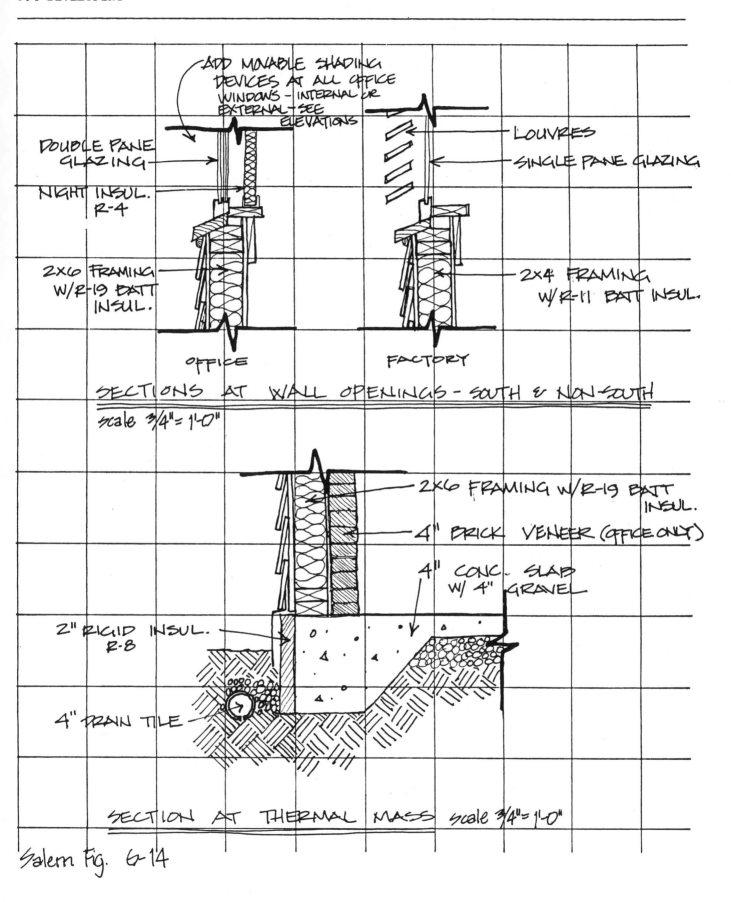

ADD MOVABLE SHADING DEVICES AT ALL OFFICE WINDOWS - INTERNAL OR EXTERNAL - SEE ELEVATIONS

DOUBLE PANE GLAZING

NIGHT INSUL. R-4

2X6 FRAMING W/R-19 BATT INSUL.

LOUVRES

SINGLE PANE GLAZING

2X4 FRAMING W/R-11 BATT INSUL.

OFFICE

FACTORY

SECTIONS AT WALL OPENINGS - SOUTH & NON-SOUTH

scale 3/4" = 1'-0"

2X6 FRAMING W/R-19 BATT INSUL.

4" BRICK VENEER (OFFICE ONLY)

4" CONC. SLAB W/ 4" GRAVEL

2" RIGID INSUL. R-8

4" DRAIN TILE

SECTION AT THERMAL MASS   scale 3/4" = 1'-0"

Salem Fig. 6-14

EVALUATION

A. Criterion A
All glazed apertures are fully shaded during overheated months. From 6.3 Criterion D:
Factory overheated months are: *All months*
                                 May through October
Office overheated months are:

Evaluation Tool for Criterion A - "Shading Masks"
1. Plot the overheated months on photocopies of the sunchart for your latitude (found in section 6.5).
a. For each building the Balance Point (from section 3.6 climate exercise) is:
  office:
  factory:

50° - office balance point
5° - factory

b. On the graph of your climate on the Bioclimatic Chart the max temperature occurs approximately 4 p.m., the min temperature approximately 4 a.m. Temperatures can be matched to the remaining hours of the day along the line drawn between the two points with 10 a.m. and 10 p.m. occurring at mid-point. Note that you will graph the temperature pattern for August in the next criterion.
c. Plot a separate sun chart for each building. For each overheated month between January and June, use <u>vertical lines</u> to shade hours of the day in which the <u>exterior</u> temperature is GREATER than that of the balance point.
d. For each overheated month between July and December use <u>horizontal lines</u> to shade in hours of the day in which the <u>exterior</u> temperature is GREATER THAN the Balance Point.

Plot indicates little if any pairing of shaded months - so fixed shading devices are inappropriate for this climate. See LOF Sun Chart at end of chapter.

When <u>both</u> months in a pair are overheating, this shows up as a cross-hatched area, signalling the appropriateness of <u>fixed</u> shading devices. Where only one of the paired months is overheated, <u>movable</u> shading devices are needed; for example, to admit sun in cold March, exclude it in warm September.

2. Shading Masks. External shading devices produce distinct zones of full shade. These patterns of sun protection are called "masks" and can be compared to the plot's overheated periods. Some masks are illustrated in Olgyay, <u>Design with Climate</u>, p. 82-83 and in Olgyay and Olgyay, <u>Solar Control and Shading Devices</u>, p. 88.
a. For all apertures not shaded at the building and site scale draw the mask of the shading device on tracing paper at the same scale as the LOF suncharts (label by aperture and key to elevations or paralines).

---

b. Evaluate the performance of each shading device by superimposing the mask over the overheated plot for the right building. Whenever the building is overheated the mask should cover the plot. Conversely when the building is not overheating the mask should not cover the plot.

c. Each of your shading devices has a "shading coefficient" (S.C.) that indicates what percent of the window's unshaded solar gain is admitted through (or around) the shading device. The S.C. is listed both in Olgyay Design with Climate p. 68-71, and in MEEB (Table 4.28 a, b).

List the shading coefficient for each of your shading devices.

Shading Devices   Olgyay, p. #
Office: East and
Northwest Windows:
(vertical) venetian
blind inside window,
white, max. S.C. =
0.56   69
North Windows:
(horiz.) venetian
blinds, as above
max. S.C. = 0.56   69
South lower windows,
and skylight
outside moveable
louvres (over the
arcade); max. S.C. =
0.15   71

Does your first trial design meet Criterion A?
Yes.
If not, what will you change for your second trial?

South clerestory
windows
outside canvas
awnings, max. S.C. =
0.25   70

B. Criterion B: Daily heat peaks are minimized during overheated months by avoiding simultaneous peaks from more than one source.

Factory: All windows:
adjustable outside
louvres, max. S.C. =
0.15   71

Evaluation Tool for Criterion B - "Heat Gain Calculation"
The following heat gain calculation is an abbreviated version of the procedure you will follow in 7.0 THERMAL FINALIZATION. This calculation looks at the daily patterns of heat gain in your buildings, and will be done for August 21st.

(The area of opaque wall tabulated for the office, 2,840 sq.ft., does not include openings for ventilation, 200 sq.ft. added in this exercise. See Salem Fig. 6-7.)

1. Roof and Wall Gains
Opaque skin elements receive and store heat from the sun as well as heat from the air. To get an approximate 24-hour average of the heat passed to the interior through these elements, heat transfer factors (HTF's) are used. Heat transfer factors combine the U-value of a construction with various temperature differences, or differentials, between inside and outside.

A more precise nonresidential calculation would use Total Equivalent Temperature Differentials (TETD's) rather than HTF's. However, most of the heat gain in this exercise is solar gain or internal gain, so HTF's provide a reasonable approximation.

Use MEEB Table 4.22(c) p. 149 to find the HTF's for your design. For all buildings, assume the lowest (20°) temperature differential.

Office:
R-19 walls, standard frame: HTF at 20° = 1.1
R-38 roof, pitched, with standard ventilation: HTF at 20° = 0.8

Factory:
R-11 walls, standard frame: HTF at 20° = 1.5
R-30 roof, pitched, with standard ventilation: HTF at 20° = 1.0

		Trial 1					Trial 2	
	Area sq.ft.	X	HTF	=	Daily average gain Btuh	Area sq.ft.	X HTF =	Daily average gain Btuh
a. OFFICE								
opaque wall	(incl. doors) 2840	x	1.1	=	3124			
opaque roof	(1700-50)	x	0.8	=	1320			
				TOTAL:	4444		TOTAL:	
b. FACTORY								
opaque wall	(incl. doors) 3400	x	1.5	=	5100			
opaque roof	3780	x	1.0	=	3780			
				TOTAL:	8880		TOTAL:	

2. Solar Gains

To look at the pattern over a day you first find the gain through east-facing glazed apertures for morning, then through south glazed apertures at noon, then west-facing for afternoon. For horizontal skylights, calculate gains at all three times. For sloped skylights, calculate for the hour corresponding to the direction it faces, and interpolate between vertical and horizontal solar gains. Shading coefficients (column 3) are from the previous evaluation tool. Solar Heat Gains for August 21 (column 2) are as follows (Mazria, 1979, Appendix I):

		32°	36°	40°	44°
	Northeast	101	91	80	70
9am	East	195	194	192	189
	Southeast	176	184	191	197
noon	South	109	127	145	151
	Southwest	165	171	175	179
4pm	West	211	210	208	206
	Northwest	137	130	122	115
Horizontal		216	211	205	197

(Salem, Or) For sloped east skylight, Salem: 189 vertical, 197 east; gain = $\frac{189 + 197}{2}$ = 19.

6.0 COOLING

6.4 DEVELOPING

---

a. OFFICE

	Col. 1	Col. 2		Col. 3		Col. 4	Col. 5
Trial 1	aperture	sq.ft. area	X	S.C.	X	Btuh/sq.ft. Heat Gain Factor =	Btuh Total
East 9 am	*Lower window*	*100*		*.56*		*189*	*10,580*
	Skylight	*50*		*.15*		*193*	*1,450* *12,030*
South noon	*Lower windows*	*200*		*.15*		*151*	*4,530*
	Clerestories	*200*		*.25*			*7,550* *12,080*
West 4 pm	*Northwest window*	*50*		*.56*		*206*	*5,770*

Trial 2							
East 9 am	*Lower window*	*100*		*.15*		*189*	*2,840*
	Skylight	*50*		*.15*		*193*	*1,450* *4,290*
South noon	*Lower windows*	*200*		*.15*		*151*	*4,530*
	Clerestories	*200*		*.25*			*7,550* *12,080*
West 4 pm	*Northwest window*	*50*		*.56*		*206*	*5,770*

b. FACTORY

	Col. 1	Col. 2		Col. 3		Col. 4	Col. 5
Trial 1	aperture	sq.ft. area	X	S.C.	X	Btuh/sq.ft. Heat Gain Factor =	Btuh Total
East 9 am	*none*						
South noon	*All* *south windows*	*950*		*.15*		*151*	*21,520*
West 4 pm	*none*						

Trial 2							
East 9 am							
South noon							
West 4 pm							

---

3. Internal Gains
Use the internal gain rate from p. 6.19 to calculate
total while occupied. (This should be the same value
found in 3.6 Climate.)

Office: _10.4_ Btuh/sf x _1,750_ sf = _18,280_ Btuh

Factory: _34.2_ Btuh/sf x _3,900_ sf = _133,440_ Btuh

Evaluation Tool for Criterion B - "24 Hour Gain Patterns"

Note: This tool will also be used to assess performance
for Criterion C. Look ahead and consider this in any
redesign.

1. Plot the gains from the preceeding tool below. Be sure
to use different colors or hatch patterns to distinguish
the gain sources. (See Salem Example as a model)
a. First, plot roof and wall gains as constant over the
24 hour period.
b. Next, add the internal gains for the hours of building
use you specified in the design presentation.
c. Finally, plot each of the solar gains at the hour for
which it was calculated. Plot zero gain at sunrise and
sunset (approx. 5:30 am, 6:30 pm on August 21). Connect
the points.

2. Below the gains plot the 24 hour temperature pattern
for August estimated for the shading masks in the
previous criterion. The high temperature generally
occurs at 4 pm, the low at 4 am.

*....Guess we'll have to clean
up our act...*

Trial 1

Trial 2

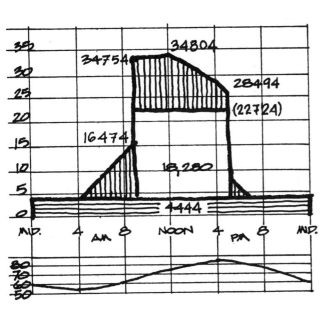

Office Heat Gain Pattern
Pattern

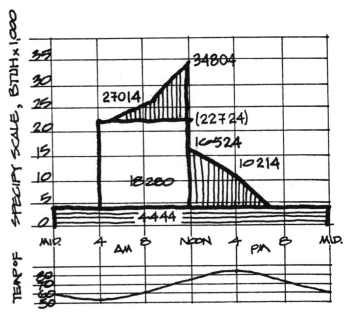

Office Heat Gain

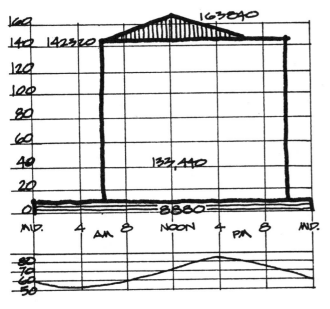

Factory Heat Gain Pattern
Pattern

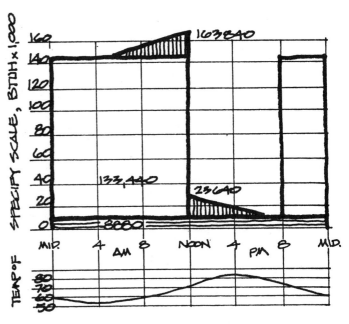

Factory Heat Gain

Does your first trial design meet Criterion B by avoiding simultaneous heat gain peaks from more than one source?
Office: No, morning and noon peaks are particularly high, and correspond with internal gains.
Factory: No, but solar peak is relatively slight.

If not, what will you change for your second trial?
Office: Increase shading coefficient for east glass; go to exterior moveable vertical louvres, S.C. = 0.15 (Olgyay p. 71)
Factory: Leave as is.

C. Criterion C: Heat gain peaks do not occur during the hours of the day it is hottest outside. Use the previous Evaluation Tool "Daily Heat Gain Patterns" to see if your design meets this criterion.

Does your first trial design meet Criterion C?
Office: Comes close; temperatures peak as it closes.
Factory: Temperatures peak at about middle of usage period.

If not, what will you change for your second trial?
Office and Factory: Considering both criteria B and C, show a changed use schedule which, although socially unrealistic, allows most building usage to take place during the colder hours of early morning.

OFFICE: Overheated months, May thru October.
Criterion A, part 1 (p.S6.32).

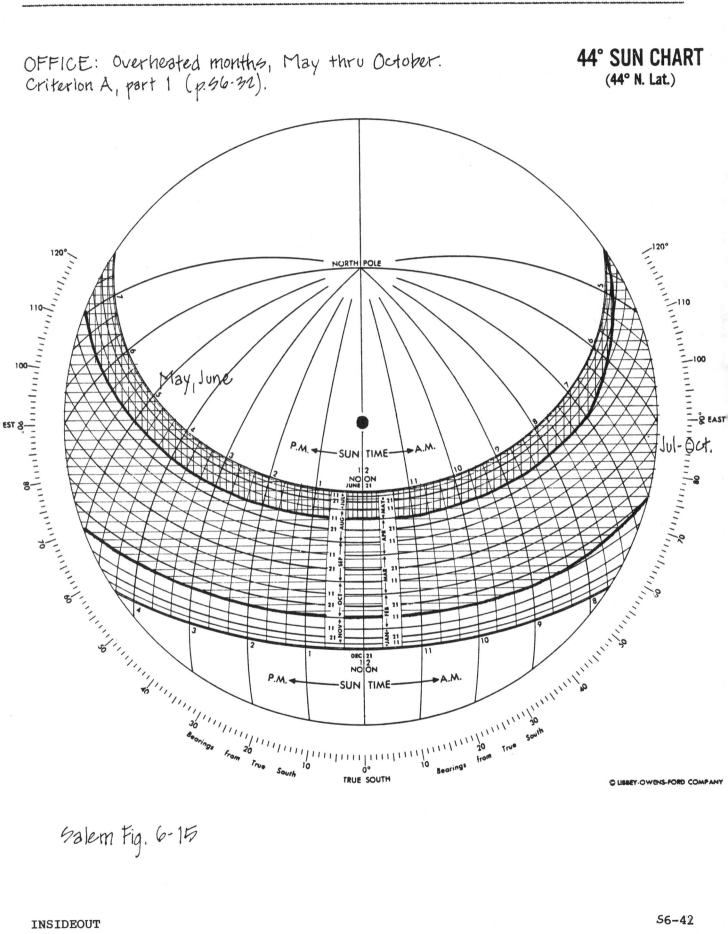

May, June

Jul-Oct.

Salem Fig. 6-15

FACTORY: Overheated months, all (internally dominated)
Criterion A, part 1 (p56.32).

44° SUN CHART
(44° N. Lat.)

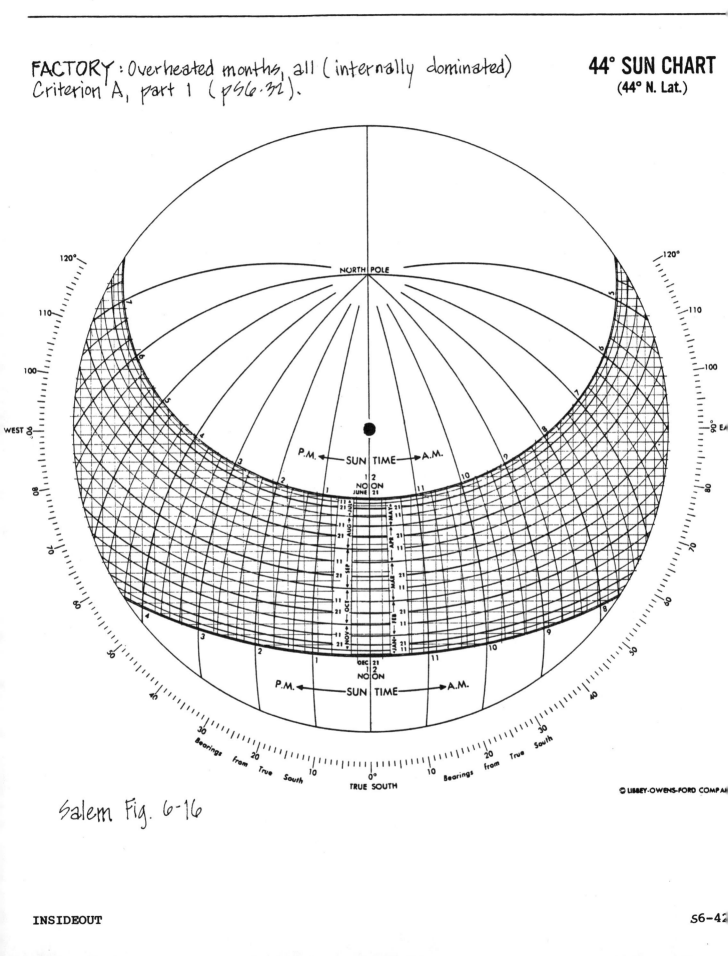

Salem Fig. 6-16

SHADING MASK: FACTORY
horizontal louvers at factory clerestory (item #4, p. 56-29);
for criterion A, part 2, p. 56-32.

44° SUN CHART
(44° N. Lat.)

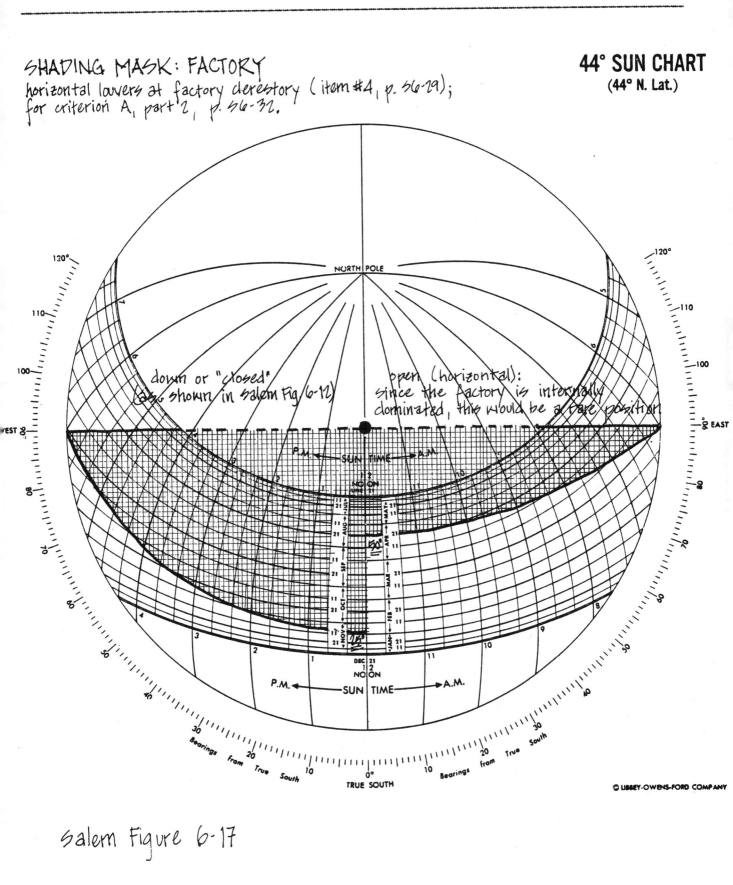

down or "closed"
(as shown in Salem Fig 6-12)

open (horizontal):
since the factory is internally
dominated, this would be a rare position

Salem Figure 6-17

7.1 INTRODUCTION

In this finalization exercise you will see how your building design will work for both cooling and heating. The first step for each section is to figure the quantity of heat that needs to be supplied in the heating season (heat loss or heating load) and needs to be removed in the cooling season (heat gain, or cooling load). The calculations may seem relatively long, but they are the only way you can evaluate your design at this level. The Btu totals you calculate are not a goal in themselves, but are used to evaluate thermal performance in ways that tell you something about the building.

Contents

7.2 HEATING PERFORMANCE

GOAL

Keep occupants thermally comfortable throughout the year by using the natural energies available in your climate, thereby minimizing the use of electrical or fossil fueled auxiliary equipment.

CRITERIA

A. Your buildings will use less fuel for heating in the winter than comparable "energy conserving" buildings in your climate.

B. The temperatures inside are between $60^{\circ}F$ and $80^{\circ}F$ when the buildings are occupied in the winter months.

C. The capacity of the heating backup system is sufficient to keep you warm when there is no sun or internal gain.

DESIGN

Fully document the design which you are evaluating in this exercise.

See Salem Example, Figures 7-5 to 7-12.

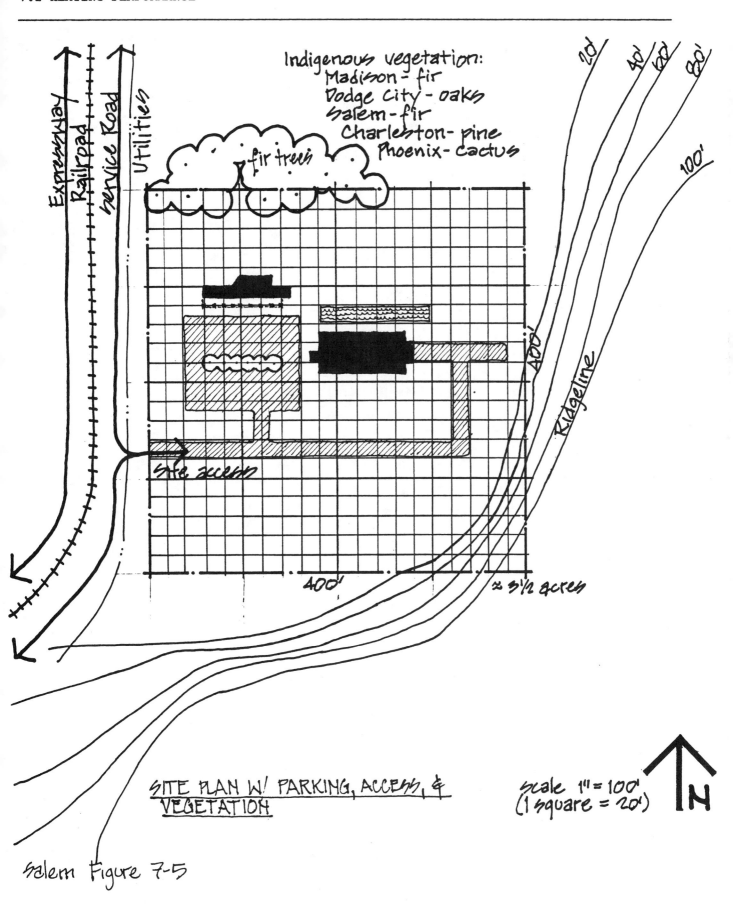

Indigenous vegetation:
Madison - fir
Dodge City - oaks
Salem - fir
Charleston - pine
Phoenix - cactus

fir trees

Expressway
Railroad
Service Road
Utilities

20' 40' 60' 80'

100'

400'

Ridgeline

site access

400'

≈ 3½ acres

SITE PLAN W/ PARKING, ACCESS, &
VEGETATION

scale 1" = 100'
(1 square = 20')

N

Salem figure 7-5

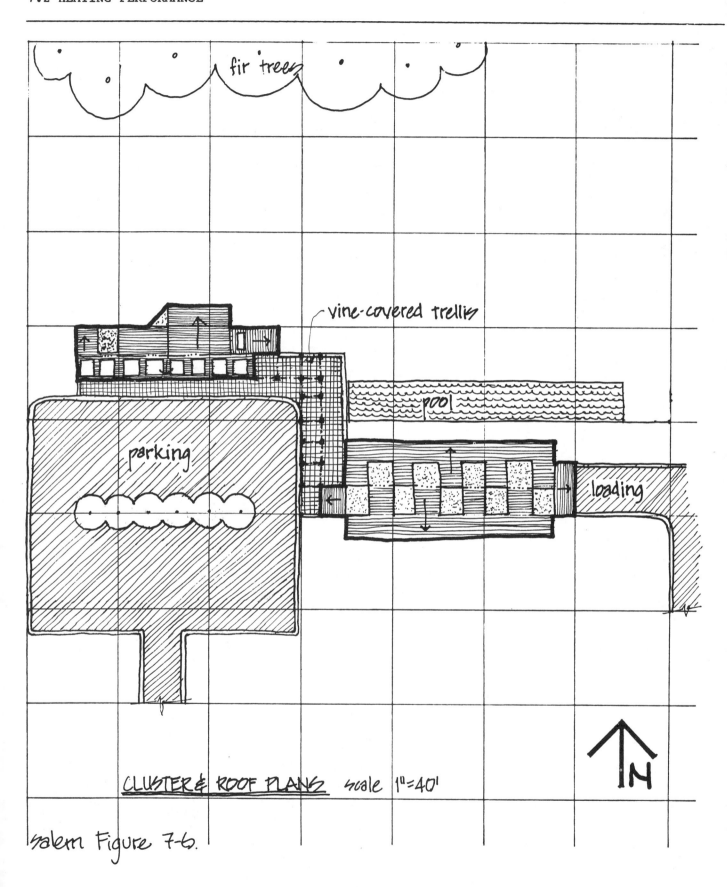

fir trees

vine-covered trellis

pool

parking

loading

CLUSTER & ROOF PLANS scale 1"=40'

salem Figure 7-6.

N

for plans, sections, & elevations :   if   1/16"=1'-0",   1 square = 10'.

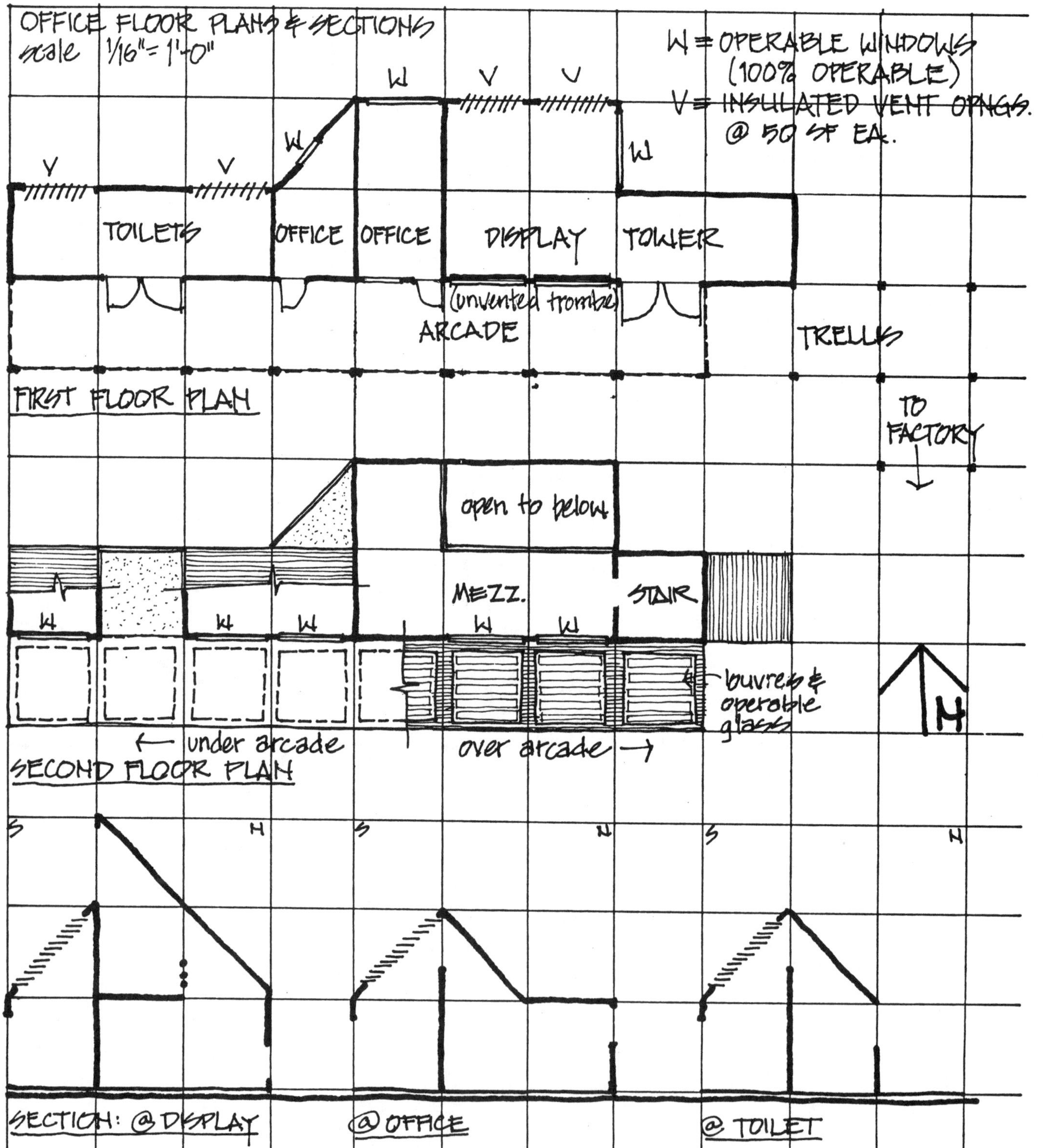

OFFICE FLOOR PLANS & SECTIONS
scale 1/16"= 1'-0"

W = OPERABLE WINDOWS
(100% OPERABLE)
V = INSULATED VENT OPNGS.
@ 50 SF EA.

V   V   W   V   V   W

TOILETS   OFFICE OFFICE   DISPLAY   TOWER

(unvented trombe)
ARCADE   TRELLIS

FIRST FLOOR PLAN

TO FACTORY

open to below

MEZZ.   STAIR

W   W   W

buvres &
operable
glass

← under arcade   over arcade →

SECOND FLOOR PLAN

S   H   S   N   S   H

SECTION: @ DISPLAY   @ OFFICE   @ TOILET

salem Figure 7-7.

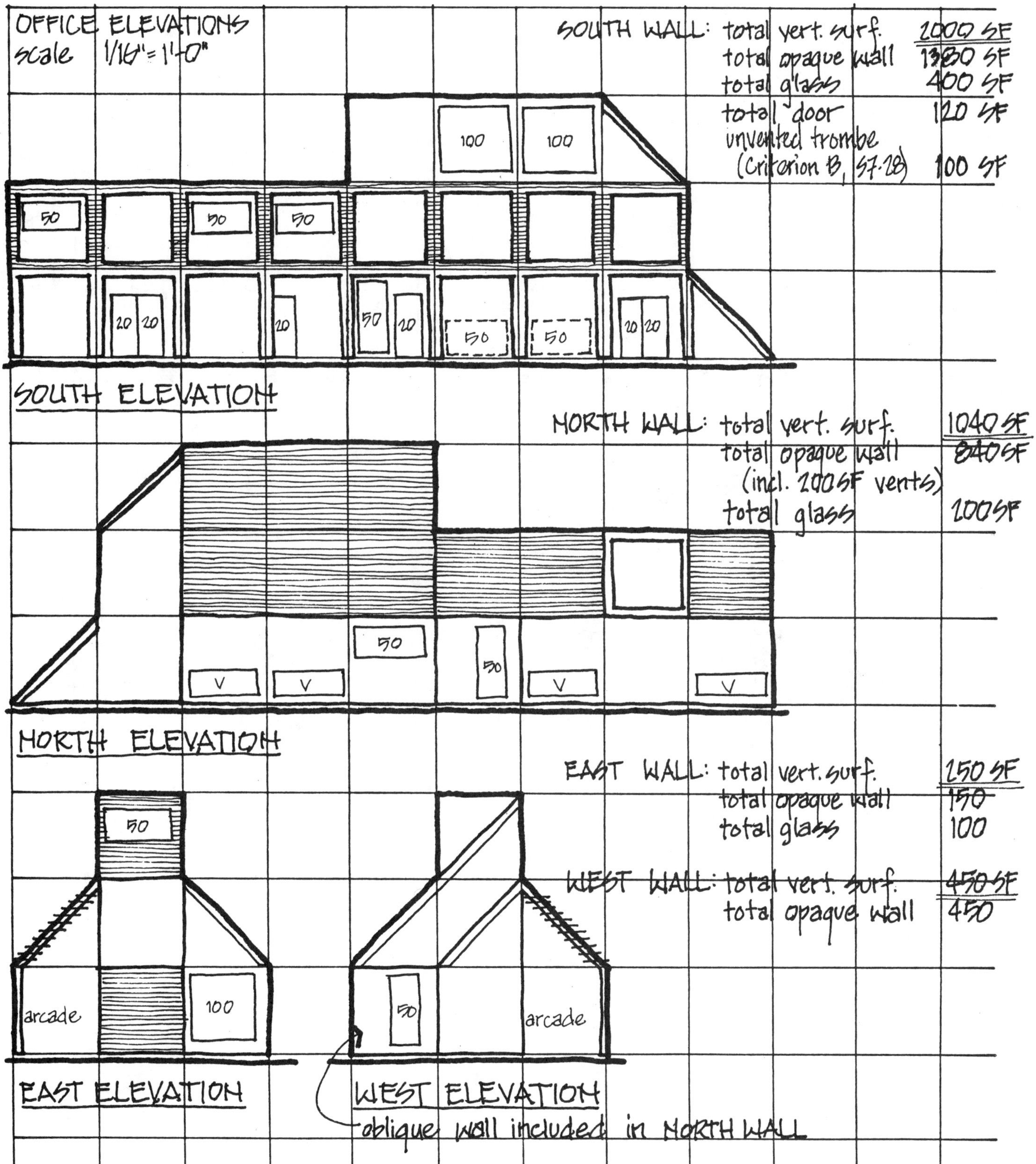

OFFICE ELEVATIONS
scale 1/16" = 1'-0"

SOUTH WALL: total vert. surf.   2000 SF
total opaque wall   1380 SF
total glass   400 SF
total door   120 SF
unvented trombe
(Criterion B, §7.28)   100 SF

100   100

50   50   50

20 20   20   50 20   50   50   20 20

SOUTH ELEVATION

MORTH WALL: total vert. surf.   1040 SF
total opaque wall   840 SF
(incl. 200 SF vents)
total glass   200 SF

50

50

V   V   V   V

NORTH ELEVATION

EAST WALL: total vert. surf.   250 SF
total opaque wall   150
total glass   100

WEST WALL: total vert. surf.   450 SF
total opaque wall   450

50

50

50

arcade   100   arcade

EAST ELEVATION   WEST ELEVATION
oblique wall included in NORTH WALL

Salem Figure 7-8

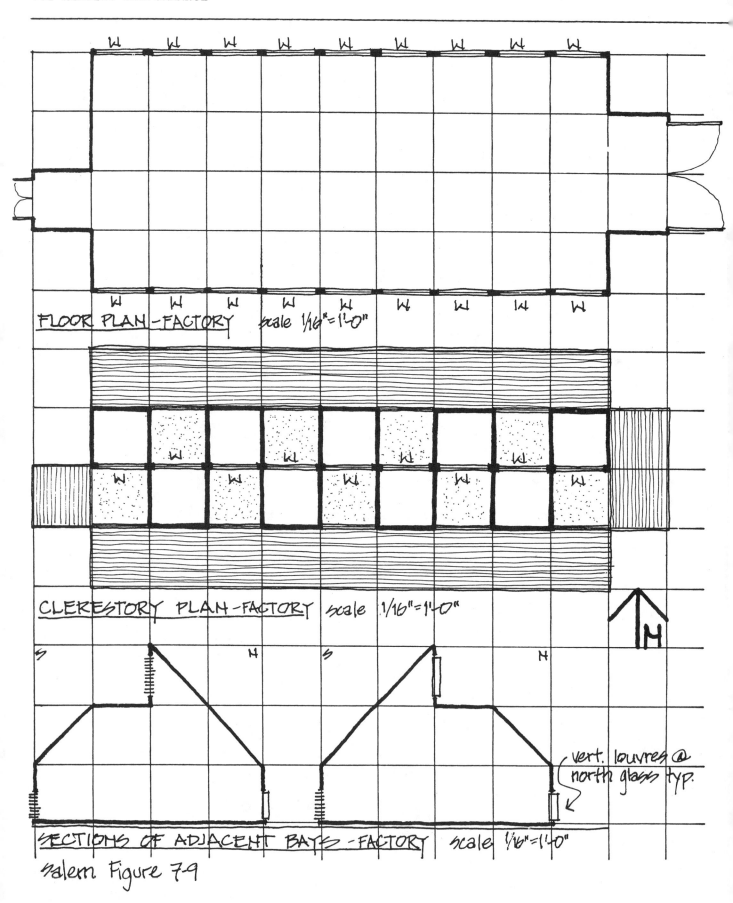

FLOOR PLAN - FACTORY   scale 1/16"=1'-0"

CLERESTORY PLAN - FACTORY   scale 1/16"=1'-0"

SECTIONS OF ADJACENT BAYS - FACTORY   scale 1/16"=1'-0"

vert. louvres @ north glass typ.

Salem Figure 7-9

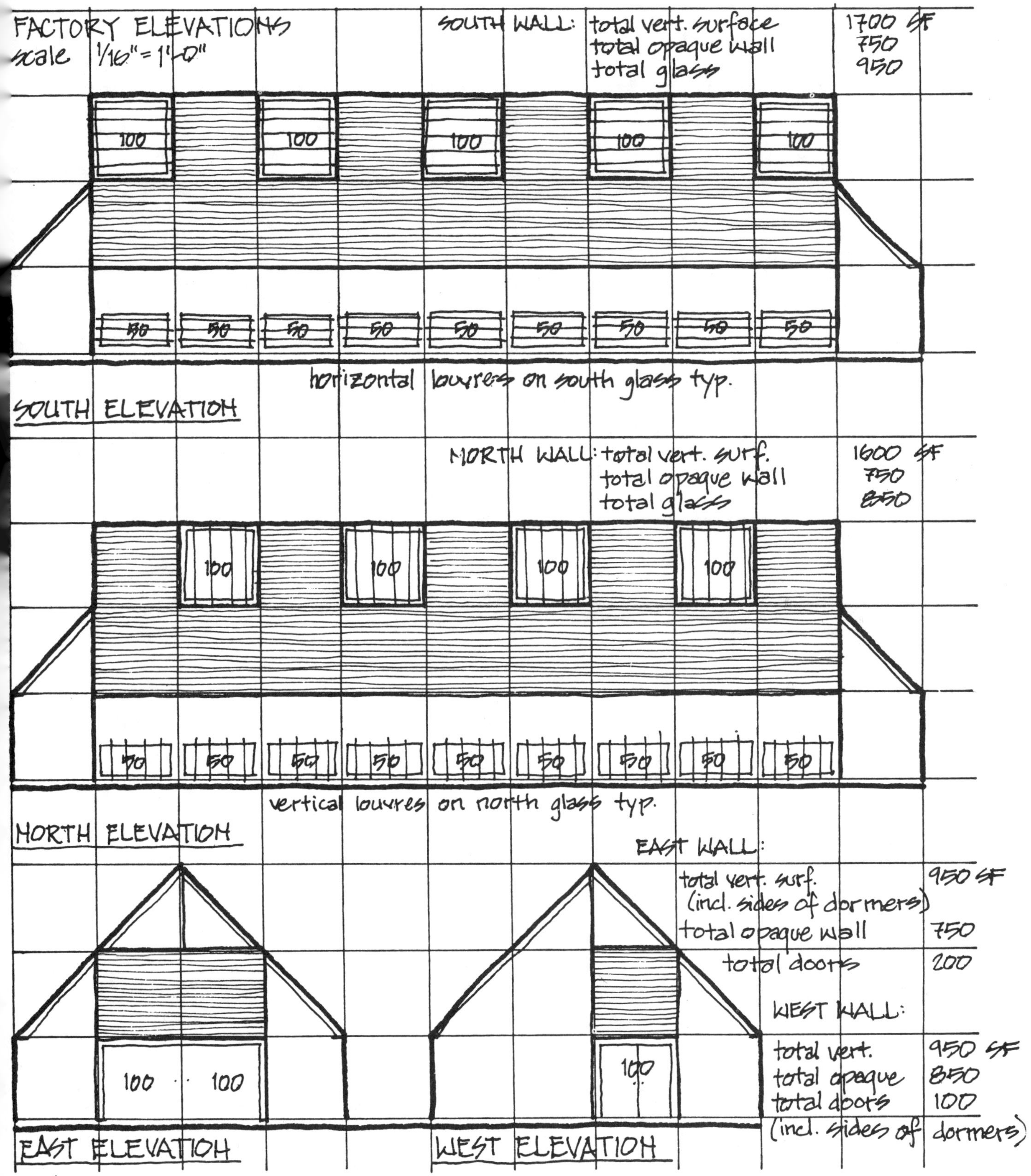

FACTORY ELEVATIONS
scale 1/16"=1'-0"

SOUTH WALL: total vert. surface   1700 SF
total opaque wall   750
total glass   950

| 100 | 100 | 100 | 100 | 100 |

| 50 | 50 | 50 | 50 | 50 | 50 | 50 | 50 | 50 |

horizontal louvres on south glass typ.

SOUTH ELEVATION

MORTH WALL: total vert. surf.   1600 SF
total opaque wall   750
total glass   850

| 100 | 100 | 100 | 100 |

| 50 | 50 | 50 | 50 | 50 | 50 | 50 | 50 | 50 |

vertical louvres on north glass typ.

NORTH ELEVATION

EAST WALL:

total vert. surf.   950 SF
(incl. sides of dormers)
total opaque wall   750
total doors   200

WEST WALL:

total vert.   950 SF
total opaque   850
total doors   100
(incl. sides of dormers)

| 100 | 100 |

| 100 |

EAST ELEVATION   WEST ELEVATION

Salem Figure 7-10

---

for plans, sections, & elevations :   if   1/16"=1'-0",   1 square = 10'.

for plans, sections, & elevations: if   1/16"=1'-0",   1 square = 10'.

DESIGN DOCUMENTATION FOR OFFICE & FACTORY.
TYPICAL DETAILS @ 3/4" = 1'-0" (cont.)

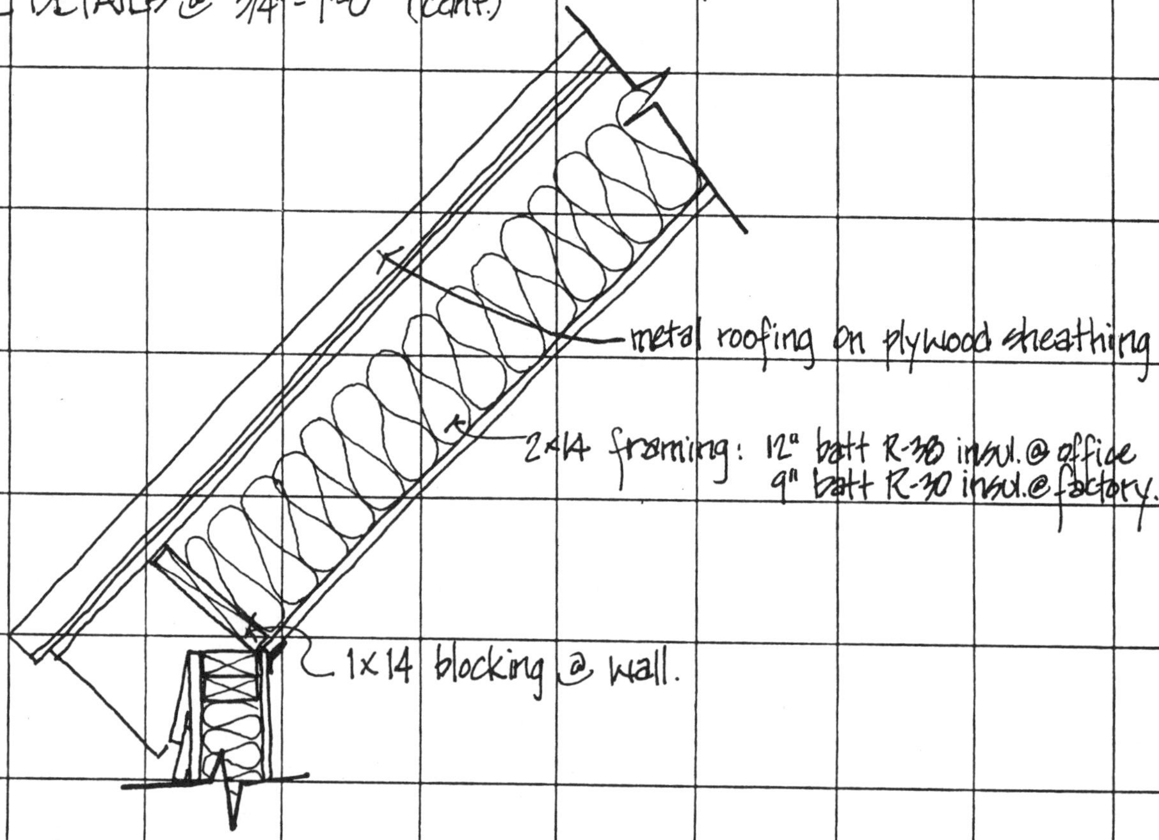

metal roofing on plywood sheathing

2x14 framing: 12" batt R-38 insul.@ office
9" batt R-30 insul.@ factory.

1x14 blocking @ wall.

SECTION @ PITCHED ROOF (flat roof similar)

salem Figure 7-11

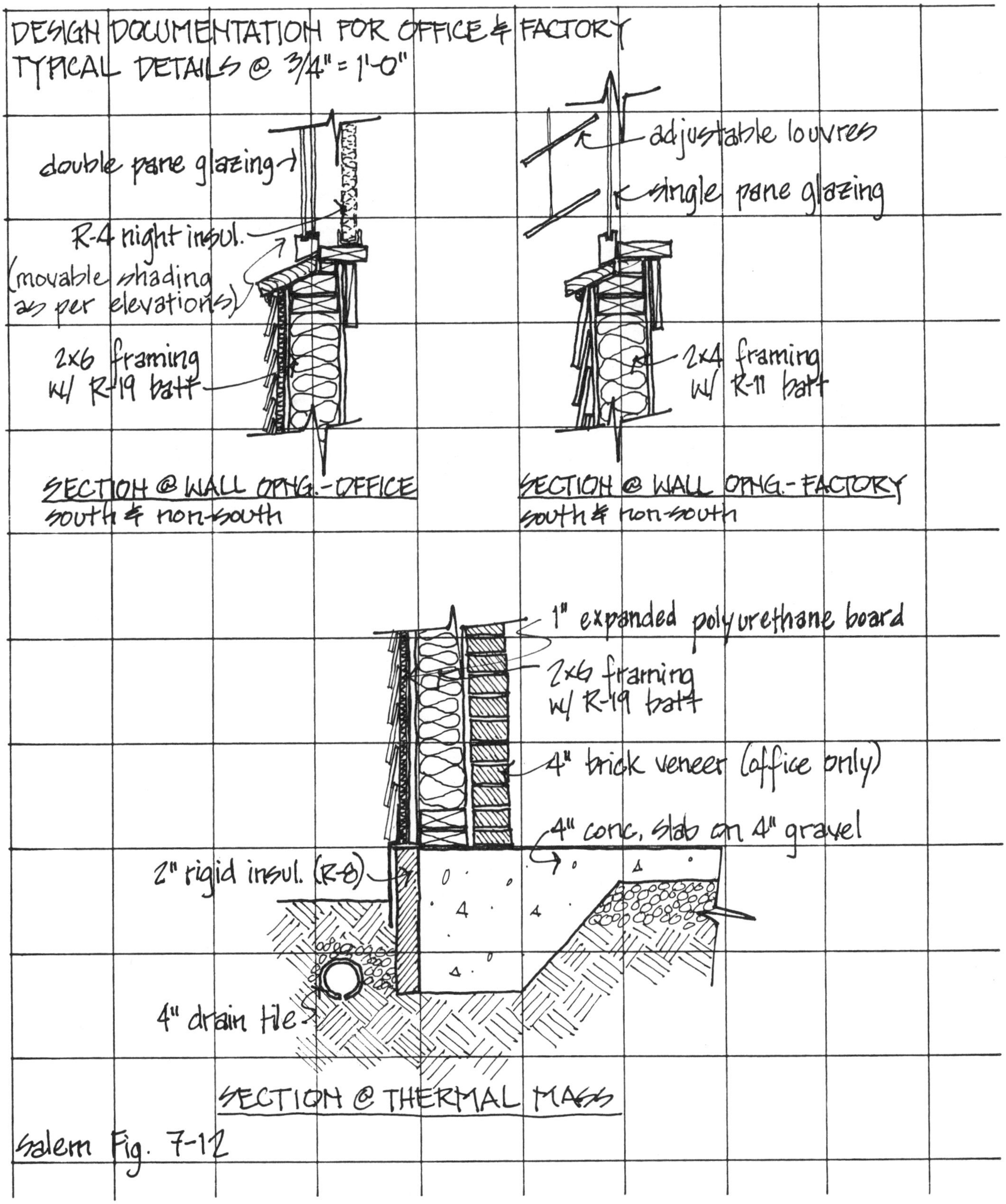

DESIGN DOCUMENTATION FOR OFFICE & FACTORY
TYPICAL DETAILS @ 3/4" = 1'-0"

double pane glazing

adjustable louvres

single pane glazing

R-4 night insul.
(movable shading
as per elevations)

2x6 framing
w/ R-19 batt

2x4 framing
w/ R-11 batt

SECTION @ WALL OPNG.-OFFICE
south & non-south

SECTION @ WALL OPNG.-FACTORY
south & non-south

1" expanded polyurethane board

2x6 framing
w/ R-19 batt

4" brick veneer (office only)

4" conc. slab on 4" gravel

2" rigid insul. (R-8)

4" drain tile

SECTION @ THERMAL MASS

Salem Fig. 7-12

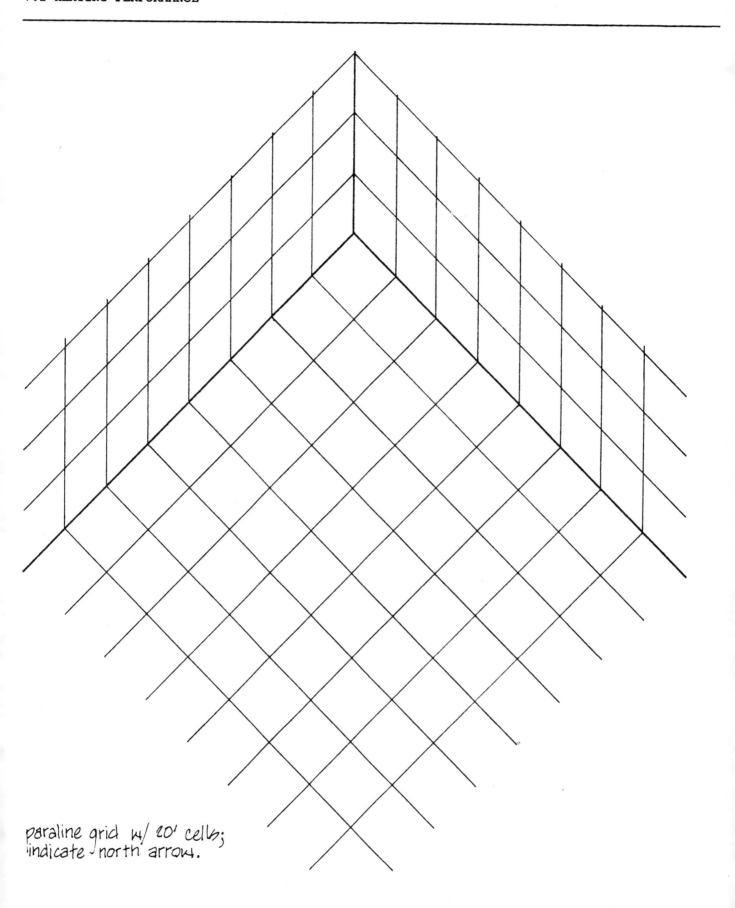

paraline grid w/ 20' cells;
indicate north arrow.

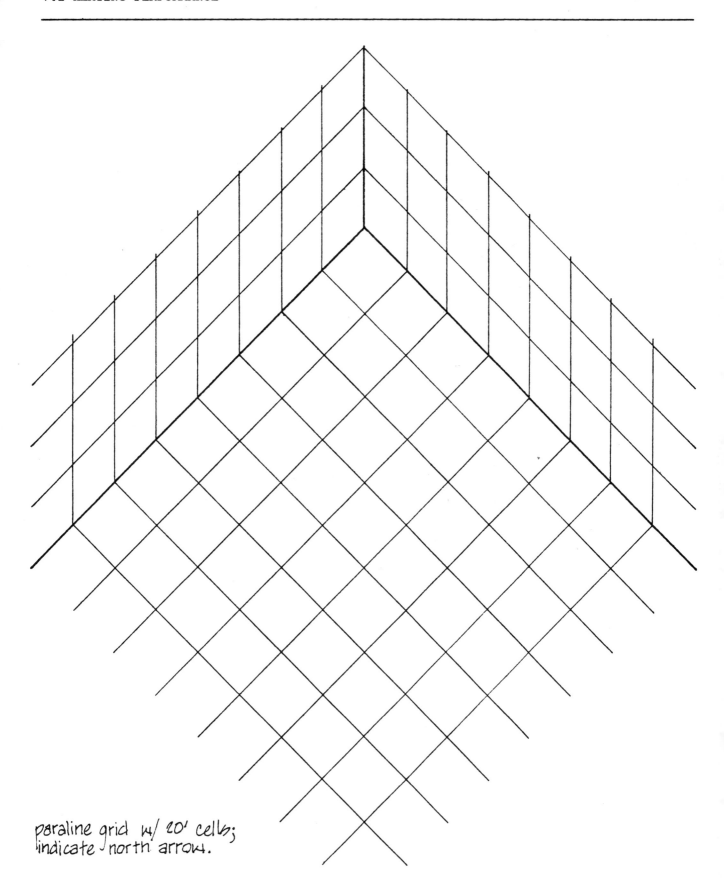

paraline grid w/ 20' cells;
indicate north arrow.

EVALUATION

A. Criterion A: Your buildings use less fuel for heating in the winter than comparable "energy conserving" buildings in your climate. (criteria from Balcomb, p. 24)

Heating Degree-days	"Energy Conserving", Conventionally fuelled	Solar Heated, not including the solar walls heat losses
less than 1,000	9 Btu/DD sq.ft.	7.6 Btu/DD sq.ft.
1,000-3,000	8	6.6
3,000-5,000	7	5.6
5,000-7,000	6	4.6
greater than 7,000	5	3.6

Because these values assume no internal gain, the criteria may be adjusted for Internally Dominated Load (IDL) buildings similarly to the adjustment for solar aperture in 5.3 Scheming:

$$\text{Adjusted Btu/DDSF} = \text{Btu/DDSF} \times \frac{65}{\text{Balance Point Temp (24 hr avg)}}$$

Office Building: *(Not internal dominated)*

Factory:

"Conventionally Fueled"

$$7 \times \frac{65^o}{5^o} = 91 \; Btu/DD \; sq.ft.$$

HEAT BALANCE ON A CAT

From "Thermal and Economic Performance..." by J. Augustyn et al, 4th National Passive Conference Proceedings, 1979. Used by permission of the author.

Evaluation Tool - "Heat Loss Calculation"

Use the format below to establish UxA for all skin elements except the south facing solar aperture. (South aperture is excluded because it is assumed that there will be no net loss through this heat-admitting surface area.)

1. Office Building Trial 1
Include all spaces that are heated.

	Materials	Area (sf.ft.)	U(Btuh/sq.ft./°F)	UA(Btuh/°F)	
Exterior opaque walls	R-19 standard frame	2920	0.06	175	include insulated vent openings as opaque wall
Roof: horizontal 45° slope	R-38, both roofs	1650	0.02	33	
Slab edge	insulated	214 lin. ft.	0.06	13	
Non-south glass: vertical	dbl. glazed R4 night insulation	200	0.30	60	
45° slope (skylight)	as above	50	0.30	15	
horizontal	none				
Exterior doors (6 at 20 SF each)	steel, with urethane core	120	0.19	23	U-value from MEEB p. 130
Infiltration	(N.A.)	volume in cu.ft. x .018 x ACH = 21,000 x .018 x 0.5ACH		189	

TOTAL UA = 508
(excluding solar aperture)

For the "Building Load Coefficient" (BLC), multiply this
UA total by 24 hours: 508 x 24 = 12,192

BLC Office =   12,192   Btu/DD

To relate to the initial criteria, divide by floor area:

$$\frac{Btu/DD}{floor\ area\ in\ sq.ft.} = \underline{\quad 7.0 \quad}\ Btu/DD\ sq.ft. \quad \frac{12,192}{1,750\ sf} = 7.0$$

Trial 1: 7.0 Btu/DD sq.ft.; maximum allowable 5.6; go to R9 insulating shades, which, by fitting more tightly, further reduce infiltration. Also, glaze roof of arcade for winter, trapping warmer air outside solar aperture to office and toilets. Assume infiltration drops to .33ACH, as a result. Also, add 1" expanded polyurethane board to R19 frame walls, R-value 6.25 (MEEB, p. 113), new U-value for wall 0.05. See Salem Figure 7-12.

Office Building Trial 2
Include all spaces that are heated.

	Materials	Area (sq.ft.)	U(Btuh/sq.ft./$^\circ$F	UA(Btuh/$^\circ$F
Exterior opaque walls	R-19 frame w/ expanded polyurethane board	2920	0.05	146
Roof: horizontal 45° slope	as before	1650	0.02	33
Slab edge	as before			13
Non-south glass: vertical	double glazed	200	.24	48
45° slope horizontal	R-9 night insulation	50	.24	12
Exterior doors	as before			23
Infiltration	(N.A.)	volume in cu.ft. x .018 x ACH = 21,000 x .018 x .33		125

TOTAL UA = 400
(excluding solar aperture)

For the "Building Load Coefficient" (BLC), multiply this
UA total by 24 hours:

BLC Office = __9,600__ Btu/DD

To relate to the initial criteria, divide by floor area:

$$\frac{\text{Btu/DD}}{\text{floor area in sq.ft.}} = \underline{\quad 5.5 \quad} \text{ Btu/DD sq.ft.}$$

This should be no more than the criteria listed for your
type of building in your climate. How does it compare?

Trial 2: 5.5 Btu/DD sq.ft., below the 5.6 allowable.

7.0 THERMAL FINALIZATION

7.2 HEATING PERFORMANCE

Factory Building Trial 1	Materials	Area (sq.ft.)	U(Btuh/sq.ft.$^{\circ}$F)	UA(Btuh/$^{\circ}$F)	
Exterior opaque walls	R-11, standard frame	3100	.08	248	
Roof - horiz: 45° slope:	R-30, both roof conditions	3780	.03	113	South glazing: assume that the factory louvres are sufficiently adjustable to admit 50% of the sunlight striking them. Thus, the other, shaded 50% south glass should be included as "non-south" glass.
Slab edge	insulated	300 lin. ft.	.06	18	
Non-south glass: vertical (include So. because 45° slope it's shaded horizontal	sgl. glazed	850	1.10	935	
		950 x .5	1.10	523	
Exterior doors	steel, mineral fibrecore	300	.59	177	U-value from MEEB, p. 130.
Infiltration	N.A.	volume cu.ft. x .018 x ACH 71,500 x .018 . 75ACH		965	

TOTAL UA = 2979
(excluding south aperture)

For the Building Load Coefficient (BLC), multiply this UA total by 24 hours:

BLC Factory = Total UA _2979_ x 24 = _71,500_____ Btu/DD

and to relate to the initial criteria, divide by the floor area:

$$\frac{Btu/DD}{floor\ area\ in\ sq.ft.} = \underline{18.3\ \ } \ Btu/DD\ s.f.$$

$$\frac{71,500}{3900\ sf} \doteq 18.3$$

7.0 THERMAL FINALIZATION

7.2 HEATING PERFORMANCE

Factory Building Trial 2	Materials	Area (sq.ft.)	U(Btuh/sq.ft.$^{\circ}$F)	UA(Btuh/$^{\circ}$F)
Exterior opaque walls				

Roof - horiz:

45° slope:

Slab edge:

Non-south glass: vertical

45° slope

horizontal

Exterior doors

---

		volume cu.ft. x .018 ACH =		
Infiltration	N.A.			

TOTAL UA =
(excluding south aperture)

For the Building Load Coefficient (BLC), multiply this UA total by 24 hours:

BLC Factory = _____ Btu/DD

and to relate to the initial criteria, divided by the floor area:

$$\frac{\text{\_\_\_\_\_ Btu/DD \_\_\_\_\_}}{\text{floor area in sq.ft.}} = \text{_____ Btu/DD sq.ft.}$$

18.3 is much more than the 9 allowed for a conventional building, but is considerably less than the 91 allowable when balance point is considered.

This should be no more than the criteria listed for your type of building in your climate. How does it compare?

Evaluation Tool for Criterion A - "Solar Savings Fraction"

Now that you know your heat loss and building load coefficient, you can see whether using passive solar heating makes much difference. The Solar Savings

---

raction, or SSF, is a means of comparing your passively
olar heated building to a non-solar, but
nergy-conserving building. (See Balcolm, 1980, p. 10).

omparing the non-solar energy needed by above buildings,
he solar building needs 25 units, while the conserving
uilding needs 70. The difference is 70-25=45, which =
4% of the conserving buildings' total. Therefore, the
olar building has a SSF (Solar Savings Fraction) of 64%.
It is, however, 75% solar heated. It needs more total
eat, because it has substantially greater glass areas
han does the energy conserving building.)

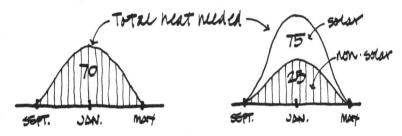

A "conserving" building. A nearly identical building
except that it is solar heated.

For office "collector", assume 10% of aperture area is blocked by the adjustable louvres, mullions, etc.

. Determine the LCR (Load-Collector Ratio)
A comparison between the buildings' need for heat, and
ts heat-admitting south aperture is called the
load-collector ratio (LCR). Small LRC's mean large
window areas and higher SSF's.

For factory "collector", assume that fully open adjustable horizontal louvres could admit 50% of the sunlight striking them:
Factory aperture: .5 x 950 s.f. = 475.

LCR = Building load coefficient
      South aperture area

	1st Trial	2nd Trial	*1st Trial*		*2nd Trial*
LCR Office	26.7	21.4	*Office:* $\frac{9,600 \ Btu/DD}{.9x400 \ sq.ft.} = 26.7$		$\frac{9,600}{.9x500} = 21.4$
LCR Factory	150		*Factory:* $\frac{71,500 \ Btu/DD}{475} = 150$		

2.a. What passive systems are you using, in what portions
of the total, for each building?

Office: *Direct gain, in all cases, all with night insulation.*

Factory: *Direct gain (if needed...)*

b. Find your SSF in the charts below, using the LCR by interpolating where necessary.

WW = water wall system
TW = trombe wall system
DG = direct gain system
NI = night insulation

7.2 HEATING PERFORMANCE

MADISON WISCONSIN	SSF =	.1	.2	.3	.4	.5	.6	.7	.8	.9
	WW	62	25	13	7	–	–	–	–	–
43.1°NL	WWNI	137	62	39	27	20	15	12	9	6
7730 DD **LCR**	TW	67	27	15	8	3	–	–	–	–
T(Jan)=17	TWNI	130	59	36	25	19	14	11	8	5
	DG	–	–	–	–	–	–	–	–	–
	DGNI	133	59	36	24	17	12	8	5	2

DODGE CITY KANSAS	SSF =	.1	.2	.3	.4	.5	.6	.7	.8	.9
	WW	140	64	39	27	19	15	10	6	–
37.8°NL	WWNI	228	107	67	48	36	29	23	18	13
5046 DD **LCR**	TW	138	62	37	25	17	11	8	5	2
T(Jan) = 31	TWNI	215	100	63	45	34	26	20	15	10
	DG	132	54	27	10	–	–	–	–	–
	DGNI	236	109	68	47	35	26	20	14	8

SALEM OREGON	SSF =	.1	.2	.3	.4	.5	.6	.7	.8	.9
	WW	137	57	34	22	15	9	5	–	–
44.9°NL	WWNI	226	101	62	43	32	24	18	14	9
4852 DD **LCR**	TW	133	56	32	20	13	8	4	–	–
T(Jan)=39	TWNI	212	95	58	40	29	22	16	11	7
	DG	X 121	41	–	–	–	–	–	–	–
	DGNI	231	102	62	42	30 X	21	15	10	5

Factory SSF looks well under .10% Office Building; SSF = about
.55, 1st Trial

PHOENIX ARIZONA	SSF =	.1	.2	.3	.4	.5	.6	.7	.8	.9

.60, 2nd Trial

PHOENIX ARIZONA	SSF =	.1	.2	.3	.4	.5	.6	.7	.8	.9
	WW	467	219	139	100	75	58	45	34	22
33.4°NL	WWNI	620	293	188	136	104	82	65	51	36
1552 DD **LCR**	TW	436	202	126	87	63	46	34	25	16
T(Jan)=51	TWNI	583	275	176	126	95	73	56	42	28
	DG	555	256	157	107	75	53	37	24	12
	DGNI	673	316	201	143	107	82	62	45	29

CHARLESTON SO. CAROLINA	SSF =	.1	.2	.3	.4	.5	.6	.7	.8	.9
	WW	252	118	72	51	38	29	22	16	9
32.9°NL	WWNI	358	173	108	76	58	46	37	29	21
2146 DD **LCR**	TW	238	110	67	46	32	23	17	11	7
T(Jan)=49	TWNI	339	161	101	71	54	42	32	24	16
	DG	276	124	73	47	30	19	10	–	–
	DGNI	384	180	113	79	59	45	34	24	16

Fig. 7-1   Solar Savings Fraction Charts (Balcomb, 1980)

7.0 THERMAL FINALIZATION

7.2 HEATING PERFORMANCE

	1st Trial	2nd Trial
SSF Office building:	55%	60%
SSF Factory:	under 10%	

Does your first trial design use less non-solar fuel for heating than a conserving building to meet Criterion A?

Office: yes. Factory: self-heating; solar minor.
If not, what will you change in your second trial?
(Due to later considerations, aperture increased on office building)

B. Criterion B: Your buildings stay between 60°F and 80°F inside while occupied in the winter months.

Note: At 60°F air temperature, for a person to be comfortable, a higher radiant temperature than 60° should be available from walls or floors which have stored solar heat.

Evaluation Tool for Criterion B - "Internal January Temperature Estimate"

The average interior temperature is the sum of three temperatures:

1. Average outside January temperature, T out
+ 2. Internal heat-generated temperature, T int
+ 3. Temperature increase due to solar energy,
     called $\triangle$T solar

= 4. Average interior January temperature

1. T out = avg. outside January temperature (from Fig. 7-1) = _39_ $^\circ$F

2. T int = $\dfrac{\text{total daily internal gains}}{\text{BLC}}$

	Office	Factory
Total Internal gains (in 24 hrs)	146,160 Btuh	2,135,040 Btuh
BLC	9,600 Btu/DD	71,500 Btu/DD
T int	13.7 $^\circ$F	29.9 $^\circ$F

Office: 6,090 Btuh x 24 hrs = 146,160
Factory: 88,960 Btuh x 24 hrs = 2,135,040

3. ($\triangle$T - solar) depends on your latitude and your Load Collector Ratio (LCR). (Balcomb, 1980, p. 45)

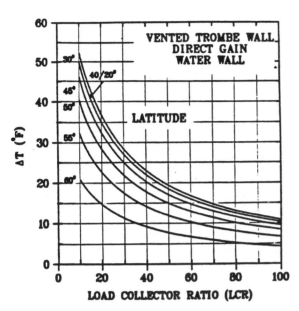

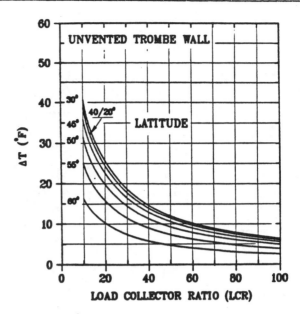

Fig. 7-2 Δ T Solar for Direct Gain (and Vented Trombe Wall.)

Fig. 7-3 Δ T Solar for Unvented Trombe Wall.

Note: These curves assume no night insulation, so your resulting ΔT solar may be higher than these conservative values. These curves are also for a clear day. If your climate rarely has clear January days, you may wish to adjust the Δ T solar value you obtain below, to account for a prevalence of cloudy weather. This approximate adjustment involves comparing your average January insolation on a vertical surface with the clear day insolation for your latitude.

7.2 HEATING PERFORMANCE

```
CHARLESTON, SOUTH CAROLINA                          LAT = 32.9      ELEV =    39
        JAN   FEB   MAR   APR   MAY   JUN   JUL   AUG   SEP   OCT   NOV   DEC  YEAR
HS      744   995  1339  1732  1860  1844  1799  1585  1394  1193   934   721  1345
VS     1058  1135  1128  1006   832   753   759   808  1010  1226  1251  1083 12049
TA     48.6  50.5  56.5  64.6  72.1  77.9  80.2  79.6  75.2  66.1  56.3  49.3  64.7
D50     120    81    23     2     0     0     0     0     0     2    24   108   360
D55     222   161    75    10     1     0     0     0     0     7    75   205   756
D60     360   275   157    36     5     0     0     0     0    26   156   339  1355
D65     521   419   300    69     5     0     0     0     0    74   271   487  2146
D70     664   547   422   192    66    16     9    10    31   165   414   642  3178


DODGE CITY, KANSAS                                  LAT = 37.8      ELEV = 2582
        JAN   FEB   MAR   APR   MAY   JUN   JUL   AUG   SEP   OCT   NOV   DEC  YEAR
HS      827  1122  1476  1886  2090  2358  2295  2055  1687  1301   894   732  1560
VS     1345  1466  1436  1273  1063  1065  1084  1212  1417  1535  1369  1258 15523
TA     30.8  35.2  41.2  54.0  64.0  73.7  79.2  78.1  68.9  57.9  42.8  33.4  54.9
D50     596   417   289    48     4     0     0     0     1    20   239   517  2132
D55     750   555   432   116    15     1     0     0     4    63   373   670  2980
D60     905   695   584   210    50     4     0     0    15   139   518   825  3945
D65    1060   834   738   344   115    21     0     0    41   247   666   980  5046
D70    1215   974   893   482   218    51    14    19   118   382   816  1135  6318


MADISON, WISCONSIN                                  LAT = 43.1      ELEV =   860
        JAN   FEB   MAR   APR   MAY   JUN   JUL   AUG   SEP   OCT   NOV   DEC  YEAR
HS      515   804  1136  1398  1743  1948  1934  1708  1299   911   504   389  1191
VS      971  1214  1285  1109  1037  1014  1061  1185  1275  1244   893   776 13064
TA     16.8  20.3  30.2  45.3  56.0  65.8  70.1  68.7  59.7  49.9  34.7  21.9  44.9
D50    1029   832   614   167    19     1     0     0     6    86   460   871  4086
D55    1184   972   769   297    70     4     1     2    26   182   609  1026  5143
D60    1339  1112   924   442   157    19     5     8    87   318   759  1181  6352
D65    1494  1252  1079   591   297    72    14    39   173   474   909  1336  7730
D70    1649  1392  1234   741   436   156    83   106   314   623  1059  1491  9284


PHOENIX, ARIZONA                                    LAT = 33.4      ELEV = 1112
        JAN   FEB   MAR   APR   MAY   JUN   JUL   AUG   SEP   OCT   NOV   DEC  YEAR
HS     1021  1374  1814  2355  2676  2739  2486  2293  2015  1576  1150   932  1869
VS     1472  1589  1552  1388  1211  1128  1059  1186  1482  1643  1561  1419 16692
TA     51.2  55.1  59.7  67.7  76.3  84.6  91.2  89.1  83.8  72.2  59.8  52.5  70.3
D50      78    29     9     1     0     0     0     0     0     0     9    60   187
D55     162    85    36     4     0     0     0     0     0     1    34   137   459
D60     285   168   100    16     0     0     0     0     0     5    96   250   919
D65     428   292   185    60     0     0     0     0     0    17   182   388  1552
D70     584   419   327   130    24     2     0     1     3    64   314   544  2411


SALEM, OREGON                                       LAT = 44.9      ELEV =   200
        JAN   FEB   MAR   APR   MAY   JUN   JUL   AUG   SEP   OCT   NOV   DEC  YEAR
HS      332   588   947  1370  1738  1849  2142  1775  1328   769   410   277  1127
VS      659   933  1126  1148  1093  1014  1240  1301  1372  1103   764   582 12336
TA     38.8  42.9  45.2  49.8  55.7  61.2  66.6  66.1  61.9  53.2  45.2  40.9  52.3
D50     348   204   163    68    10     1     0     0     1    26   158   285  1265
D55     502   340   306   168    57     8     1     1     6    97   296   437  2220
D60     657   479   459   308   151    48     7     9    39   217   444   592  3411
D65     812   619   614   456   295   133    43    53   120   366   594   747  4852
D70     967   759   769   606   444   267   130   141   247   521   744   902  6496
```

Salem: VS = 659 for January

HS = Normal daily value of total hemispheric solar radiation on a horizontal surface (Btu/ft^2 day)

VS = Normal daily value of total solar radiation on a vertical south-facing surface (Btu/ft^2 day)

TA = (T min. + T max.)/2 where T min. and T max. are monthly (or annual) normals of daily minimum and maximum ambient temperature ($^{\circ}$F)

Dxx = monthly (or annual) normal of heating degree-days below the base temperature xx ($^{\circ}$F days)
Base 65° is usually used in calculating heating loads.

Figure 7-4 Monthly Average Solar Radiation, Temperature, and Degree-Days

Source: Balcomb, 1980, Appendix A

Average insolation: see "VS" data, Fig. 7-4.
Clear Day insolation: approximate values may be obtained by using MEEB table 4.27 for 40°N.Lat., pages 170-171. (ASHRAE Fundamentals also lists these for 24°, 32°, 48° and 56°N.Lat.)

Salem, Or: January: average 3 clear days!
Fig. 7-4 average day = 659 Btu/sf day;
MEEB p. 170, clear Jan. day, south windows: 815 Btu/sf, half day x 2 = 1630 Btu/sf day.

Adjusted Δ T Solar = clear day Δ T solar x $\dfrac{\text{average insolation}}{\text{clear day insolation}}$

So Δ T solar is adjusted by
$\dfrac{659}{1630} = .40$

Trial 1

	LCR	Δ T solar	Adjusted Δ T solar, if applicable
Office	26.7	27°	27 x .4 = 10.8°
Factory	150	approx. 5°	5 x .4 = 2°

Trial 2

Office	21.4	30°	30 x .4 = 12°
Factory			

Mixture of 1/5 unvented trombe (22°), 4/5 direct gain (32°); (.2 x 22) + (.8 x 32) = 30°

4. Average interior January temperature = T out + T int + ΔT solar

	Trial 1:		Trial 2:
	average	*clear (rare)*	
Office:	63.5 °F	79.7°	64.7° (82.7° clear)
Factory:	70.9 °F	73.9°	_____

Trial 1: office
average day
39 + 13.7 + 10.8 = 63.5
clear day
39 + 13.7 + 27 = 79.7
Factory
average day
39 + 29.9 + 2 = 70.9
clear day
39 + 29.9 + 5 = 73.9

Trial 2: office
average day
39 + 13.7 + 12 = 64.7
clear day
39 + 13.7 + 30 = 82.7

Evaluation Tool for Criterion B – "January Temperature Swing Estimate"
The temperature swing estimates how far your building's temperature will fluctuate above and below that average temperature on a clear January day. The total range from high to low is roughly proportional to the T solar, as follows for various systems (remember, cloudy climates might use an adjusted T solar):

Direct gain	T swing = 0.74 x Δ T solar
Vented trombe wall	T swing = 0.65 x Δ T solar
Water wall	T swing = 0.39 x Δ T solar
Unvented trombe wall	T swing = 0.13 x Δ T solar

High temperature = Av. Jan. interior temp. + $\dfrac{\text{T swing}}{2}$

Low temperature = Av. Jan. interior temp. - $\dfrac{\text{T swing}}{2}$

Average Salem:

	Trial 1		Trial 2	
	Office	Factory	Office	Factory
high temperature:	66.7°F	71.7°F	68.4°F	___°F
low temperature:	58.7°F	70.2°F	61.0°F	___°F

If you assume high temperature occurs at 4 p.m. and low temperature at 4 a.m., what are the approximate high and low temperatures during operating hours?

	Trial 1		Trial 2	
	Office	Factory	Office	Factory
high temperature:	67°F	72°F	68°F	___°F
low temperature:	63°F	71°F	64°F	___°F

Does your first trial design meet the 60°F-80°F criteria range?

Yes, in office on average days if internal gains materialize; could get too hot on clear days if all those people are in the display room all day. Factory is steady, and adequately warm.

If not, is more or less solar aperture, or another passive system, appropriate?

Due to unreliability of internal gains in office building, (33 people in conference room all day is rare), and to rather high temperature swing, consider more aperture, in unvented trombe wall.

What will you change on your second trial design?

Office: convert 100 s.f. of lower conference south opaque wall to unvented trombe; this becomes additional aperture.

C. Criterion C: The capacity of the heating backup system is sufficient to heat the building both when there is no sun, and when there is no sun nor gains from people, lights or machines.

Trial 1: office
average day
.74 x 10.8 = 8
62.7 + 8/2 = 66.7
62.7 - 8/2 = 58.7
clear day
.74 x 27 = 20
77.7 + 20/2 = 87.7, hot!
77.7 - 20/2 = 67.7
Factory
average day
.74 x 2 = 1.5
70.9 + 1.5/2 = 71.7
70.9 - 1.5/2 = 70.2
clear day
.74 x 5 = 3.7
73.9 + 3.7/2 = 75.8
73.9 - 3.7/2 = 72.1
Trial 2: office
mixed passive system constant:
1/5(.13) + 4/5(.74) = .62
average day
.62 x 12 = 7.4
64.7 + 7.4/2 = 68.4
64.7 - 7.4/2 = 61.0
clear day
.62 x 30 = 18.6
82.7 + 18.6/2 = 92.0
82.7 - 18.6/2 = 73.4
Although 92° is even hotter. the average conditions are less dependent on high internal gains to maintain comfort.

Assume winter:
office 8:30 am - 4:30 pm
factory 6 am - 10 pm

Evaluation Tool for Criterion C - "Heating Backup System Sizing"

1. Figure your building heat loss in Btuh, <u>including</u> all south facing solar aperture. If you have no solar gains, the south facing glazing becomes an additional area which loses heat.

	Office	Factory
		950 x .5 = 475
a. Area of south aperture	*500* sq.ft.	___ sq.ft.
b. U value of south aperture	*.24*	*1.10*
c. Is this using insulating shades?	*yes*	*no*
d. UA for south aperture	*120* Btuh/°F	*523* Btuh/°F
e. UA total excluding south aperture (from Criterion A)	*400* Btuh/°F	*2979* Btuh/°F
f. Total building UA (d + e = f)	*520* Btuh/°f	*3502* Btuh/°f

2. The "Conservative" approach assumes no contributions from either solar or internal gains. This may be highly unrealistic, but quite "safe" in supplying lots of auxiliary heat in extremely cold and cloudy conditions. Use the maximum heat loss directly in sizing heater(s). Such heating units will be vastly oversized, and may never be used to full capacity.
Determine maximum hourly heat loss, under the coldest hour usually encountered in your climate. This "design temperature" can be determined from MEEB Appendix C, Table C-2a, Winter DB temperature column.

Maximum Heat Loss = U x AΔT
Where ΔT is the difference between interior temeperature   *Salem Or, MEEB p. 1301*
and exterior winter DB, and UA have been determined in the previous step.

<u>Office</u> maximum hourly heat loss:

UAΔT = *520* Btuh/°F x *50* °F = *26,000* Btuh

<u>Factory</u> maximum hourly heat loss:

UAΔT = *3502* Btuh/°F x *50* °F = *175,100* Btuh

---

3. The "Reasonable Risk" approach reduces the maximum heat loss by the internal gains (averaged over 24 hours). We can call this the "Assumed Heat Loss."

Assumed Heat Loss = Maximum Heat Loss - Average Internal Heat Gain

	Office:	Factory:
Average Internal Heat Gain =	6090 Btuh	88,960 Btuh

(as found in Heating p. 5-7.)

Office assumed heat loss = (____ Btuh)-(____ Btuh)= ____ Btuh   26,000 - 6,090 = 19,910

Factory assumed heat loss = Maximum hourly loss - average   175,100 - 88,960 = 86,410
internal heat gain = (____ Btuh)-(____ Btuh)= ____ Btuh

4. Auxiliary heat sources are specified by Btuh delivered to the space. You have just calculated two different needs above. What capacity will your backup system be? Does this meet Criterion C?

Office: 26,000   Factory: 175,100

The conservative approach meets Criterion C.

With the Btuh heat needs determined by either of these approaches, you now can find catalogs of manufacturers, and choose the auxiliary equipment. This isn't required here, but it's a useful experience. Additional information about choosing these systems:

Several decisions are necessary when choosing a means for supplying the non-solar portion of your building's space heat need. One central heat source or several zones? The more separate spaces in your building, the more likely that each space will need heat in different quantities, and at different times. Each space could be called a "zone" for heating. Small heaters in each room, each controlled by its own thermostat, are an appropriate way to respond to several zones.

The disadvantage of this approach is that electric resistance heaters are almost always the economical choice for this task, and electrical resistance heat mismatches a very high grade energy source with a low grade energy task. (See MEEB, chapter 1.)

One central heat source is particularly appropriate for simple, one-space buildings, and can utilize a lower grade energy source more appropriate for space heating. Since it is controlled by only one thermostat, the location of this control device is particularly important: when the thermostat calls for heat, the majority of the building should also need heat.

If you choose one central auxiliary heat source:
Natural gas, oil, wood or other combustion fuel heaters can be chosen by the Btuh delivered to the space. (The total Btuh burned includes some heat that is wasted, up the flue.) Btuh delivered should be approximately equal to the Maximum (or assumed, if you take risks) heat loss in Btuh.

Electric resistance heaters can be chosen by converting the Btuh needed into Kw:

$$\frac{Btuh}{3,412} = Kw$$

Heat pumps, while an attractive option for non-solar buildings, are more rarely an economic choice for passively solar heated buildings. The most likely time for auxiliary heaters to be utilized is during the coldest weather; in most climates where auxiliary heaters are important, this cold weather is not conducive to efficient heat pump operation, because there is less heat available in the outdoor air for the heat pump to deliver to the building.

After determining the Btuh (or Kw) to be delivered by your heater, consult manufacturers' literature (such as found in Sweets Files) to determine the range of choices available.

If you chose individual zone heaters:
An approximate way to size the heaters in each zone is simply use a ratio of its floor area to total floor area:

$$\text{Maximum Heat Loss (or assumed, if you take risks)} \times \frac{\text{Zone floor area}}{\text{Total floor area}} = \text{Zone Heat Loss}$$

A more exact method would be to break down the UxA calculation by zone instead of for the whole building. Choice of heaters then proceeds as outlined for one central source, above.

A TYPICAL ACTIVE SOLAR THERMAL CAT SYSTEM

From J. R. Augustyn et al, 1979.

7.3 COOLING PERFORMANCE

GOAL

Keep occupants thermally comfortable throughout the year by using the natural energies available in your climate, thereby minimizing the use of electrical or fossil fueled auxiliary equipment.

CRITERIA

A. The internal temperature in your buildings is no greater than 3° hotter than outside during the overheated months.

B. The capacity of your cooling backup system is sufficient to keep you cool during overheated months.

DESIGN

Use the design documentation from 7.2 for your evaluation in this section.

A. **Criterion A:** The internal temperature in your buildings is no greater than 3° hotter than outside during the overheated months. You must of course make certain that this internal temperature is within the comfort zone. Many overheated months have an air temperature low enough to maintain human comfort.

Evaluation Tool for Criterion A - "Heat Gain Calculation"
Heat gain is calculated for the worst condition - people, lights and machines producing heat inside and sun striking glazed openings. Calculation procedures below are for both open buildings (such as naturally ventilated buildings) and for closed buildings (buildings with minimal untreated outdoor air, relying upon thermal mass or evaporative cooling while occupied).

The open building procedure will include internal heat sources, direct solar gains through glass, and some "stored solar" gains through roof and wall surfaces.

The closed building procedure will include all of the above, plus gains due to heat transfer from hot outside air to colder inside air, such as through glass, and by infiltration.

You should use the procedure for open or closed depending on which condition describes each of your buildings while occupied. It must be emphasized that these are still approximate calculation procedures. Much more exact and lengthy calculations can be done, which incorporate the hourly change in solar load on buildings. The ASHRAE Handbook of Fundamentals describes these in detail.

We will illustrate both closed and open procedures for the office building. Salem climate allows open procedures, all summer, for both buildings.

The calculations have three parts:

Heat gain through skin, regardless of its orientation, where differences between open and closed building procedures are accounted for;

Instantaneous solar gain through glass (same procedure for both open and closed buildings), which is highly dependent upon the choice of date and time for which calculation is performed, and is also greatly influenced by the shading devices which accompany windows;

Internal gain (same procedure for both open and closed buildings), as previously calculated.

Information necessary to do the calculations will be found as follows:

Heat Transfer Factors: if your construction hasn't changed since the developing phase of Cooling, you can use those. If it has, go back to MEEB Table 4.22c, p. 149 and use the lowest temperature differential of 20°, as previously.

Solar gains through glass area are listed in Table 4.27, MEEB, for 40° N. Latitude. (ASHRAE Handbook of Fundamentals also lists these for 24°, 32°, 48° and 56° N. Latitude.) Your choice of date and time can greatly change the resulting hourly Btu to be removed. After considering your climate and your function, aim for a date/time which approaches the <u>worst hour</u> (greatest hourly gain), condition. A fairly common date/time choice is late summer afternoon (4 pm August 21) when relatively low altitude sun can penetrate unshaded west glass at the time when peak daily air temperatures are occuring. On the other hand, a building with mostly unshaded south facing glass would experience highest solar gains near noon. Obviously, you may choose different days/times for the office building as compared to the factory.

<u>Shading coefficients</u> are from Cooling Exercise 6.4 in the "Shading Masks" Evaluation Tool.

ΔT for air changes is the same as that specified for the HTF choice.

For the "closed" example, we assume extreme Salem heat of 90° outdoors; we allow 80° indoors. $\Delta T = 10°$.

1. OFFICE BUILDING Specify the date and hour you are calculating for: *Noon, August 21*

Why is this your "worst" hour? *South glass is by far the predominant solar load in the office building, and August sun is low enough to potentially cause problems.*

a. <u>Gains Through Skin Trial 1</u>

All spaces combined:	Closed Building			Open Building			
	Area (sq.ft.)	HTF (UΔT)	Hourly Gain (Btuh)	Area (sq.ft.)	HTF (UΔT)	Hourly gain (Btuh)	
Exterior opaque walls	2920	1.1	3210	2920	1.1	3,210	Ventilating openings not included;
Roof: horizontal 45° sloped	1650	0.8	1320	1650	0.8	1,320	new trombe wall included (stores heat)
All glass (insulating u=0.61)	non-s 250 s 400	6.1 (ΔT=10°)	3970	(note #1 below)			
Exterior doors	120	1.9	240	120	1.9	240	Use HTF for insulated door, MEEB p. 149.
Infiltration	21,000 cu.ft.	.018x10x.33 .018xΔTxACH	1250	(note #2 below)			
	Subtotal		9990 Btuh		Subtotal	4,770 Btuh	

Notes to above calculations:
#1 open buildings are hotter inside than outside, so the heat will flow, if at all, from inside to outside through the glass, therefore contributing no gain. Opaque skin contributes a small gain because it can store and later release heat.
#2 the air will be cooler outside than inside so it contributes no heat.

Gains Through Skin, Trial 2

All spaces combined:	Closed Building			Open Building		
	Area (sq.ft.)	HTF (UΔT)	Hourly Gain (Btuh)	Area (sq.ft.)	HTF (UΔT)	Hourly Gain (Btuh)
Exterior opaque walls						
Roof: horizontal 45° sloped						
All glass				(note #1 above)		
Exterior doors						
Infiltration	Volume c.f.	.018 x ΔT x ACH		(note #2 above)		
		Subtotal_____ Btuh			Subtotal_____ Btuh	

---

b. Instantaneous Solar Gains Through Glass, Trial 1

For 40°N Latitude, noon August 21; MEEB p. 171

All Spaces Combined	Area (sq.ft.)	Solar Heat Gain Factor	Shading Coefficient	Hourly Gain (Btuh)	
Horiz. glass	*none*				
Vertical glass: N	*250*	*35*	.51	4460	North Area: Include ventilating
NE	--				openings of 100 sf in toilets
E	*100*	*38*	.15	570	and 100 sf in display rooms,
SE	--				and consider shaded by the
S	*clerest.200*	*149*	.25	7450	ventilating cover.
SW	*lower 200*	*149*	.15	4470	
W	--				
NW	*50*	*35*	.51	890	

45° sloped glass: interpolate (between horizontal, vertical values)

	50	143	.15	1070	

Horiz = 247
Vert (E facing) = 38
Interpolation = 143

Subtotal **18,910** Btuh *(Compare to approx. total of 12,000 from Cooling, Section 6.4)*

Instantaneous Solar Gains Through Glass, Trial 2

All Spaces Combined	Area (sq.ft.)	Solar Heat Gain Factor	Shading Coefficient	Hourly Gain (Btuh)
Horiz. glass				
Vertical glass: N				
NE				
E				
SE				
S				
SW				
W				
NW				

45° sloped glass: (interpolate between horizontal, vertical values)

Subtotal _____ Btuh

---

c. Internal Gains

You may use the hourly gains you found in the Climate Exercise, section 3.6. However, the electric lighting assumptions may be reduced, either based on your available daylight in this "worst month" compared to needed fc illumination for the tasks, or if you have completed Lighting Finalization 10.0. If you adjust your lighting gains, show your procedure and assumptions. Ordinarily you will be calculating gains for a time when the building is occupied, so you should use the higher internal gains while occupied, rather than the lower averaged internal gains (over a 24 hour period).

Subtotal: _____ Btuh internal gains while occupied.

d. Approximate Maximum Hourly Heat Gain, Trial 1

		Closed	Open
PART	I. Skin	9,990 Btuh	4,770 Btuh
	II. Solar	18,910	18,910
	III. Internal	10,610 (occupied)	10,610 (occupied)

Office TOTAL Gains 39,510 Btuh   34,290 Btuh

Approximate Maximum Hourly Heat Gain, Trial 2

		Closed	Open
PART	I. Skin	_____ Btuh	_____ Btuh
	II. Solar	_____	_____
	III. Internal	_____	_____

Office TOTAL Gains _____ Btuh   _____ Btuh

How does this compare to the estimated peak gain from the Cooling Exercise 6.4 "Heat Gain" calculation?

Cooling, Ex. 6.4: peak estimated at 34,760 Btuh. Finalization calculation for open building: 34,290 Btuh, is very close; daylighting total heat gain about equalled the reduction in electric lighting. For closed building: 39,510 Btuh calculated is 14% greater than Ex. 6.4 estimated 34,760 Btuh.

2. FACTORY BUILDING Specify the date and hour you are calculating for: *Noon, August 21*
Why is this your "worst" hour?
South openings dominate, as in office building.

Office floor area = 1750 sf
Total openings = 800 sf, or 46% of floor area. Even with shading devices, we should have at least 10% of floor area.
From Climate Section 3.5, this would require 500 ft candles outdoors to fully light the most demanding tasks. To check available light on August 21, see MEEB p. 776, clear day table, Salem at 44° N:
June 21 noon = 950 on north glass, more on south
Sep 21 noon = 800 on north glass, more on south
Therefore, assume no gain from electric lighting.
People = 9,250 Btuh (Climate, Section 3.6)
Equipment = 1,360
Total   10,610

----

a. Gains Through Skin, Trial 1

All spaces combined:	Closed Building *(not using)*			Open Building		
	Area (sq.ft.)	HTF (U△T)	Hourly Gain (Btuh)	Area (sq.ft.)	HTF (U△T)	Hourly Gain (Btuh)
Exterior Opaque walls				3100	1.5	4650
Roof: horizontal 45° sloped				3780	1.0	3780
All glass				(note #1 below)		
Exterior Doors *(use thermal equivalent of solid core; MEEB p. 149)*				300	12.0	3600
Infiltration	Volume c.f.	.018 x △T x ACH		(note #2 below)		

Subtotal _____ Btu/h

Subtotal 12,030 Btu/h

Notes to above calculations:
#1 open buildings are hotter inside than outside so that the heat will flow, if at all, from inside to outside through the glass, therefore contributing no gain. Opaque skin contributes a small gain because it can store and later release heat.
#2 the air will be cooler outside than inside so it contributes no heat.

Gains Through Skin, Trial 2

All spaces combined:	Closed Building			Open Building		
	Area (sq.ft.)	HTF (U$\triangle$T)	Hourly Gain (Btuh)	Area (sq.ft.)	HTF (U$\triangle$T)	Hourly Gain (Btuh)
Exterior Opaque walls						
Roof: horizontal 45° sloped						
All glass				(note #1 above)		
Exterior Doors						
Infiltration	Volume c.f.	.018 x $\triangle$T x ACH		(note #2 above)		
		Subtotal	Btuh		Subtotal	Btuh

b. Instantaneous Solar Gains Through Glass, Trial 1

All spaces combined	Area (sq.ft.)	Solar Heat Gain Factor	Shading Coefficient	Hourly Gain (Btuh)
Horiz. glass				
Vertical glass: N	850	35	0.15	4,460
NE				
E				
SE				
S	950	149	0.15	21,230
SW				
W				
NW				

45° sloped
glass: (interpolate between horizontal, vertical values)

Subtotal   25,690 Btuh

Instantaneous Solar Gains Through Glass, Trial 2

All spaces combined	Area (sq.ft.)	Solar Heat Gain Factor	Shading Coefficient	Hourly Gain (Btuh)
Horizontal glass				
Vertical glass: N				
NE				
E				
SE				
S				
SW				
W				
NW				

45° sloped
glass: (interpolate between horizontal, verticaal values)

Subtotal _____ Btuh

c. Internal Gains
See the discussion in the office calculation above about Internal Gains. If you adjust your lighting gains show your procedure and assumptions.

Subtotal: _____ Btuh internal gains while occupied.

d. Approximate Maximum Hourly Heat Gain, Trial 1

	Closed	Open
Part I. Skin	_____ Btuh	12,030 Btuh
II. Solar	_____	25,690
III. Internal	_____	106,160
Factory TOTAL GAINS	_____ Btuh	144,150 Btuh

Approximate Maximum Hourly Heat Gain, Trial 2

	Closed	Open
Part I. Skin	_____ Btuh	_____ Btuh
II. Solar	_____	_____
III. Internal	_____	_____
Factory TOTAL GAINS	_____ Btuh	_____ Btuh

Again, daylighting alone seems adequate to light the factory; 1800 sf window is 46% of the floor area. Therefore, assume no electric light gain at noon on August 21. From Climate, Section 3.6:

People	*4,400*
Machines	*101,760*
	106,160 Btuh

How does this compare to the <u>estimated</u> peak gain from the calculation in the Cooling Exercise 6.4?
Here, daylighting replaces electric light, with some savings in cooling load.

Cooling Exercise: 163,840
Finalization: 144,150

Evaluation Tool for Criterion A - "Mass Sizing for Closed Buildings"
Use this tool for buildings closed by day and cooled with mass storage, but open at night for ventilation. For buildings cooled by natural ventilation only ("open"), go ahead to the following Evaluation Tool "Aperture Sizing for Cooling", p. 7-47 (Procedure adapted from Crowther, "Night Ventilation Cooling of Mass", 1980)

Closed Building: We'll return to a more socially acceptable "9 to 5" for this example (4am to noon doesn't require a closed strategy). We'll also assume a more extreme 90° day, for peak summer conditions (MEEB appendix C, for Salem). This is 8° above the average high of 82°. Raise low temp of 51°, accordingly, to 59°.

1. Size Mass for Heat Storage
a. Combine your final heat gain pattern from the Cooling Exercise, p. 6-37 with a graph of your building schedule showing the hours the building is closed and storing heat and the hours the building will be open to ventilate the stored heat:
Ideally your building will close at the lowest temperature of the day and open again after the outside temperature has peaked and dropped to a bearable level, such as 82°F when it crosses into the comfort zone.

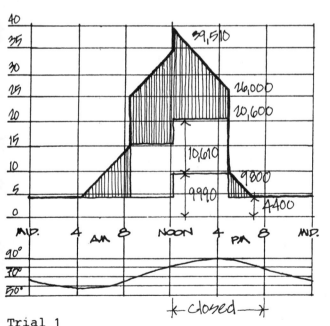

Trial 1
Office 24-hour Pattern

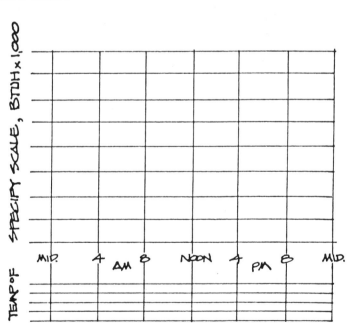

Trial 2
Office 24-hour Pattern

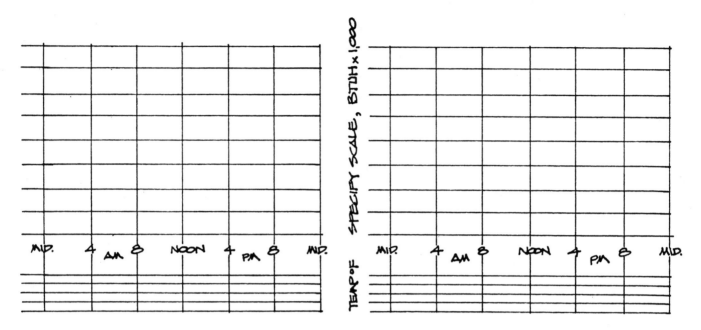

Trial 1
Factory 24-hour Pattern

Trial 2
Factory 24-hour Pattern

b. Find the total number of BTU's to be stored during the hours the building will be closed. This total is the sum of BTU's stored during each closed hour.

Office _182,350 Btu_

Factory _(open strategy)_

5 hrs x (39,500 @ noon + 26,500 @ 5)/2 = 163,750 Btu

2 hrs x (9,800 @ 5   +   4,400 @ 7)/2 = 14,200

1 hr x   4,400   = __4,400__

182,350 Btu

c. Determine the amount of mass you have provided in your design from the Heating Exercise, p. 5-34ff

See S5-37; S7-28, 100 sf trombe added.

	Office	Factory			
sf of mass exposed surface	2300	____	950	1250	100
material	total*	____	brick*	conc. slab*	conc. trombe*
thickness	____	____	.33 ft	.33 ft	1.0 ft
cubic feet	____	____	314	413	100

---

d. Mass heat
storage capacity = (mass volume) x (density) x (specific heat)

Material	Density #/cuft	Specific Heat Btu/#/^oF		
Ordinary Concrete	144	0.156	*conc. 513cf x 144 #/cf x 0.156*	*= 11,525*
Common Brick	123	0.2	*brick 314 x 123 x 0.2*	*= 7,725*
Gypsum	78	0.259		*19,250*
Limestone	103	0.217		
Sand	95	0.191		
Wood:				
Softwoods	27	0.45 approx.		
Hardwoods	45	0.55 approx.		
Water	62	1.0		

Fig. 7-4 Density and Specific Heat of Materials. From
The Passive Solar Energy Book by Ed Mazria, 1979,
Appendix A.
(Other tables of density and specific heat can be found
in Olgyay "Design With Climate".)

Office mass capacity 19,250   Btu/^oF

Factory mass capacity_____ Btu/^oF

Although you have only calculated the storage in mass   *Considerable gypsum wall*
specifically provided for that purpose, some heat will   *and ceiling surface avail-*
also be stored in the rest of the building materials.   *able as a safety factor.*
You may consider this additional cooling potential as a
"safety factor" for extreme conditions, particularly if
there is a lot of wall and floor area you have not
included as mass above.

e. In column 2 below list the August hourly outside   *We're illustrating a worst*
temperatures, from the Cooling Exercise 6.4, p. 6.37.   If   *case: 90° high, 59° low.*
average figures are used, then average performance will
be calculated.   If worst case figures are used, then
worst case performance will be calculated.

f. Begin calculating column 3 at the time when the
outside temperature goes below 82°:

Cooling BTU = $\left(\begin{array}{c}\text{mass temp. col. 4} \\ \text{preceeding hour}\end{array} - \begin{array}{c}\text{outside temp.} \\ \text{col. 2}\end{array}\right)$ x $\begin{array}{c}\text{mass} \\ \text{surface area}\end{array}$

Assume the 1st mass temperature preceeding hour to be 80°

g. For column 4,

Mass temp. = mass temp. from $-$ $\dfrac{\text{(cooling from col. 3)}}{\text{(mass heat capacity)}}$
preceeding hour

h. Continue calculating columns 3 and 4. STOP when the mass temperature, col. 4, is lower than the outside temperature, col. 2

i. Add up the hourly Cooling Btu, in Column 3.

j. Very rough ventilation check: there must be apertures large enough to admit the outdoor air that is being used to cool down the storage mass. A rough approximation:

Inlet areas <u>and</u> outlet areas $\dfrac{\text{each}}{} = \dfrac{1}{30}$ mass surface area

(Technical note: refer to the window sizing formula on p. 7-48. Assume that only 1 mph (88 fpm) of wind is available, and that it strikes the window diagonally (E=0.3) rather than head-on. With these conservative assumptions,

Inlet Area (sq.ft.) = $\dfrac{\text{Cooling Btu per hour}}{1.08 \times 0.3 \times \Delta T \times 88 \text{ fpm}}$

and since Cooling Btu = $\Delta T \times$ Mass Surface Area,
Inlet Area (sq.ft.) =

$\dfrac{\text{Mass Surface Area}}{1.08 \times 0.3 \times 88}$, or about $\dfrac{\text{Mass Surface Area})}{30}$

Rough Office Inlet Area Rqd:
$\dfrac{2300 \text{ sf}}{30}$ = 77 sq.ft.
Inlet + outlet = 154 sq.ft.;
300 sq.ft. provided. Ok.

This procedure, steps c through i, is taken from Karen Crowther "Night Ventilation Cooling of Mass", in <u>Passive Cooling Handbook</u>, prepared for the Fifth National Cooling Conference, Amherst, MA, 1980.

7.0 THERMAL FINALIZATION

7.3 COOLING PERFORMANCE

Trial 1

(1) Hour	(2) Outside Air Temp. ($^{\circ}$F)	OFFICE (3) Cooling Btu	(4) Mass Temp. ($^{\circ}$F)	FACTORY (3) Cooling Btu	(4) Mass Temp. ($^{\circ}$F)
8 pm	80	no storage	80		(calculations)
9	77	6,900	80 - .36 = 79.6		(80 - 77)2300 = 6900
10	74	12,880	79.6 - .67 = 78.9		(79.6 - 74)(2300) = 12,880
11	71	18,170	78.9 - .94 = 78		(78.9 - 71)(2300) = 18,170
12	68	23,000	78 - 1.2 = 76.8		(78 - 68)(2300) = 23,000
1 am	65	27,140	76.8 - 1.4 = 75.4		(76.8 - 65)(2300) = 27,140
2	63	28,520	75.4 - 1.5 = 73.9		(75.4 - 63)(2300) = 28,520
3	61	29,710	73.9 - 1.5 = 72.4		(73.9 - 61)(2300) = 29,710
4	59	30,820	72.4 - 1.6 = 70.8		(72.4 - 59)(2300) = 30,820
5	61	22,540	70.8 - 1.2 = 69.6		(70.8 - 61)(2300) = 22,540
6	63	15,180	69.6 - .8 = 68.8		(69.6 - 63)(2300) = 15,180
7	66	6,440	68.8 - .3 = 68.5		(68.8 - 66)(2300) = 6,440
8	69	stop here	69 > 68.5		
9	71				

	Office		Factory	
Total Btu's that can be cooled by mass: (sum of col. 3)	221,300 Btu		_____ Btu	
Final mass temperature (last temp. in col. 4)	68.5^{o} $^{\circ}$F		_____ $^{\circ}$F	

Trial 2

(1) Hour	(2) Outside Air Temp. (^{O}F)	OFFICE		FACTORY	
		(3) Cooling Btu	(4) Mass Temp. (^{O}F)	(3) Cooling Btu	(4) Mass Temp. (^{O}F)
8 pm	_____	_____	_____	_____	_____
9	_____	_____	_____	_____	_____
10	_____	_____	_____	_____	_____
11	_____	_____	_____	_____	_____
12	_____	_____	_____	_____	_____
1 am	_____	_____	_____	_____	_____
2	_____	_____	_____	_____	_____
3	_____	_____	_____	_____	_____
4	_____	_____	_____	_____	_____
5	_____	_____	_____	_____	_____
6	_____	_____	_____	_____	_____
7	_____	_____	_____	_____	_____
8	_____	_____	_____	_____	_____
9	_____	_____	_____	_____	_____

	Office	Factory
Total Btu's that can be cooled by mass: (sum of col. 3)	_____ Btu	_____ Btu
Final mass temperature (last temp. in col. 4)	_____ ^{O}F	_____ ^{O}F

Check now to see that your mass is sized to meet Criterion A before continuing to size apertures:

Can you cool all the Btu's you need to with your mass?

INSIDEOUT

Is your mass cooled down so the building can be closed at
the coolest hour of the day?   *Yes. Mass down to comfortable 68.5^{o} by opening hour.*
182,350 Btus need to be stored; 221,300 Btus can be stored; this is more than required.

If not, what will you change for your second trial?

Evaluation Tool for Criterion A - "Aperture Sizing for Cooling"
To check the size of openings you have provided for
ventilation, complete the following table and then use
either procedure 1. cross-ventilation or procedure 2.
stack ventilation for each building.

"Heat gain to be removed" for open buildings will be the   *We will return to the open*
total Btuh hourly heat gain from the "Heat Gain"   *building for this evaluation*
Evaluation Tool above.   *tool.*

"Heat gain to be removed" for closed buildings will be
the hour with the largest total heat gain to be removed,
including both stored BTU's (from column 3, from the
preceeding evaluation tool), and internal, skin, and
solar gains if any of these occur while the building is
being ventilated in this worst hour:

Specify what is	Office	Factory
Worst Hour:	*Noon, Aug 21*	*Noon, Aug 21*
Stored Btu's:	*none*	*none*
Skin gains:	*4,770*	*12,030*
Solar gains:	*18,910*	*25,690*
Internal gains:	*10,610*	*106,690*
Total Btuh to be removed:	*34,290*	*144,150*

7.0 THERMAL FINALIZATION

7.3 COOLING PERFORMANCE

	Source of Information	Office	Factory
Heat gain (Btuh to be removed	See above	34,290	144,150
Inlet area (sq.ft.) provided	your design	300	850 *Cross ventilation*
			(450 lower openings only, for stack ventilation)
ΔT	3° maximum (Criteria A)	3°	3°
Average wind speed (mph)	Climate exercise	6.3	6.3 *August in Salem: 6.3, NNW*
Q (cu ft of air per minute required to remove heat gain)	$Q = \dfrac{Btuh}{1.08 \times \Delta T}$	10,590	44,510

(Note: 1.08 = .018 x $\dfrac{60\ min.}{1\ hour}$ where .018 is air constant, used with cubic feet per hour)

1. Procedure for cross-ventilation. Use the following equation:

$$A = \frac{Q}{EV}$$

Where:
A = inlet area required to remove heat gain, in sq.ft.
Q = cu ft air per minute from preceeding table.
E = effectiveness of opening due to angle of wind; for winds perpendicular to openings E = 0.5, for winds diagonal to openings E = 0.3
(ASHRAE Fundamentals, 1972, p. 344)
V = wind speed, in f.p.m. = mph x 88

Use 0.3 because NNW wind, combined with trees, may deflect wind further away from true-north openings.

A for office = __64__ sq.ft. $A = \dfrac{10,590}{0.3 \times (6.3 \times 88)}$

A for factory = __268__ sq.ft. $A = \dfrac{44,510}{0.3 \times (6.3 \times 88)}$

2. Procedure for stack ventilation. Use the following equation:

$$A = \frac{Q}{9.4\ \ h(\Delta T)}$$

Office: h = 22' at display room, 12' at toilet/offices; use 17'

Factory: h = 22'
Where:
A = inlet area required to remove heat gain, in sq.ft.
h = height between centers of inlet and outlet openings
Q and $\triangle$T are from the table above.

A for office = <u>158</u> sq.ft.

A for factory =<u>58</u> sq.ft.

Office:
$$A = \frac{10,590}{9.4 \quad 17x3} = 158 \ sf$$
Factory:
$$A = \frac{44,510}{9.4 \quad 22x3} = 593 \ sf$$

Have you provided enough inlet area to keep your open buildings within the temperature range for Criterion A? If not, what will you change for your second trial?
Decide to live with more than 3° T for factory: 3° calc. x $\frac{583 \ s.f. \ calc.}{450 \ s.f. \ actual}$ = 3.9° actual T.
If it is 3° hotter inside your buildings than outside, what is the inside temperature?

Office: yes, both in cross and stack ventilation. Factory: yes for cross ventilation; but for stack ventilation, 450 actually is less than 583 required.

In a climate where average maximum temperature is 81.3, this produces an average maximum indoor = 85.2. At this ventilation rate (450 s.f. of lower openings), this should still be within the comfort zone for faster-moving air (see Climate section, Bioclimatic Chart).

Remember - there is adequate opening area for cross-ventilations. This 85° interior only occurs at dead calm conditions (10% of August).

B. Criterion B: The capacity of your cooling backup system is sufficient to keep you cool during your worst overheated conditions.

Evaluation Tool for Criterion B - "Cooling Backup System Sizing"
If a building is using a mechanical system for cooling the air itself is being cooled to a lower temperature, rather than the heat moving to a cooler sink such as mass or outside air. This cooling process also involves removing moisture in the air to prevent an uncomfortable increase in humidity at the lower temperature. Doing so means more Btu's must be removed. (See MEEB 5.12 Psychrometry p. 226-234) These additional Btu's due to water vapor are termed "latent gains" and can be approximated as 0.3 x sensible heat gains, (MEEB 1980, p. 160). Sensible heat gains are those you calculated for Criterion A above.

7.0 THERMAL FINALIZATION

7.3 COOLING PERFORMANCE

Total cooling load = sensible heat gain + latent heat gain = 1.3 x sensible heat gain for mechanical system.

Total cooling load:

office = _____ Btuh

factory = _____ Btuh

office, closed condition:
1.3 x 39,510 = 51,363 Btuh

Mechanical cooling systems can be sized using the above total (such as heat pumps) or in tons of refrigeration (Flynn & Segil, 1970, p. 224).

1 ton refrigeration capacity = 12,000 Btuh.

What capacity will your refrigeration system be for the office?   4   tons.

The factory? _____ tons

Does this meet Criterion B?
Yes, if we use refrigeration, which isn't the answer we're looking for...

9.0 DAYLIGHTING

9.1 INTRODUCTION

9.1 INTRODUCTION

.

This exercise begins with exploring organizational attitudes and then achieving the necessary quantity and
distribution of daylight (enough light in the right
places) through manipulation of building, room and window
sizes and locations. This design is then developed
through consideration of reflectances, obstructions and
transmission properties. In Exercise 10 you will evaluate the annual performance of the design.

It is very important in starting this exercise to know if
your climate is generally sunny or cloudy, or if it
changes seasonally. In lower latitudes the light is
brighter and the direct sun both more intense and less
desirable from a thermal standpoint than farther north.
These factors should be recognized in your design from
the beginning. Vernacular buildings in your climate will
provide good clues about where to start.

At this point, review the initial daylight explorations
you made in the Climate Exercise, sections 3.4 and 3.5.

Contents

9.0 DAYLIGHTING

9.2 CONCEPTUALIZING

9.2 CONCEPTUALIZING

GOALS

A. Use available daylight for lighting.
B. Prevent discomfort from excessive brightness or contrast.

DESIGN STRATEGIES

A. Site Scale
Locate buildings and spaces on site to either utilize or avoid obstructions of the skydome.
B. Cluster Scale
Recognize that buildings and ground surfaces can reflect light.
C. Building Scale
Consider zoning work by the visual difficulty of tasks. Put spaces which need light near openings in the skin.
D. Component Scale
Orient openings to recognize the brightest part of the skydome: zenith for overcast sky, horizon for clear sky.

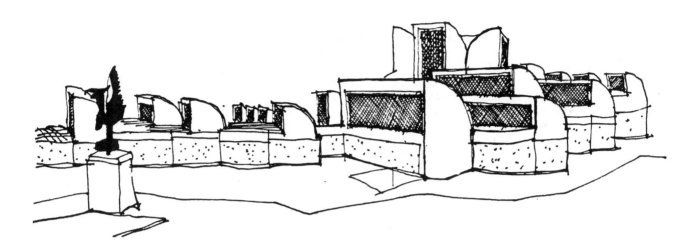

DAE Architects, Tehran Museum of Contemporary Art.

Consider openings for illumination as potentially different from openings for view, thermal gain, or ventilation.

DESIGN

Choose an existing building/site which has a clear conceptual approach, at any scale, to achieving these goals. Document your choice as follows:

A. Identify the location, program, architect (if any), and the source of your information.

B. Include photo copies or drawings (whatever is quick and easy for you) to explain the design.

EVALUATION

Evaluation Tool - "Building Response Diagrams"

Diagram how this design is organized to achieve these goals. If the typical sky condition for this building is different from yours, speculate on how the design responses might change your climate.

(This more general part of these sections is not included in the Salem, Or example solution.)

9.3 SCHEMING

GOAL

A. Open yourself to the sky in proportion to the brightness of the sky and the amount of light you need.

CRITERIA

A. The daylighting is distributed so there is light where people need it.
B. There are gradual changes in lighting levels between inside and out, and within rooms, to prevent glare problems.
C. Recommended Daylight Factors (DF) for tasks are achieved.
D. Solar collecting and ventilating apertures are maintained as required for heating and cooling.

DESIGN STRATEGIES

A. Cluster Scale

Place light-colored walls to the north of north-facing spaces to increase available reflected light.
Recognize the potential of thermal buffer zones (arbors, vestibules, courtyards, atria, greenhouses, etc.) to control glare and heat gain.
Snow is a highly effective reflector of light, but can also cause glare.

B. Building Scale
Consider which one of these strategies is most
appropriate to your climate and function:
1. Design for adequate daylight under average sky
conditions; below this average, supplementary lighting
will be necessary.
2. Design for minimum acceptable daylighting under
minimum daylight available conditions (such as 9 am in
December).
3. Design for adequate daylight in the great majority of
typical working hours.
4. Use supplementary artificial lighting for dark areas
rather than oversupplying daylight in lighter areas to
achieve the minimum everywhere.

*...We chose this approach, to
minimize electric lighting
usage and its accompanying
internal gains.*

Proportion rooms and/or locate tasks to use outer 15'
band of floor area where maximum light from side windows
is available.

Avoid over-daylighting your spaces in hot climates, where
darkest places are associated with cooler temperatures.

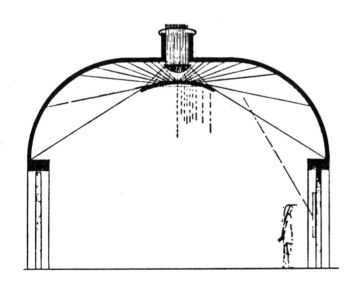

Kahn, Light Study for Kimball Art Museum (Ronner, 1977,
p. 345)

The resulting smaller daylight openings can be placed
near the ceiling, where ground reflected light can
illuminate a white ceiling without causing glare at eye
level.

C. Component Scale
Top lighting is brightest under the skylight and darkens in all directions.

Side lighting is brightest near the window and darkens toward the back wall.

Specular surfaces reflect light at an angle equal to the angle they receive it.

Tasks can be moved to use available light rather than moving light to tasks.

For standard lighting (office, lobby, circulation) the area of glazing = approximately 5%-10% floor area.

For intensive lighting (display, drafting, typing, factory work) area of glazing = approximately 25% floor area.

Proportion window height to equal 1/2 the depth of the room for good daylight penetration and distribution. (Hopkinson, 1966, p. 435)

...This causes us to move the factory windows up, midway in the outer wall, rather than leaving their sill at floor level.

Arrange shading devices so that the sun reflecting off their surfaces will not cause glare at eye level in the shaded spaces beyond.

DESIGN

Design the office, factory and outside areas according to program requirements in the 2.1 Building/site Description.
Document your design as follows using the grids provided:

A. State which sky condition (clear or overcast) you are designing for. Is this a typical or a worst condition?

Overcast, typical and worst.

B. Site plan, including parking and access drive(s). C. Cluster plan, including outdoor spaces and vegetation. Identify surfaces and their approximate light reflectances (see Olgyay, 1963, p. 33).
D. Floor plans of office and factory. Label the signifi-cant task areas on these plans.
E. Section(s) of office and factory.
F. Roof plan and elevations (or paralines) clearly showing all sides and the top of the buildings.

See Salem example Figures 9-8 to 9-13.

Important Note: Your design may change in the course of the following evaluation. The final schematic design which you show to meet all criteria should be fully documented here.

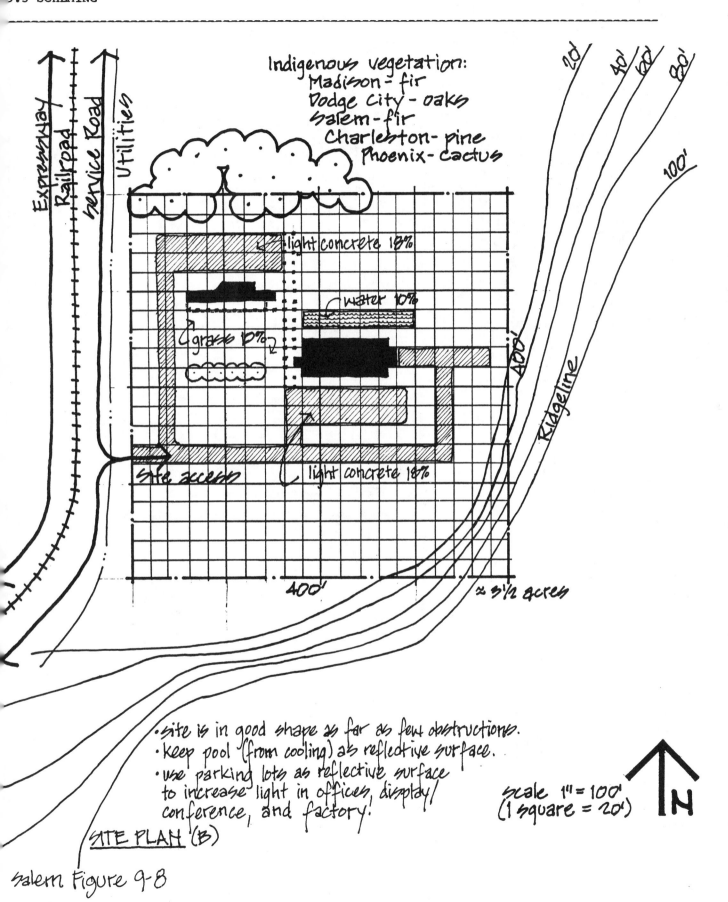

Indigenous vegetation:
Madison = fir
Dodge City - oaks
Salem - fir
Charleston - pine
Phoenix - cactus

light concrete 18%

water 10%

grass 10%

light concrete 18%

site access

400'

≈ 3½ acres

Expressway
Railroad
Service Road
Utilities

20' 40' 60' 80'

100'

400'

Ridgeline

• site is in good shape as far as few obstructions.
• keep pool (from cooling) as reflective surface.
• use parking lots as reflective surface
  to increase light in offices, display/
  conference, and factory!

scale 1" = 100'
(1 square = 20')

N

SITE PLAN (B)

Salem Figure 9-8

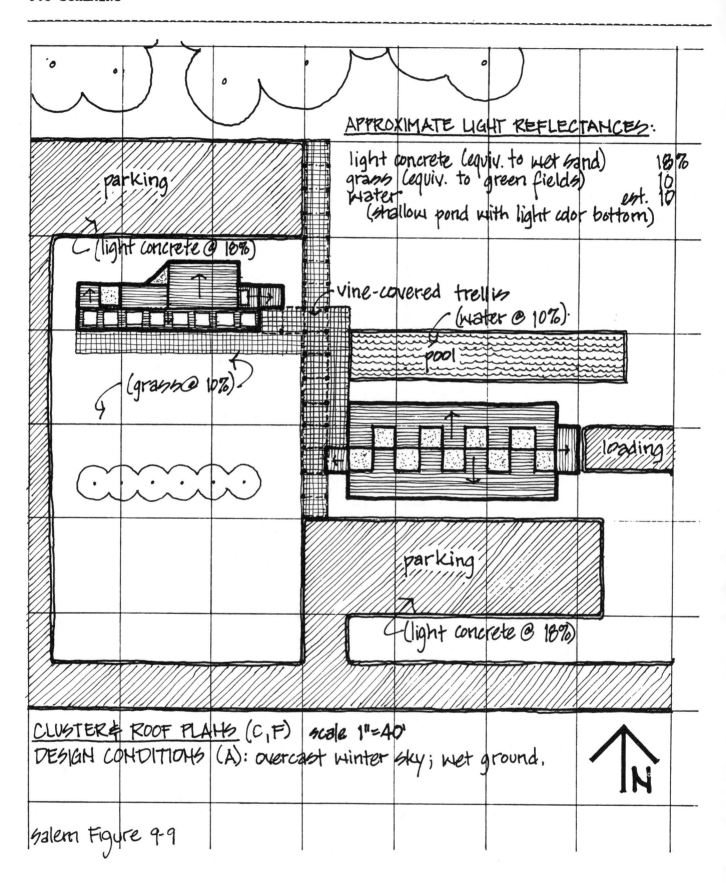

APPROXIMATE LIGHT REFLECTANCES:

light concrete (equiv. to wet sand)   18%
grass (equiv. to green fields)   10
water   est. 10
  (shallow pond with light color bottom)

parking

(light concrete @ 18%)

vine-covered trellis
(water @ 10%)

pool

(grass @ 10%)

loading

parking

(light concrete @ 18%)

CLUSTER & ROOF PLANS (C,F) scale 1"=40'
DESIGN CONDITIONS (A): overcast winter sky; wet ground.

N

salem Figure 9-9

for plans, sections, & elevations :   if   1/16"=1'-0",   1 square = 10'.

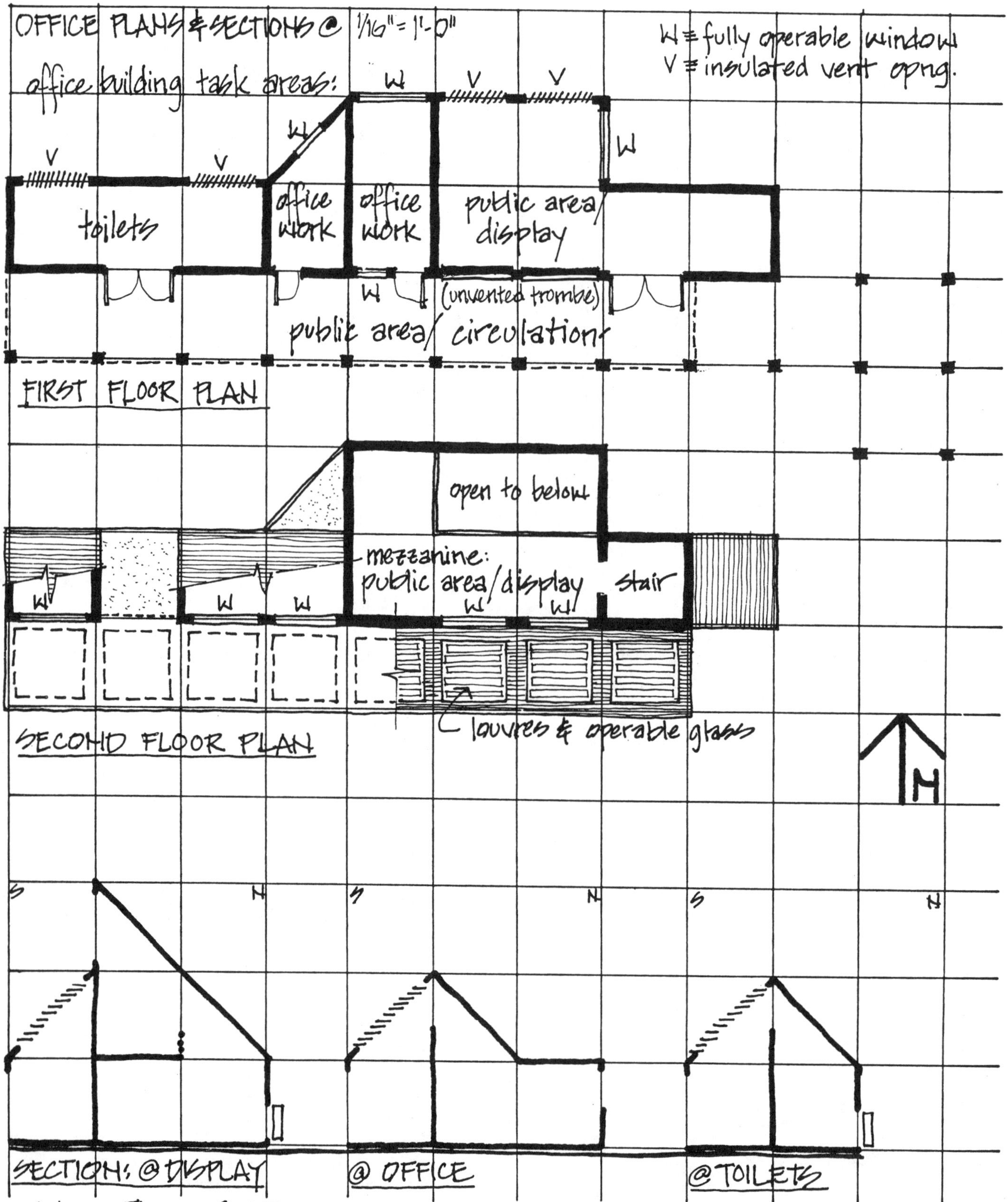

OFFICE PLANS & SECTIONS @ 1/16" = 1'-0"   W ≡ fully operable window
office building task areas: V ≡ insulated vent opng.

W = fully operable window
V = insulated vent opng.

toilets

office work   office work   public area/ display

(unvented trombe)

public area/ circulation

FIRST FLOOR PLAN

open to below

mezzanine:
public area/display   stair

louvres & operable glass

SECOND FLOOR PLAN

N

SECTION: @ DISPLAY   @ OFFICE   @ TOILETS

Salem Figure 9-10

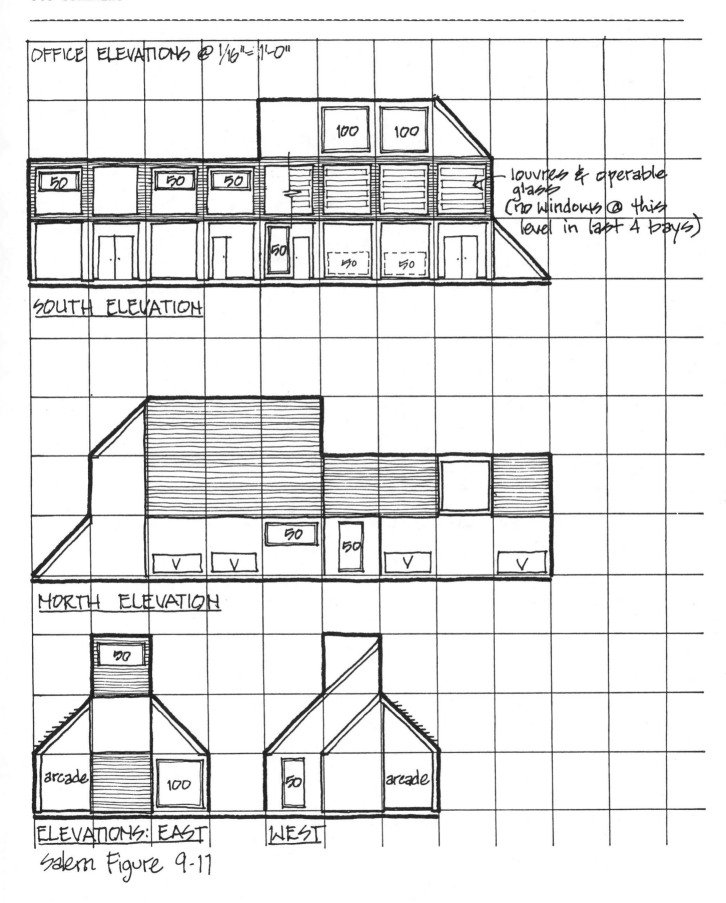

OFFICE ELEVATIONS @ 1/16" = 1'-0"

louvres & operable glass (no windows @ this level in last 4 bays)

SOUTH ELEVATION

NORTH ELEVATION

ELEVATIONS: EAST   WEST

arcade   100   50   arcade

salem Figure 9-11

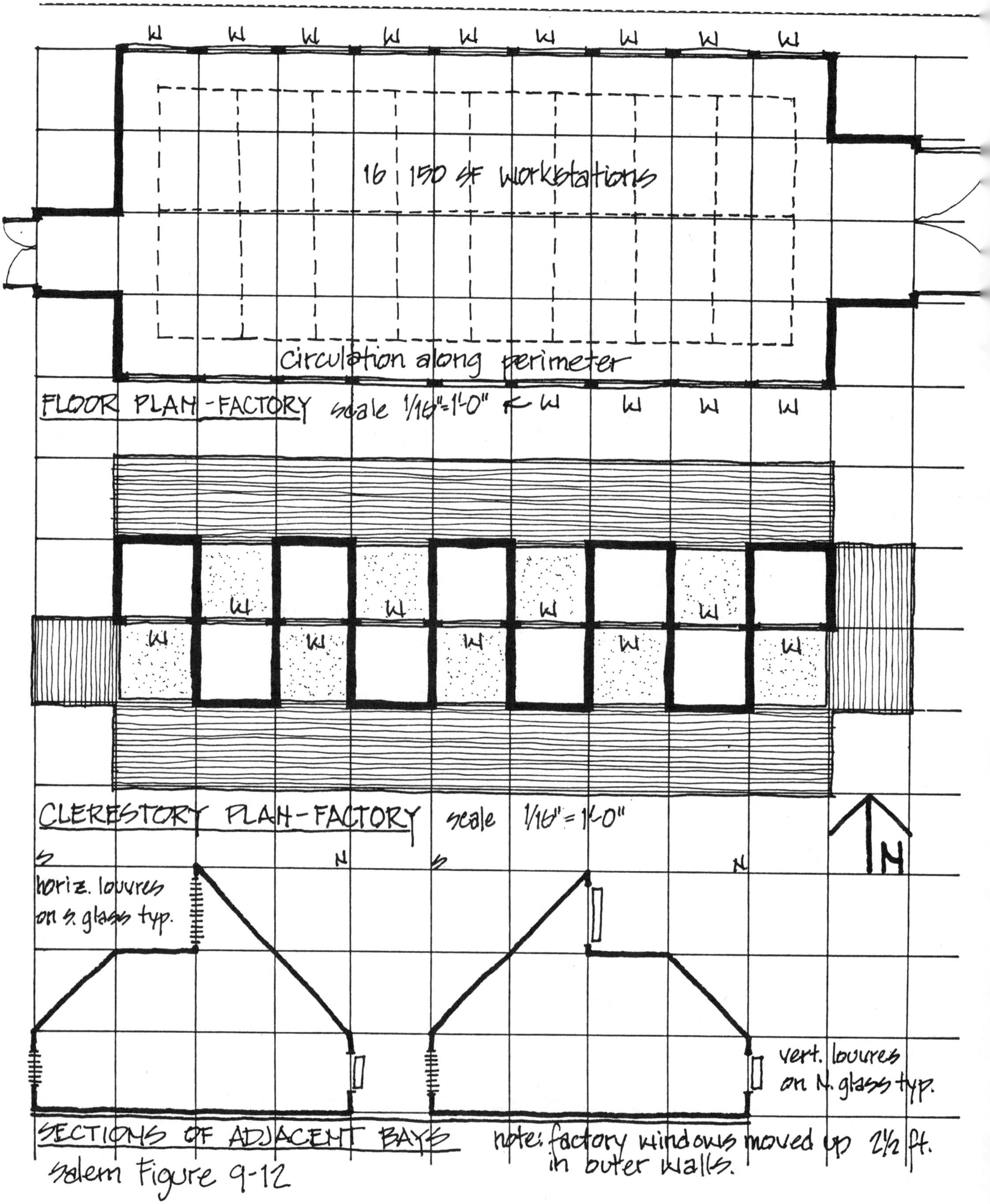

16 150 SF workstations

Circulation along perimeter

FLOOR PLAN - FACTORY   scale 1/16"=1'-0"

CLERESTORY PLAN - FACTORY   scale 1/16"=1'-0"

horiz. louvres
on s. glass typ.

vert. louvres
on N. glass typ.

SECTIONS OF ADJACENT BAYS   note: factory windows moved up 2½ ft.
in outer walls.

salem figure 9-12

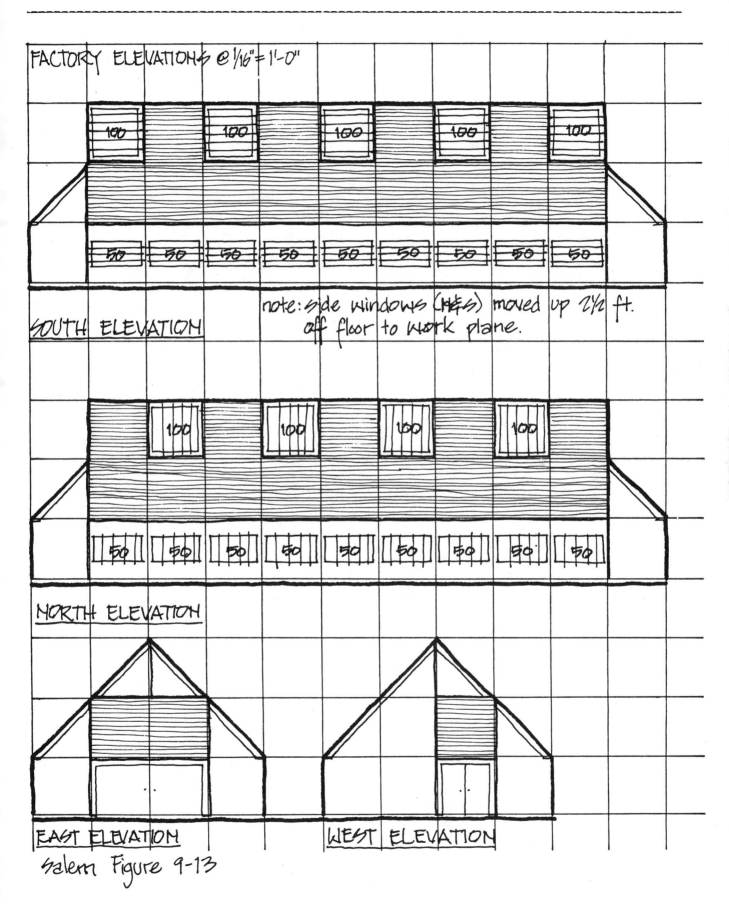

FACTORY ELEVATIONS @ 1/16" = 1'-0"

note: side windows (N&S) moved up 2½ ft.
off floor to work plane.

SOUTH ELEVATION

NORTH ELEVATION

EAST ELEVATION

WEST ELEVATION

Salem Figure 9-13

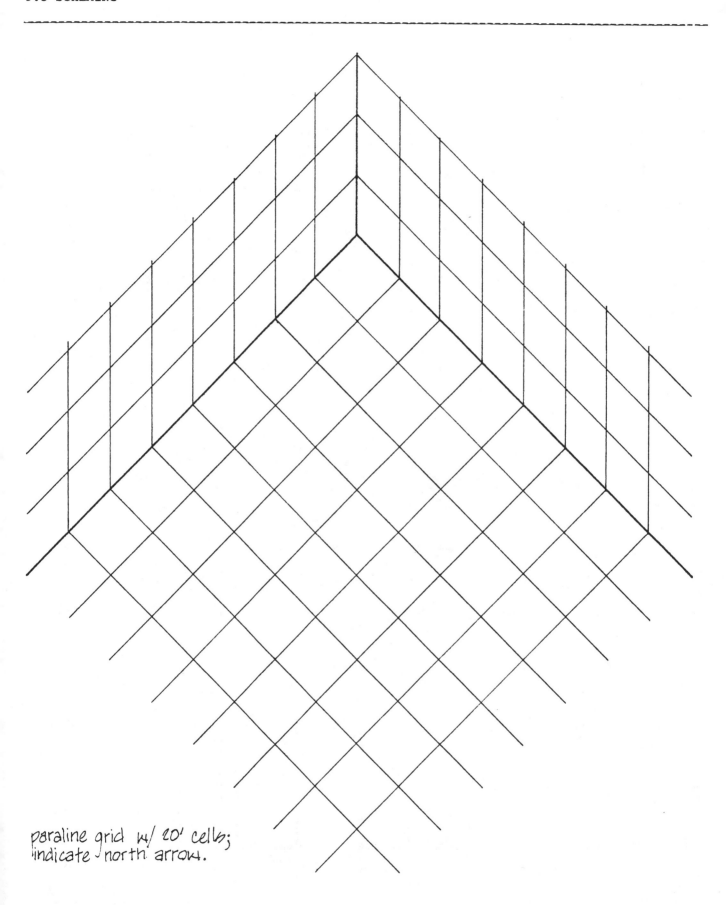

paraline grid w/ 20' cells;
indicate north arrow.

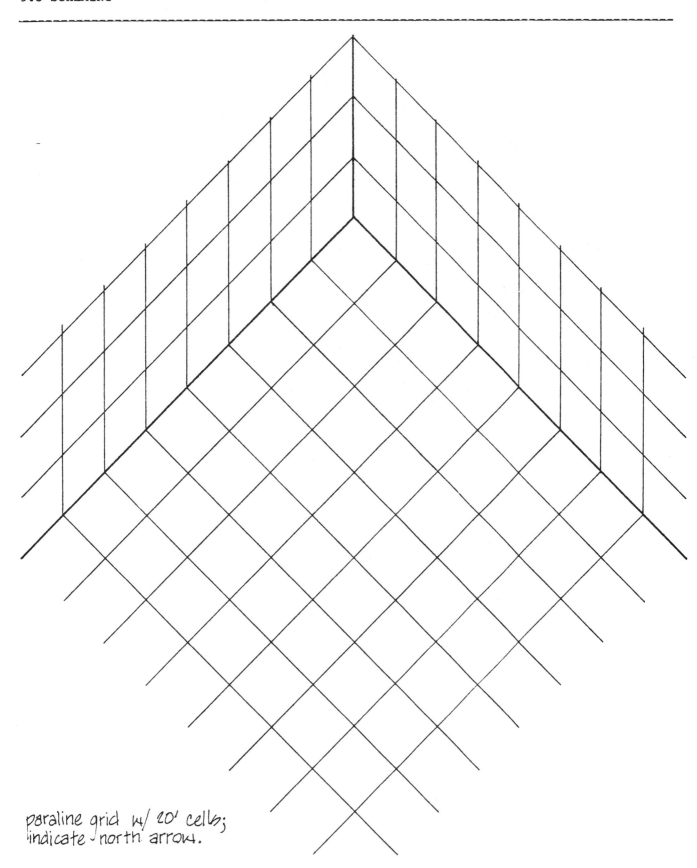

paraline grid w/ 20' cells;
indicate north arrow.

EVALUATION

A. Criterion A: Daylighting is distributed so there is light where people need it.

Evaluation Tool for Criterion A - "Light Distribution Rendering"

1. For the predominant sky condition in your climate, sketch shaded plans and sections of your office and factory buildings. Use media that allows you to quickly approximate the relative patterns of light and dark that result from your climate's daylight conditions and your methods of admitting that daylight to your buildings. (Pencil or charcoal on tracing paper is a common approach.) Each plan should have a corresponding section drawn on the same page for easy reference.
2. Clearly identify task areas, so that the light/task relationship is obvious.   *See Salem example Figure 9-14.*
3. Show how you have designed for gradual changes in brightness, especially at openings from inside to outside. Feel free to include a second trial evaluation.

Does your first trial design meet the criterion?
Yes.
If not, what will you change for your second trial?

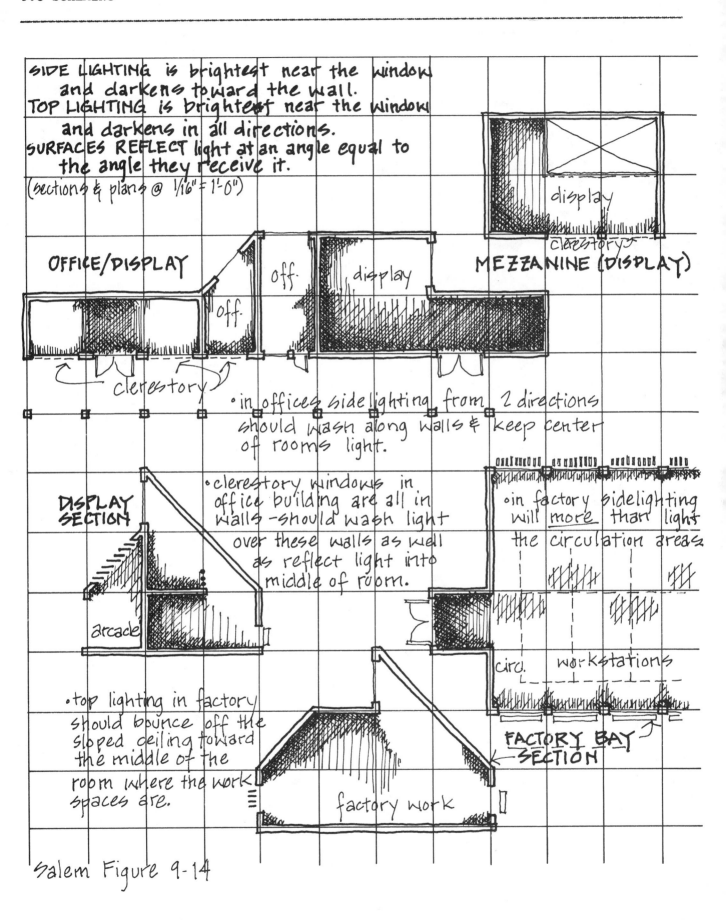

SIDE LIGHTING is brightest near the window and darkens toward the wall.
TOP LIGHTING is brightest near the window and darkens in all directions.
SURFACES REFLECT light at an angle equal to the angle they receive it.

(sections & plans @ 1/16" = 1'-0")

OFFICE/DISPLAY

off. off. display

MEZZANINE (DISPLAY)

display

clerestory

clerestory

• in offices sidelighting from 2 directions should wash along walls & keep center of rooms light.

DISPLAY SECTION

arcade

• clerestory windows in office building are all in walls - should wash light over these walls as well as reflect light into middle of room.

• in factory sidelighting will more than light the circulation areas

circ. workstations

• top lighting in factory should bounce off the sloped ceiling toward the middle of the room where the work spaces are.

factory work

FACTORY BAY SECTION

Salem Figure 9-14

B. Criterion B: There are gradual changes in lighting levels between inside and out, and within rooms, to prevent glare problems.

Evaluation Tool for Criterion B - "Light Distribution Rendering"
Use the studies done for criterion A above to show you have also satisfied criterion B. If you have designed to achieve contrast rather than gradual changes in illumination levels, explain your design goals and show that glare will not be a problem.

Does your first trial design meet the criteria?

Yes.
If not, what will you change in your second trial design?

C. Criterion C: Recommended Daylight Factors for tasks are achieved.

For overcast skies:

In Salem, Or, overcast skies predominate under the winter minimum daylight design conditions.

Task	Min. DF	Comments
office work	1%	With side lighting, min. 12' from window;
	2%	With top lighting, over whole area
typing, computing	4%	Over whole area
factory work	5%	General recommendation
public areas	1%	Depending on function, may be greater
drafting	6%	On drawing boards
	2%	Over rest of working area

Figure 9-1   Recommended Daylight Factors (Millet and Bedrick, 1978)

For clear skies, use the % from Climate section 3.5C., Summary questions A and B:

office   ____ %
factory   ____ %
circulation   ____ %

_____

Evaluation Tool for Criterion C - "Aperture Sizing for Daylighting"
1. The following table gives you recommended aperture-to-floor-area ratios to achieve average Daylight Factors with various kinds of daylight openings.

This is included as a schematic step to help you further organize your basic building form. While still rough, it is a more precise step than you took back in the Climate Exercise when you first explored daylighting.

Daylight Factor (%) = $\frac{Ag}{Af}$ x 100   where Ag = area glazing, sf
                                                     Af = area floor, sf

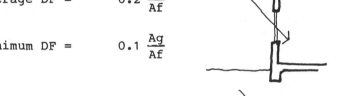

Sidelighting   average DF = $0.2 \frac{Ag}{Af}$

                    minimum DF = $0.1 \frac{Ag}{Af}$

Toplighting

Vertical monitors: average DF = $0.2 \frac{Ag}{Af}$

North-facing
sawtooth openings: average DF = $0.33 \frac{Ag}{Af}$

Horizontal
skylights: average DF = $0.5 \frac{Ag}{Af}$

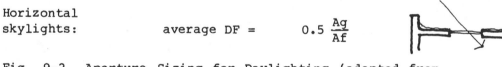

Fig. 9-2 Aperture Sizing for Daylighting (adapted from Hopkinson, 1966, p. 432)

2. Evaluate your office building and factory by completing the table below.

If more than one type of opening is used for a given space, first find the average DF for each, then add to obtain total average DF.

Room		Col. 1 Af	Col. 2 Type of opening	Col. 3 Ag	Col. 4 Applicable rule of thumb Average DF =	Col. 5 Average DF	Col. 6 Compare to recommended minimum DF
Office #1	Trial 1	200	N. window 50 S. window 50		$0.2 \dfrac{Ag}{Af}$	$0.2 \times \dfrac{100}{200} = 0.1$ (10%)	4%, ol (typi
	Trial 2						
Office #2	Trial 1	150	N. window 50 S. window 50		$0.2 \dfrac{Ag}{Af}$	$0.2 \times \dfrac{100}{150} = 0.13$ (13%)	4%, ol (typi
	Trial 2						
Display/Conf (excluding stair tower)	Trial 1	800	NE window 100 S. clerst. 200		$0.2 \dfrac{Ag}{Af}$	$0.2 \times \dfrac{300}{800} = .075$ (7.5%)	1%, ok (publi
	Trial 2						
Factory	Trial 1	3900	N. windows 850 S. windows 950		$0.2 \dfrac{Ag}{Af}$	$0.2 \times \dfrac{1800}{3900} = .09$ (9%)	5%, ok (facto
	Trial 2						

Does your first trial provide DF (in col. 5) equal or
greater than the recommended in Col. 6?
Yes.
If not, what will you change for your second trial?

D. Criterion D: Apertures needed for solar collecting and
ventilating are maintained.

If you have already done the heating and cooling
exercises, your design must still meet the criteria for
heating and cooling.   If you have not yet done these
exercises, but you know that either heating or cooling
(or both) will be concern in either building in your
climate, check the following criteria now so that you
don't make things impossible in those exercises!

For solar heating: Aperture sizing, section 5.3.

*New, raised factory window
position will not change cross
ventilation abilities, but it
will reduce stack effect. From
Section 7.3, Criterion A
Evaluation Tool: new stack h =
20' (instead of 22').*

$A\ rqd = \dfrac{44,510}{9.4\ \ 20x3} = 611\ s.f.$

Actual A = 450 s.f.
Actual T =

$3^o\ calc.\ x\ \dfrac{611\ calc.}{400\ actual} = 4.6^o$

*Considering this occurs only
under dead calm conditions (10%
August), ok.*

For cooling: Aperture sizing, section 6.3.

Does your first trial maintain adequate heating and cooling performance? *Yes; it doesn't change area of glazing or openings from the heating and cooling exercises.* If not, what will you change for your second trial?

9.4 DEVELOPING

GOALS

A. Admit daylight to achieve approximately uniform lighting within task areas, without sacrificing desirable views.
B. Use appropriate shading devices and interior surface finishes, together with the openings, to achieve recommended daylight factors.

CRITERIA

A. Provide recommended DF for each task area.
B. The range of DF within a task area should not exceed 3 to 1.

DESIGN STRATEGIES

A. Site Scale
Utilize a variety of landscape surfaces for their reflective characteristics. These will change seasonally in climates with snow. Remember that snow can cause glare, especially on sunny days.

(Snow is very rare in Salem, Or)

B. Cluster Scale
Ground reflectance is potentially the major source of daylighting in hot, sunny conditions, when shading devices may be blocking direct sun and greatly reducing light from the skydome (including summertime in cold climates).

Exterior obstructions to daylight in overcast climates (such as other buildings or hills, etc.) will not have a major effect on distribution of daylight within spaces, unless they obscure a major portion of the glazing.

Exterior obstructions are the fir trees, which stand about 60 feet from north glass. The light colored parking lot north of the office building is a response to this blocking, which is not really "major".

C. Building Scale
Proportion rooms to recognize the effect of overhangs:

the room effectively begins at the edge of the overhang, not where the glazing is located.

Use side lighting for task lighting to minimize veiling reflections and to supply useful modeling.

Use top lighting for more uniform, ambient lighting and general illumination.

D. Component Scale

Be careful of glare problems from narrow vertical windows, which create strong patterns of light and dark contrasts.

Use light colors and higher reflectances near openings to reduce glare by reducing the contrast between inside and outside.

Choose internal surface reflectances to recognize the major source of incoming light:

In temperate climates - direct and reflected skylight hits the floor first. *...This is Salem, Or typical condition.*

In tropical climates - reflected sunlight hits the ceiling first.

Use light colors/high reflectances on surfaces farthest from openings to increase light in dimmer areas.

DESIGN

Develop your design to achieve these goals for either a typical bay of your factory or the display conference room in the office building.

Document the design:

A. Plan and section of the space, at 1/8"=1'0". (State which room you are designing.) *We chose the factory. Since clerestories alternate their orientation every 10', we chose a 20' wide section to catch both orientations.*

B. Sectional perspective with finishes, materials, colors, and areas (sf) of each called out. Include 1 person for scale.

C. State for which sky condition (clear or overcast) you are designing.

D. Draw all daylight apertures in elevation and section to show mullions; label the glazing material, and show the shading devices in position for your chosen sky condition. *Overcast for Salem, Or, under winter minimum daylight conditions; also figure for summer, clear.*

See Salem example Figures 9-15 to 9-17, 9-21 and 9-22.

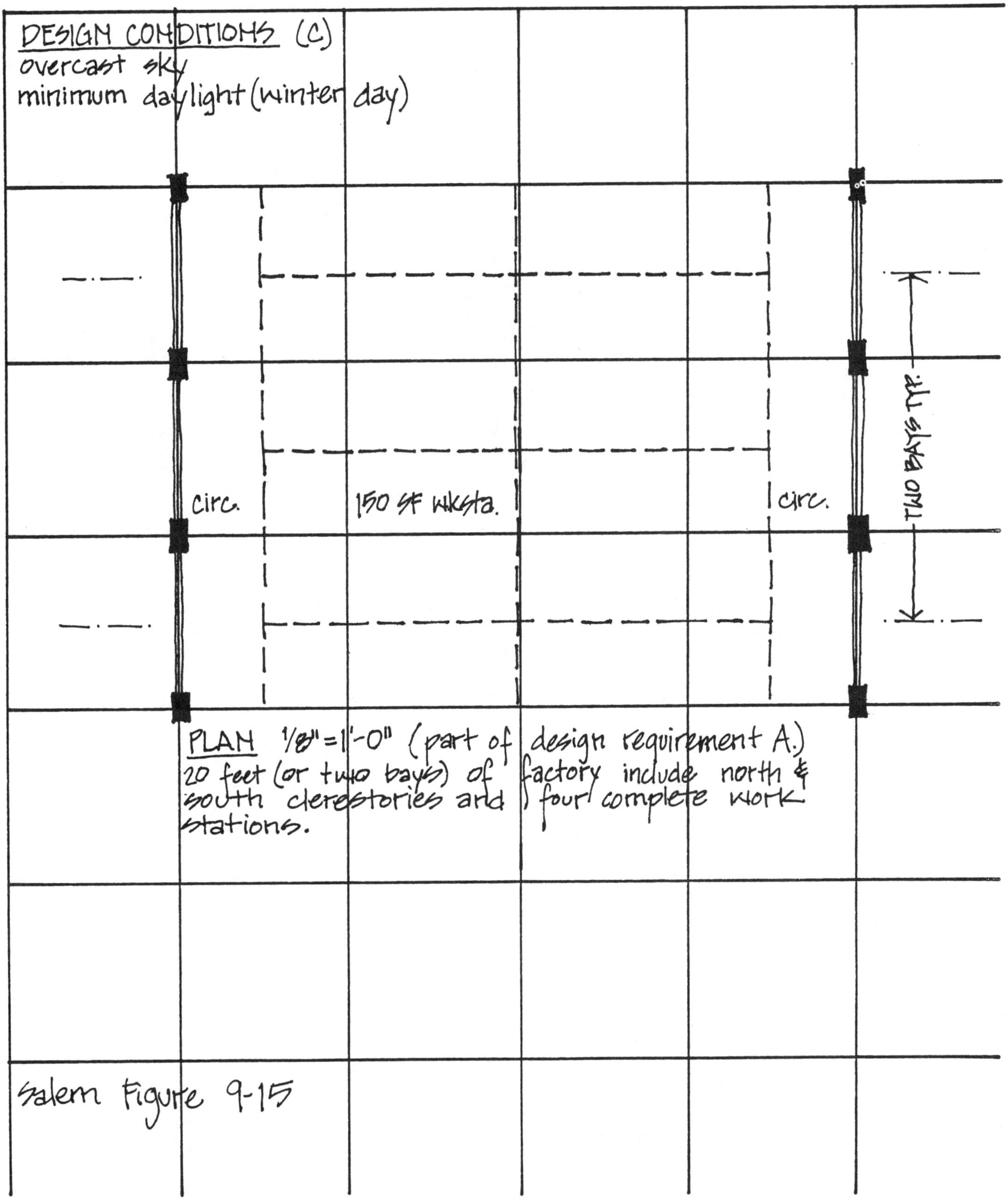

DESIGN CONDITIONS (C)
overcast sky
minimum daylight (winter day)

circ.

150 SF wksta.

circ.

TWO BAYS TYP.

PLAN 1/8"=1'-0" (part of design requirement A.)
20 feet (or two bays) of factory include north &
south clerestories and four complete work
stations.

salem figure 9-15

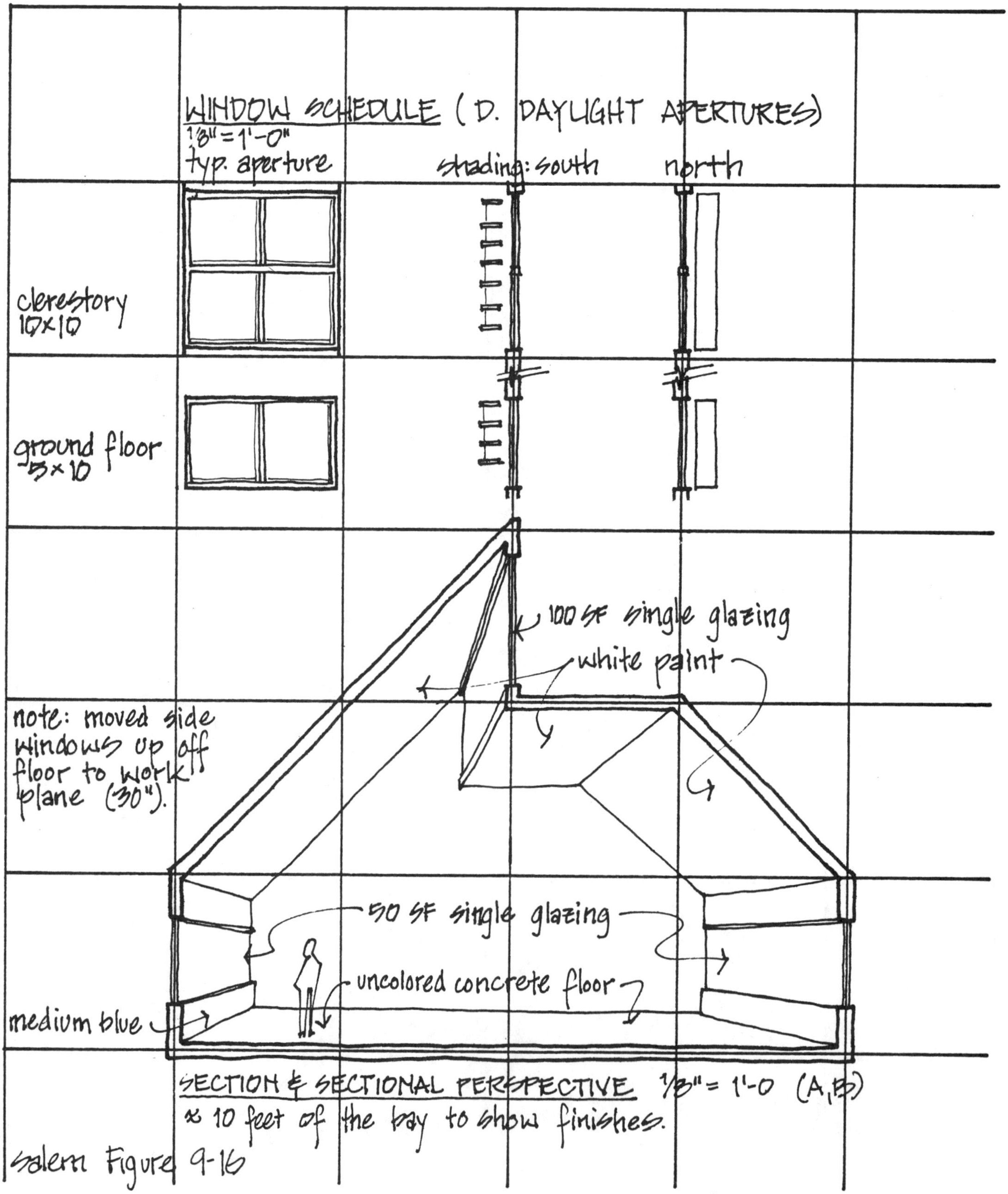

WINDOW SCHEDULE (D. DAYLIGHT APERTURES)

1/8" = 1'-0"
typ. aperture   shading: south   north

clerestory
10×10

ground floor
5×10

100 SF single glazing

white paint

note: moved side
windows up off
floor to work
plane (30").

50 SF single glazing

uncolored concrete floor

medium blue

SECTION & SECTIONAL PERSPECTIVE 1/8" = 1'-0 (A,B)
× 10 feet of the bay to show finishes.

salem Figure 9-16

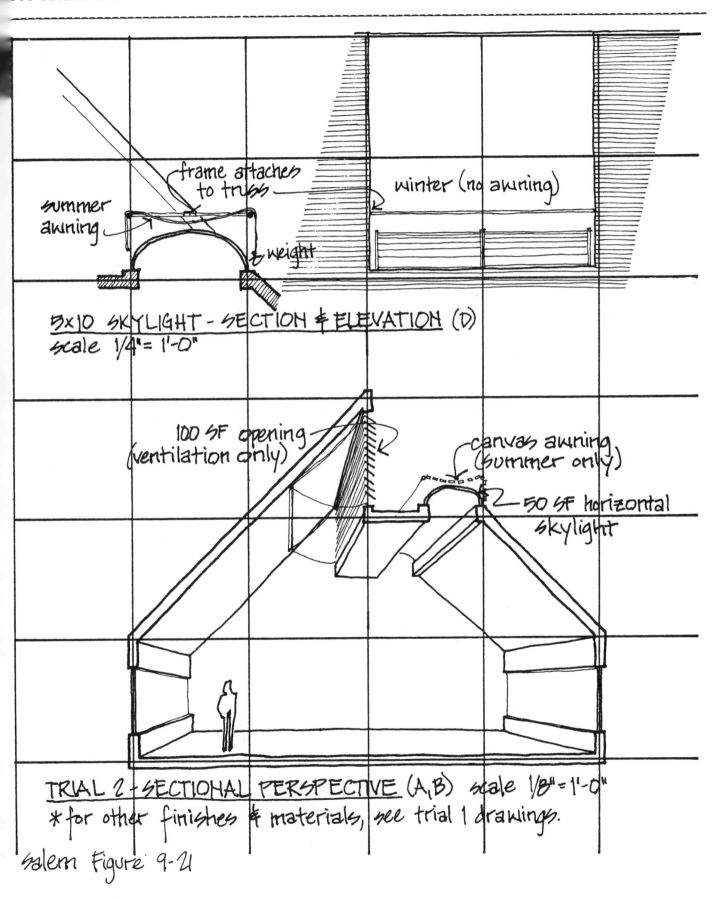

summer awning

frame attaches to truss

winter (no awning)

weight

5x10 SKYLIGHT - SECTION & ELEVATION (D)
scale 1/4"= 1'-0"

100 SF opening
(ventilation only)

canvas awning
(summer only)

50 SF horizontal
skylight

TRIAL 2 - SECTIONAL PERSPECTIVE (A,B) scale 1/8"=1'-0"
* for other finishes & materials, see trial 1 drawings.

salern Figure 9-21

----

for daylighting contours, enlarged plans: if 1/8"= 1'-0", 1 square = 10'.

EVALUATION

A. Criterion A: Provide recommended D.F. for each task area.
B. Criterion B: The range of D.F. will be less than 3 to 1 within a given task area.

Evaluation Tools for Criteria A & B - For overcast sky conditions: "Footprints"; for either clear sky or overcast: "Daylight Model" (model instructions start on p. 9-34.

"Footprints"

First determine whether the footprint prediction technique will work for your design. This technique is applicable to most spaces, however there are complications in using it with non-orthogonal openings, or in rooms of complex shape or abnormal proportions. It also does not evaluate the localized effect of interior or exterior room reflectances, the effect of exterior obstructions, or fixed-shading devices. (Fixed overhangs can be accommodated by considering the room as extending to the edge of the overhang; moveable shading devices, such as ventilation blinds, can be assumed to be clear of the opening on the darkest days you are evaluating).

If your design has one of these complications that makes the footprint method awkward, we urge you to use the model and light meter to evaluate your design. Believe it or not, you will <u>save</u> time by not trying to use the wrong technique to evaluate your design.

If you would like to use footprints, and fixed shading devices are your <u>only</u> complication, simply <u>estimate</u> the % daylight that gets through them. Do <u>not</u> use "shading coefficients" that are used in heat gain estimations, since these refer to <u>direct sun</u> more than to daylight from all sources. Thus, the % daylight will be somewhat higher than these "shading coefficients" would indicate.

OK! We estimate 60% in winter, given the reflections possible off of their surfaces, and assuming they are adjustable so as to be wide enough open to admit 50% of the <u>direct sun</u> striking them in winter (see Section 7.2 assumptions). For <u>summer</u>, cut this in half, to 30%.

The footprints included at the end of this exercise can be used <u>as is</u> for windows 5' x 10', 10' x 10', 10' x 5' <u>above</u> the workplane. For other sizes and shapes they must be re-scaled and or combined so the solid rectangle is the same size as your window. Any portion of the window <u>BELOW</u> the workplane will not contribute illumination to the workplane so "your" window for footprint purposes will be: _5' x 10'_ .

...Good thing we moved those windows up!

This technique involves three basic steps:
1. Matching your openings with available footprints;
2. Drawing isolux contours that describe light distribution and basic DF range in your space.
3. Correcting DF's applied to contours:
<u>increasing DF</u> due to factors which increase light in the space (principally interior reflectances)
<u>reducing DF</u> due to factors which reduce light in the space (glass transmission and dirt, and shading devices.)

The results give you an estimation of the distribution of light and the quantity of light (in Daylight Factors) in your space.

Note: The information and tables in this procedure are from the two sources listed below. For more information, distribution patterns, refinement steps, and procedure for dealing with obstructions, see:
Millet and Bedrick, "Architectural Daylighting Design Procedure", June 1978
Millet and Bedrick, "Manual: Graphic Daylighting Design Method," 1981.

1. Match opening and footprints.
a. List your windows and skylights and find the appropriate footprint in Fig. 9-4 on the following page.

H = height of window itself
W = width of window or skylight
L = length of skylight itself
S = distance from <u>workplane</u> to <u>sill</u>

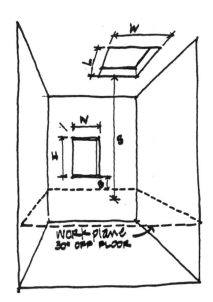

(footprint # from next page)

Clerestory windows:
S=17, because of 20' sill height less 3' workplane height

Trial 1:

Opening	H/W window; L/W skylight	S/H window; S/W skylight	Footprint #
Clerestory windows (a, Fig. 9-17)	$\frac{10}{10} = 1$	$\frac{17}{10} = 1.7$ or about 2	A-16
Sidewall windows (b, Fig. 9-17)	$\frac{5}{10} = .5$	$\frac{0}{5} = 0$	A-7

Trial 2:

Skylight (a, Fig. 9-22)	$\frac{10}{5} = 2$	$\frac{17}{5} = 3.4$ or about 3	A-36
Sidewalls (a, Fig. 9-22)	as before......................		

Openings of shapes other than those below can be used by combining shapes that are catalogued. For example, a sloping skylight:

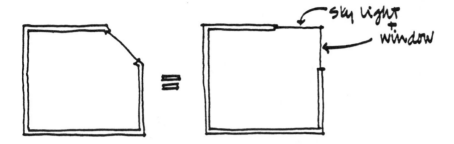

Figure 9-3 Sloped Skylight Simulation for Footprint Procedure. If a 45° skylight is square and has an area of 141 sf, the resulting window and horizontal skylight will each be 10 x 10. When estimating the IRC, use the actual area (141 sf) and Fig. 9-5, TOP-LIT Rooms.

Specify any assumptions you made to deal with your design:

No odd shapes or sizes; footprints should be ok to use.

Windows			Skylights		
H/W	S/H	# Footprint	L/W	S/W	#Footprint
.5	0	A-7 ...*Factory*	1	1	A-30
.5	.5	A-8 *sidewall*	1	2	A-31
.5	1	A-9 *windows*	1	3	A-32
.5	2	A-10	1	4	A-33
.5	3	A-11			
.5	4	A-12	2	1	A-34
			2	2	A-35
1	0	A-13	2	3	A-36 ...*Factory skylights, trial 2*
1	.5	A-14	2	4	A-37
1	1	A-15			
1	2	A-16...*Factory clerestories, trial 1*			
1	3	A-17			
1	4	A-18			
2	0	A-19			
2	.5	A-20			
2	1	A-21			
2	2	A-22			
2	3	A-23			
2	4	A-24			

Figure 9-4 Footprint Catalogue. See Section 9.5 for Footprints.

2. Drawing the contours (this will be easiest if you have your plan on a paper you can trace through):

a. Sketch the openings on your plan.
Skylights: draw the outline of the skylights as for a reflected ceiling plan (as if they were dropped straight down on to working plane).
Windows: using the section, rotate the windows 90° on to the workplane as though they were hinged at the intersection of the workplane and the vertical plane. This locates them so you can then sketch the outline on the plan.
b. Sketch the contours for each opening. Position the plan with openings sketched on it over the appropriate footprint for each opening and trace the contours for the opening. (To use the pattern, the <u>solid rectangle should be lined up with the outline of the opening on the plan.</u> The <u>solid line at the bottom of the pattern</u> indicates the position of the window plane, so you don't get them upside down.) Be sure to note values of DF contours on your plan. The values of the contour lines are % available daylight, or Daylight Factors. The dotted contour is = 0.5% DF and they increase by 1% each contour. It will probably be most useful in the evaluation if you sketch all the side lighting contours on one sheet, and all the top lighting contours on a second sheet.
c. Combine the contours. With an overlay of tracing paper, mark all the points where the contours cross, and note the sum of the values of the crossing contours. Now draw the resulting contour lines by connecting points of equal value. Outside the area of overlap, trace the original contours directly. If you have side lighting and top lighting on separate sheets, combine contours on each sheet to get total side lighting and total top lighting, and then repeat the procedure to end up with contour lines of the overall, combined pattern. Be sure to note the DF values of the contours. This is the final isolux contour drawing for your space.

At this point, things get more complicated; so if you don't like the look of your DF contour distribution, change the design <u>now</u>, and start again.

Trial 1: See Salem example Figures 9-17 to 9-20.

...We <u>don't</u> like them! Looks like very little light in the middle, where the tasks are; clerestory light appears "wasted". (See Salem Fig. 9-20) So, <u>change</u> to clerestories for ventilating only, and add 50 sf <u>horizontal skylights</u> in the flat roof sections.
With insulated hatches over the 100 sf ventilating openings, 50 sf skylights that replace them should not increase heat loss in winter. With shading devices over skylights in summer, heat gain will be increased; however, if <u>they also</u> can ventilate, the stack effect will be increased.

Trial 2: See Salem example Figures 9-22 to 9-26.

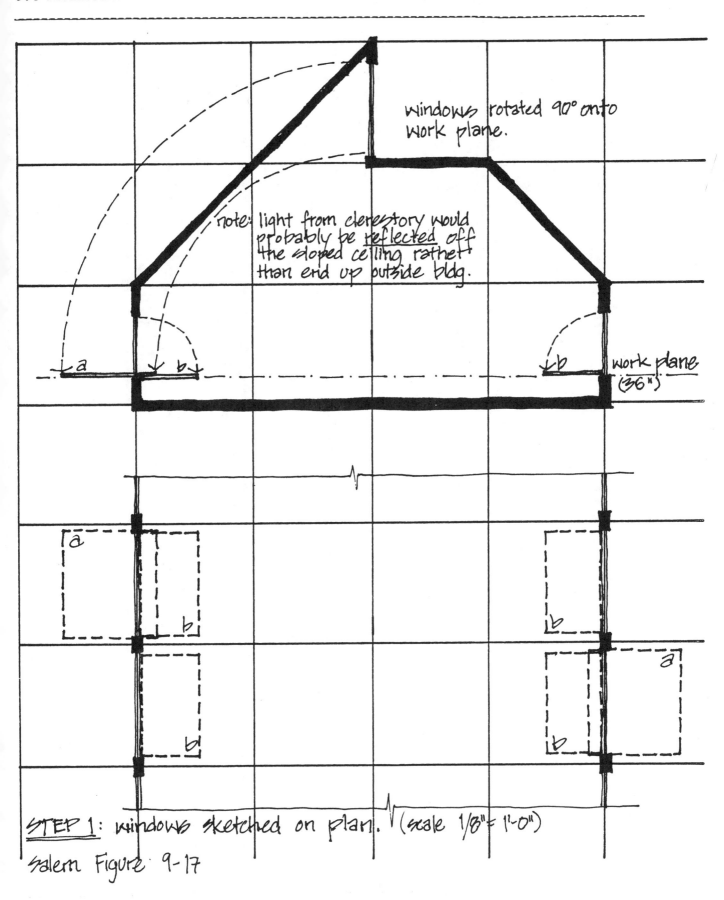

windows rotated 90° onto work plane.

note: light from clerestory would probably be reflected off the sloped ceiling rather than end up outside bldg.

work plane (36")

STEP 1: windows sketched on plan. (scale 1/8"= 1'-0")

Salem Figure 9-17

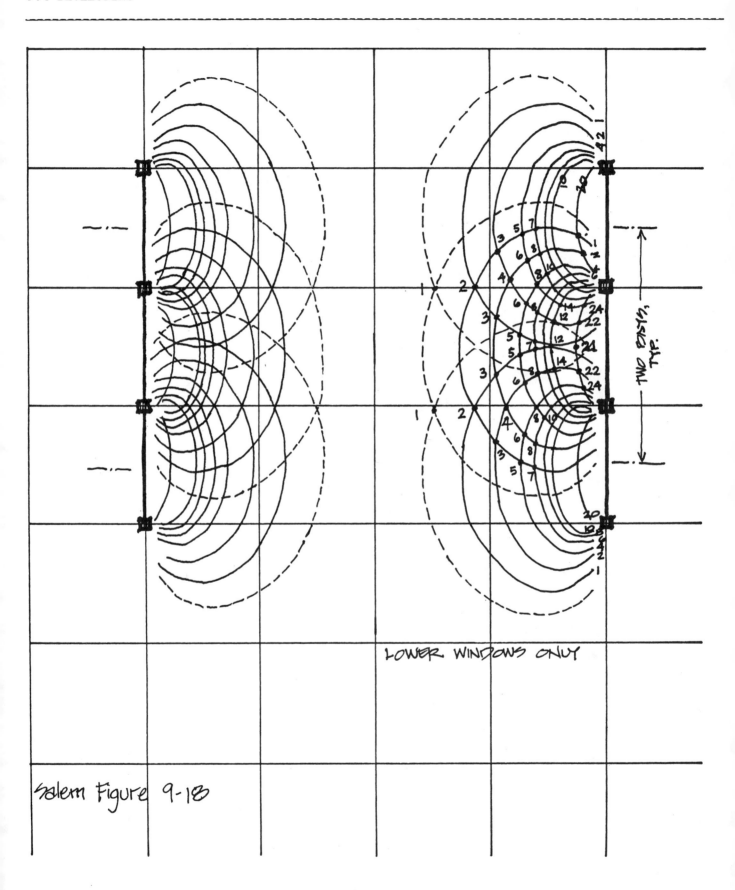

TWO BLDGS. TYP.

LOWER WINDOWS ONLY

Salem Figure 9-18

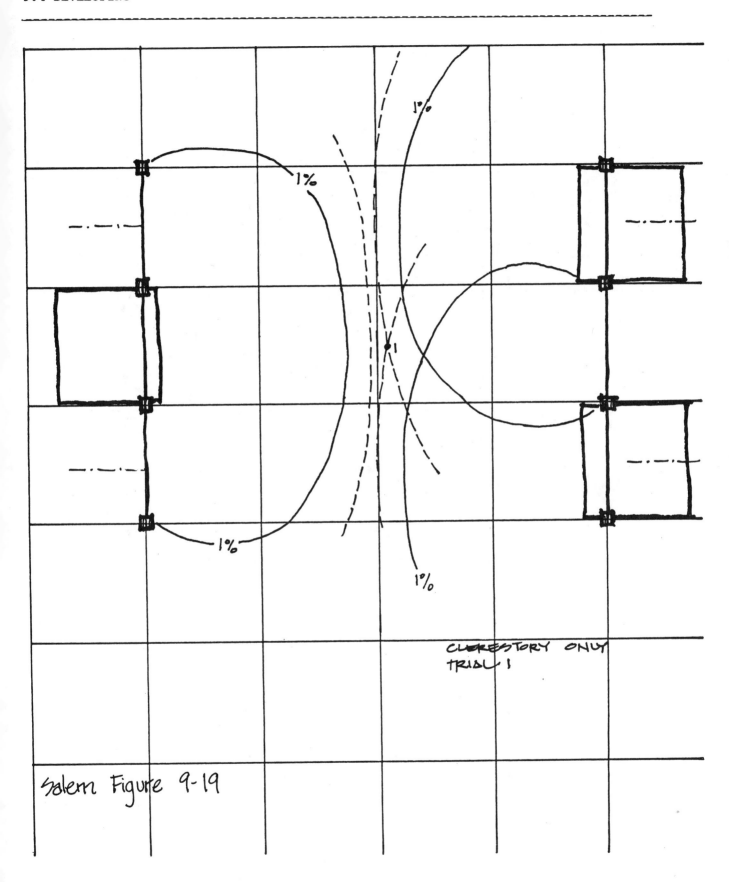

Salem Figure 9-19

1%

1%

1%

1%

CLERESTORY ONLY
TRIAL 1

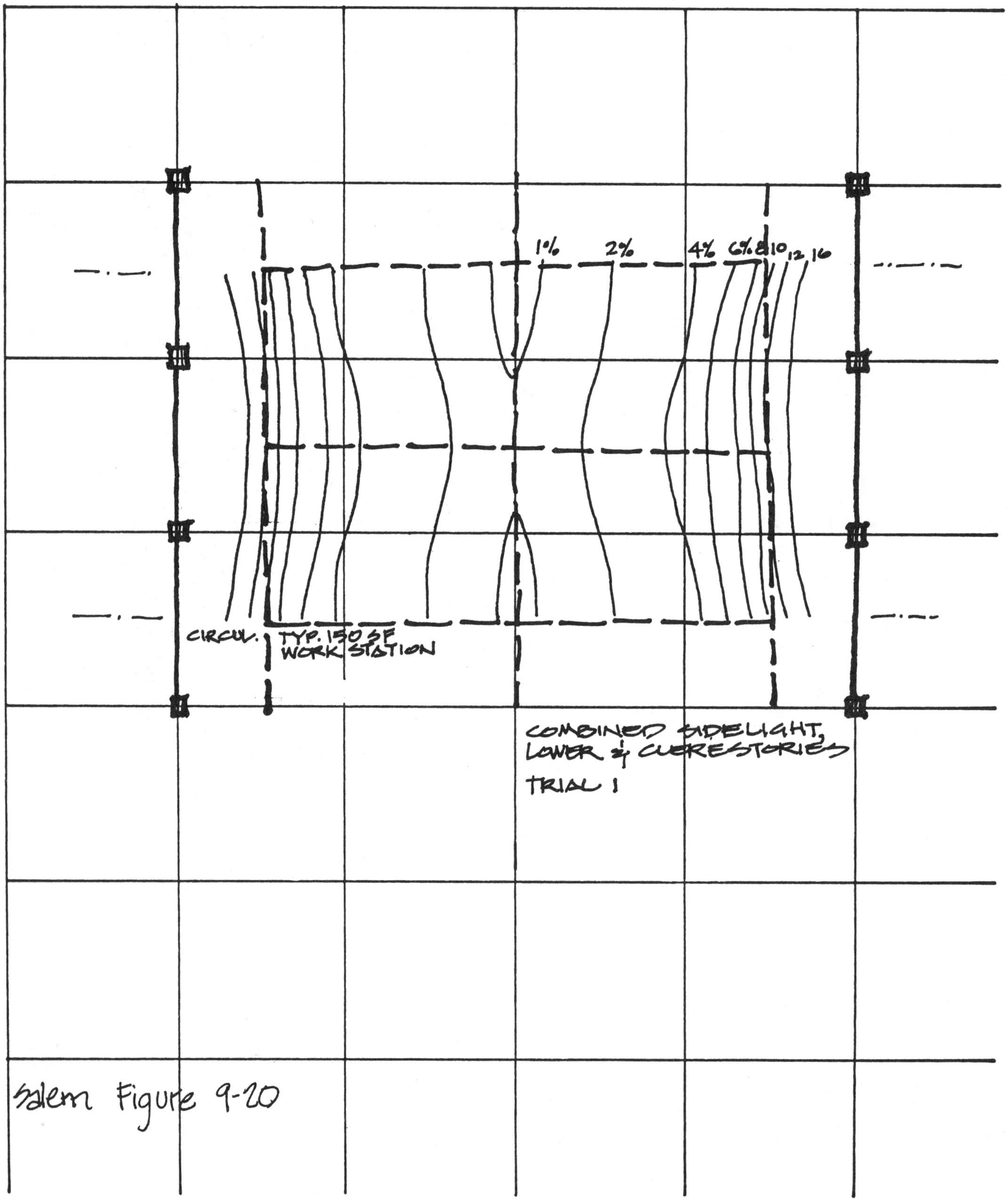

1% 2% 4% 6% 8 10 12 16

CIRCUL. TYP. 150 SF
WORK STATION

COMBINED SIDELIGHT,
LOWER & CLERESTORIES

TRIAL 1

Salem Figure 9-20

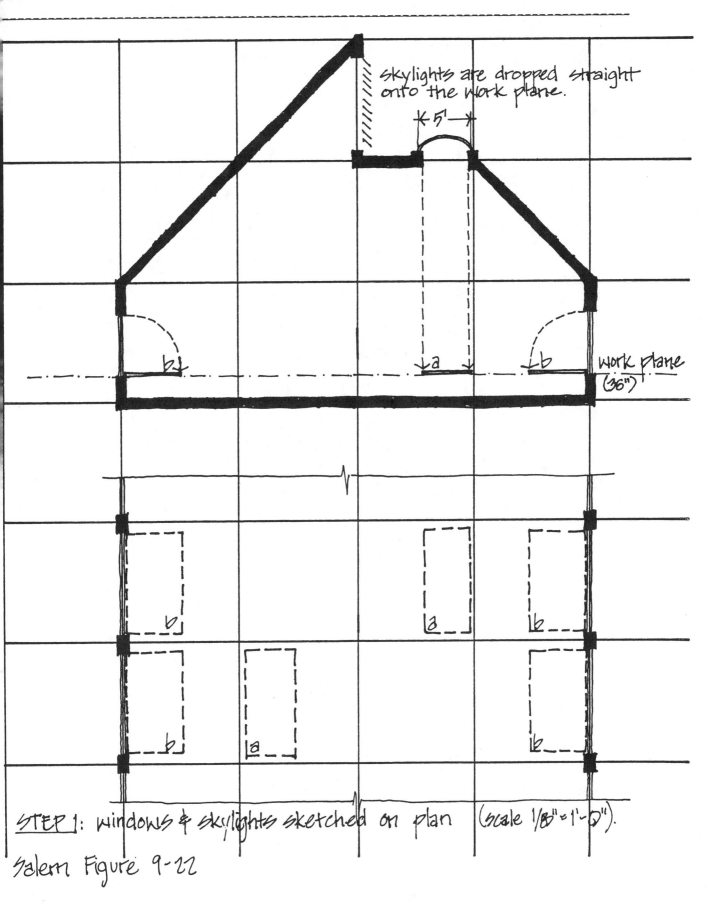

skylights are dropped straight onto the work plane.

←5'→

work plane (36")

STEP 1: windows & skylights sketched on plan (scale 1/8"=1'-0").

Salem Figure 9-22

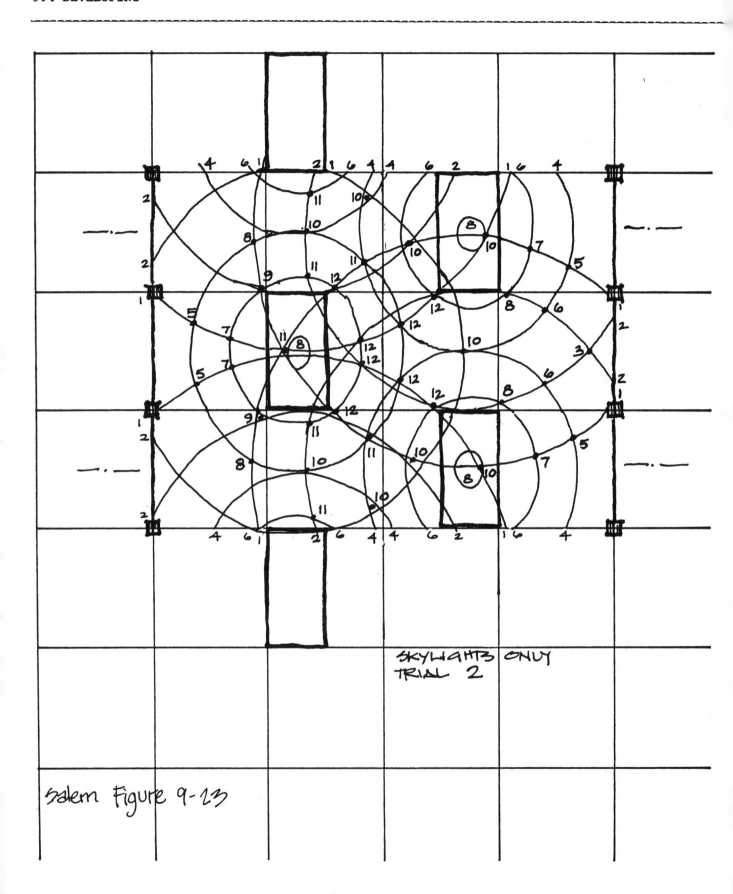

SKYLIGHTS ONLY
TRIAL 2

Salem Figure 9-23

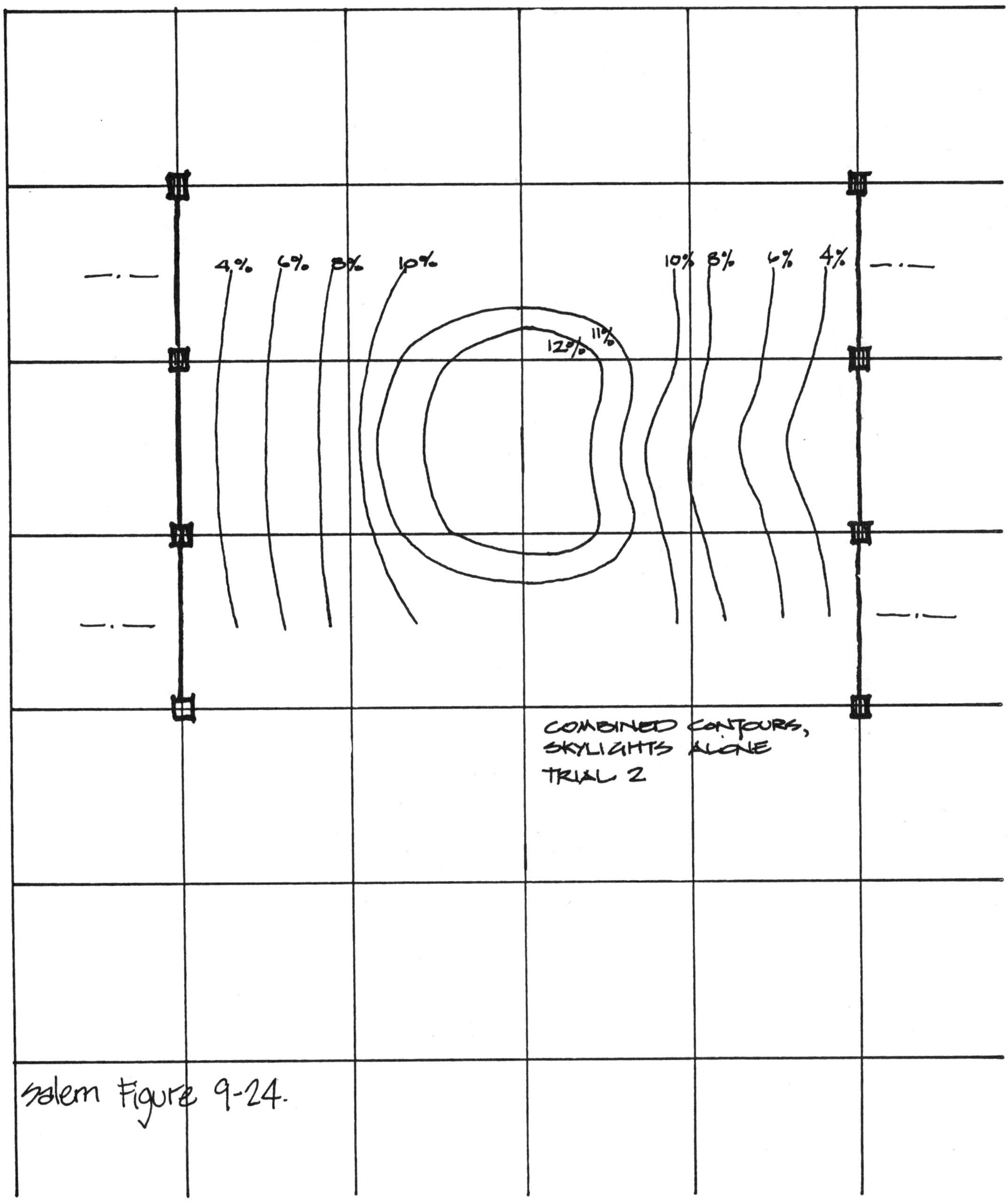

4% 6% 8% 10%   10% 8% 6% 4%

12% 11%

COMBINED CONTOURS,
SKYLIGHTS ALONE
TRIAL 2

salem Figure 9-24.

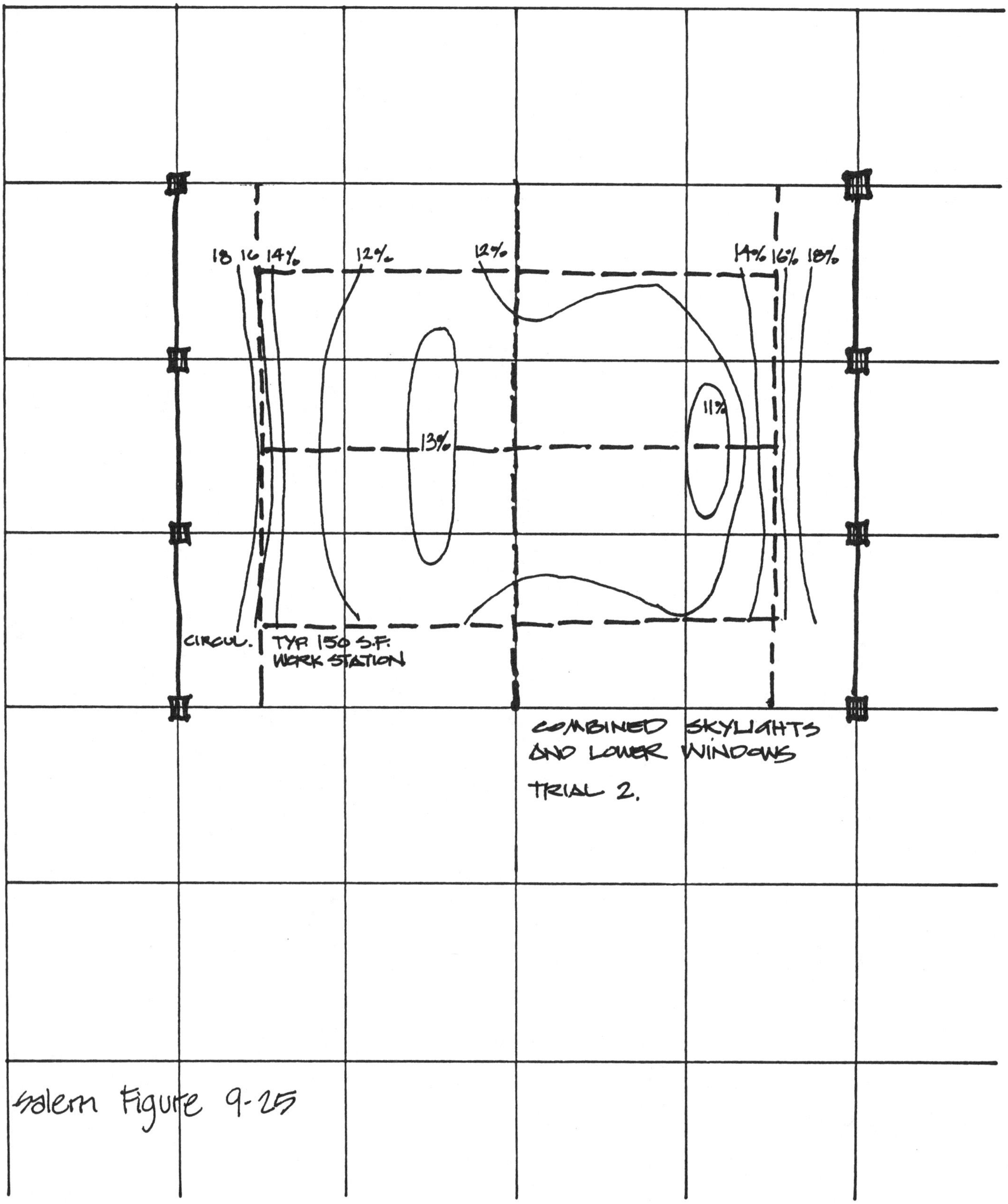

18 16 14%   12%   12%   14% 16% 18%

13%

11%

CIRCUL. TYP 150 S.F.
WORK STATION

COMBINED SKYLIGHTS
AND LOWER WINDOWS
TRIAL 2.

salern Figure 9-25

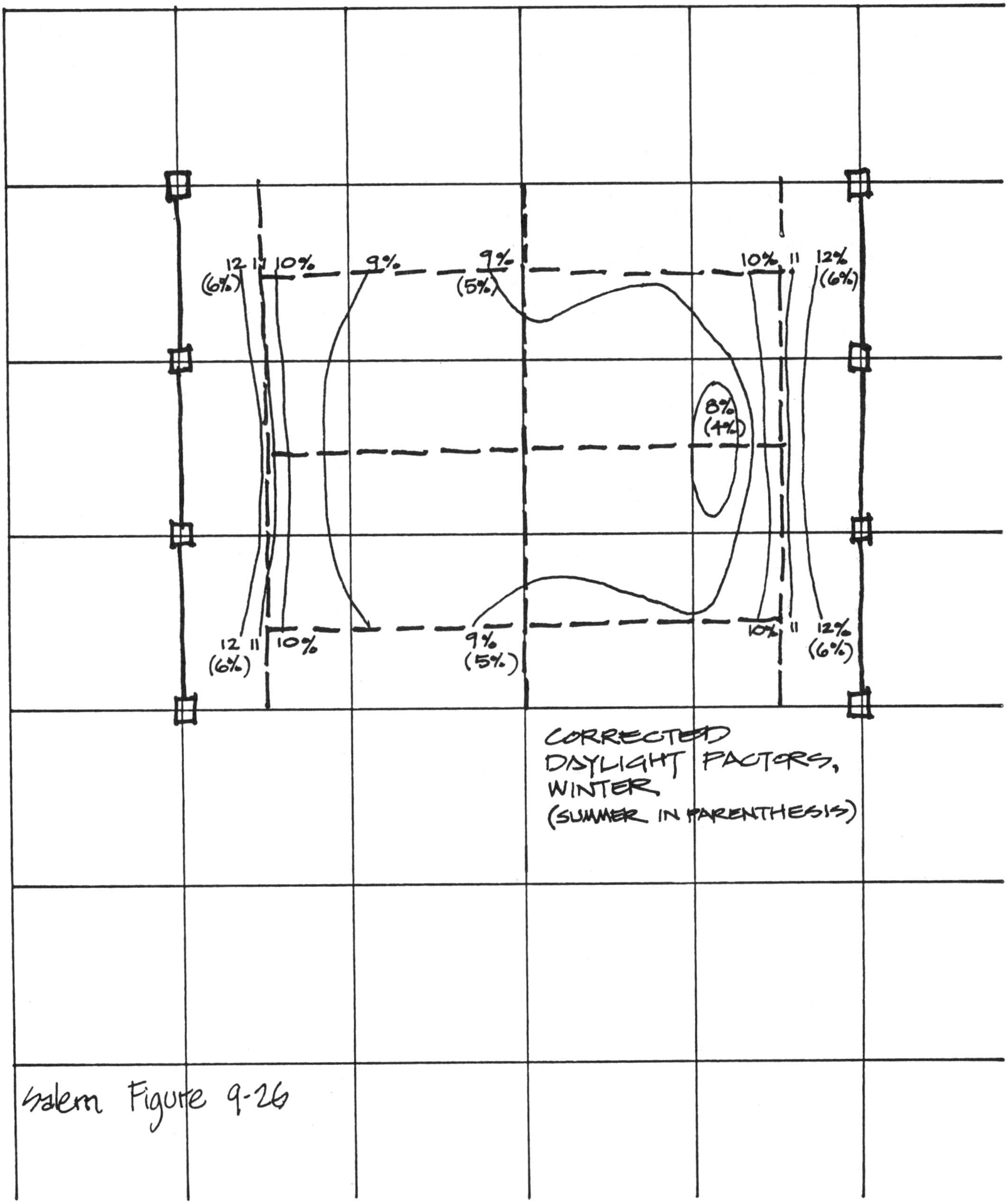

12 11 10%
(6%)
9%
9%
(5%)
10% 11 12%
(6%)

8%
(4%)

12 11 10%
(6%)
9%
(5%)
10% 11 12%
(6%)

CORRECTED
DAYLIGHT FACTORS,
WINTER
(SUMMER IN PARENTHESIS)

salem Figure 9-26

---

3. Correcting the Daylight Factors

a. Increases due to IRC are the product of the internally reflected component and the interior maintenance factor (IMF).   *Salem IMF = 0.9*

      adjusted DF = DF + (IRC x IMF)

Values of IMF for clear work in non-industrial areas; clean work in dirty-industrial areas; dirty-industrial work in non-industrial areas; and dirty-industrial work in dirty-industrial areas are 0.9, 0.8, 0.7, and 0.6 respectively (Millet and Bedrick, 1978).

Document the characteristics of your room by completing the table below (if you made assumptions about your design for the footprints use them for this section, too).

Reflectances: Flynn & Segil, Table 2-17, p. 126. MEEB, Table 20.3, p. 864.

Glass Transmission & Reflectance: MEEB, Table 19.5, p. 797. Flynn & Segil, Table 2-18, p. 127. Time Saver Standards, p. 925 (1974). ASHRAE Fundamentals. IES-Recommended Practice of Daylighting.

Surface	Material or Finish	Surface Area (sf)	Reflectance	Transmission of Glazing (and shading devices)
Floor Flat	*concrete*	*800*	*30%*	
ceiling Sloped	*white*	*100*	*85%*	
ceiling	*white*	*560 + 280*	*85%*	
Sidewalls	*blue*	*200*	*50%*	
"Imaginary sidewalls"		*740 + 740*	*50%*	*...because this is only two bays of a larger room, assume neutral "side" walls of 50% reflectance.*
Vents	*white*	*200*	*85%*	
Skylights	*translucent plastic*	*100*	*10%*	*80% winter, 40% summer*
Sidewalls	*glass*	*200*	*10%*	*60% winter, 30% summer*
	Total Room Surface Area:	*3920*		*Glass reflectances from Flynn and Segil*

9.0 DAYLIGHTING

9.4 DEVELOPING

Figure your net glazed areas for horizontal and vertical
openings separately.   Measure the actual glazed area
(omitting mullions, glazing bars, etc.) in your design,
or multiply the total opening size by a typical factor.

Window type	Net glazed factor	
all metal windows	.80	*We're assuming all-metal windows.*
metal windows in wood frames	.75	
all wood windows	.65-.70	

net glazed area, side lighting   *.8 x 200*   _160_ Trial 1   ____ Trial 2

---

net glazed area, top lighting   *.8 x 100*   _80_ Trial 1   ____ Trial 2

---

net glazed area in vertical surfaces (side lighting) =   _.04_ Trial 1   ____ Trial 2
    total room surface area

---

net glazed area in horizontal surfaces (top lighting) = _.02_ Trial 1   ____ Trial 2
    total room surface area

Figure your average room reflectance.
(If you are working with one typical bay or a large space
such as the factory, assume 50% reflectance from the
"sidewalls" that represent adjacent bays, and include
their area in the total surface area for a typical bay.
It is not necessary to calculate the entire factory.)

Average room reflectance = $\dfrac{[(\text{surface}^1)\ (\%\ \text{reflectance}^1)] + [(\text{surface}^2)\ (\%\ \text{reflectance}^2)]}{\text{total room surface area}}$

Trial 1 _53_ %   Trial 2 _____ %

Find the TOP-LIT IRC values (if you have top lighting)
from figures 9-5 and 9-6.

Find the SIDE-LIT IRC values (if you have side lighting)
from figure 9-7.

Avg. Room
Reflectance =

$$\frac{(800)(30\%)+(100)(85\%)+(840)(85\%)+(200)(50\%)+(1480)(50\%)+(200)(85\%)+(100)(10\%)+(200)(10\%)}{3920}$$

$$= \frac{240 + 85 + 714 + 100 + 740 + 170 + 10 + 20}{3920}$$

$$= .53$$

Salem Example 53%

Net Glazing Area	Average Reflectance (I.R.C.) of Room Surfaces								
Total Room Surface Area	20%	25%	30%	35%	40%	45%	50%	55%	60%
.01	.22	.32	.40	.49	.60	.72	.88	1.09	1.40
.02 ...Salem example.....	.46	.62	.78	.90	1.20	1.50	1.80	2.20	2.60
.03	.60	.90	1.15	1.40	1.80	2.15	2.70	3.40	4.10
.04	.90	1.20	1.65	1.90	2.40	3.00	3.70	4.40	5.30
.05	1.15	1.50	1.90	2.40	3.00	3.70	4.30	5.40	6.60
.06	1.40	1.85	2.30	2.90	3.70	4.40	5.40	6.60	8.00
.07	1.65	2.10	2.30	3.80	4.20	5.00	6.20	7.40	8.60
.08	1.35	2.90	3.20	3.50	4.70	5.90	7.20	8.60	10.50
.09	2.00	2.80	3.60	4.40	5.40	6.60	8.00	9.60	11.70
.10	2.20	3.10	4.10	4.50	5.90	7.30	8.80	10.70	13.00

Figure 9-5   Internally Reflected Component for TOP-LIT Rooms (glazing from horizontal up to 60°) (Millet and Bedrick, adapted from Hopkinson Daylighting)

Interpolating above,
2.04 x .9 IMF = 1.84

Net Glazing Area	Average Reflectance (I.R.C.) of Room Surfaces								
Total Room Surface Area	20%	25%	30%	35%	40%	45%	50%	55%	60%
.01	.09	.13	.16	.20	.25	.31	.37	.45	.54
.02	.19	.25	.32	.41	.50	.62	.76	.90	1.10
.03	.27	.38	.48	.60	.75	.90	1.20	1.30	1.65
.04	.38	.50	.69	.80	1.00	1.20	1.48	1.80	2.20
.05	.47	.63	.80	.99	1.22	1.60	1.85	2.20	2.70
.06	.56	.77	.95	1.20	1.50	1.80	2.20	2.70	3.40
.07	.66	.88	.96	1.40	1.72	2.10	2.55	3.20	3.70
.08	.77	.99	1.25	1.60	1.95	2.40	3.00	3.70	4.40
.09	.86	1.15	1.42	1.80	2.20	2.70	3.45	4.10	3.00
.10	.94	1.20	1.65	2.00	2.40	3.10	3.80	4.50	5.50

Figure 9-6   Internally Reflected Component for TOP-LIT Rooms (glazing vertical) (Millet and Bedrick, adapted from Hopkinson Daylighting)

| Net Glazing Area | Average Reflectance (I.R.C.) of Room Surfaces | | | | | | | *Salem example 53%* | |
Total Room Surface Area	20%	25%	30%	35%	40%	45%	50%	55%	60%
.01	.05	.07	.10	.13	.18	.25	.32	.40	.50
	.02	.03	.05	.08	.12	.17	.25	.33	.43
.02	.09	.13	.20	.27	.36	.48	.62	.78	.96
	.03	.06	.10	.16	.24	.34	.47	.64	.84
.03	.13	.20	.29	.40	.54	.74	.94	1.17	1.45
	.03	.09	.15	.24	.36	.52	.72	.98	1.24
.04 ...Salem example.......	.18	.27	.39	.53	.72	.96	1.25	1.60	1.95
	.07	.12	.20	.31	.48	.70	.96	1.27	1.66
.05	.23	.34	.48	.66	.85	1.18	1.55	1.95	2.40
	.08	.15	.26	.39	.60	.86	1.18	1.60	2.08
.06	.27	.41	.58	.80	1.07	1.43	1.80	2.35	2.92
	.10	.18	.30	.46	.72	1.03	1.43	1.95	2.50
.07	.32	.47	.68	.92	1.25	1.68	2.15	2.70	3.40
	.11	.21	.35	.56	.84	1.20	1.67	2.22	2.90
.08	.36	.54	.78	1.07	1.42	1.95	2.50	3.15	3.90
	.13	.24	.41	.64	.96	1.38	1.85	2.60	3.33
.09	.41	.60	.87	1.18	1.61	2.15	2.80	3.52	4.39
	.15	.27	.45	.72	1.06	1.55	2.12	2.92	3.70
.10	.45	.68	.96	1.30	1.74	2.40	3.10	3.90	4.80
	.16	.29	.50	.79	1.18	1.71	2.39	3.20	4.15
.12	.54	.80	1.15	1.58	2.10	2.85	3.70	4.65	5.80
	.20	.35	.60	.94	1.40	2.00	2.70	3.80	4.90
.14	.64	.94	1.35	1.85	2.50	3.33	4.30	5.40	6.90
	.24	.42	.71	1.10	1.65	2.40	3.32	4.50	5.80
.16	.72	1.07	1.55	2.10	2.80	3.80	4.90	6.55	7.70
	.27	.47	.82	1.25	1.90	2.72	3.80	5.10	6.60
.18	.81	1.20	1.70	2.35	3.20	4.30	5.50	6.90	8.30
	.29	.53	.92	1.38	2.10	3.10	4.25	5.70	7.40
.20	.90	1.30	1.80	2.60	3.50	4.70	6.00	7.70	9.70
	.33	.60	1.00	1.55	2.35	3.80	4.70	6.40	8.10

Figure 9-7   Internally Reflected Component of SIDE-LIT Rooms
(Millet and Bedrick, adapted from Hopkinson, Daylighting)

Note: Values in each box: top = average IRC
                             bottom = minimum IRC

Interpolating: above sidelight, Salem
average IRC = 1.46 x .9 IMF=1.31
minimum IRC = 1.15 x .9 IMF=1.04

b. Reductions in Daylight Factors.
Reductions due to glass transmission loss and shading devices. Glazing-shading transmission factor __60__ % *sidelights, 80% skylights in winter* Figure your reduction or maintenance factor for dirt from *(cut in half for summer,* the table below or MEEB Table 19.5 (b), p. 797. *40%) With Oregon's envi-*
ronmental reputation, we
had better assume a "clean
industrial" area....

Building Location	Inclination of Glazing	Interior Conditions	
		No industry Clean	Dirty Industry
Non-industrial area	Vertical	.9	.8
	Sloping	.8	.7
	Horizontal	.7	.6
Dirty industrial area	Vertical	.8	.7
	Sloping	.7	.6
	Horizontal	.6	.5

side windows

skylights

Figure 9-8 Maintenance Factors

Multiply all reduction factors to get total:

Total reduction factor = transmission × maintenance
loss factor   factor

Total reduction factor = __.55__ Trial 1 _____ Trial 2

Side windows:
.6 x .9 = .65 winter
skylights:
.8 x .7 = .56 winter
Compromise at .55 in winter.

c. Correct all Daylight Factors
For all contours with the lowest Daylight Factors and those in dimmer parts of the space:

Skylights =
.4 x .7 = .28 summer
Sidewindows =
.3 x .9 = .27 summer
Use .28 in summer

corrected DF = (DF + toplit IRC + min sidelit IRC) x (total reduction factor)
(show your complete work in space below)

Lowest winter corrected DF = (DF + 1.84 + 1.04) x .55   *Lowest summer corrected DF*
= .55 (DF + 2.88)   = .28 (DF + 2.88)

Lowest Footprint DF	Lowest winter Corrected DF	Lowest summer Corrected DF	
11%	8%	4%	(See Salem Figure 9-25)

For all other contours:

corrected DF = (DF + toplit IRC + <u>avg</u> sidelit IRC) x (total reduction factor)

All other contours: winter corrected DF = (DF + 1.84 + 1.31) x .55   *Summer,*
                                   = .55 (DF + 3.15)   *= .28 (DF + 3.15)*

Footprint DF	Winter Corrected DF	Summer Corrected DF	
18	12	6	
16	11	5	*Maximum within work station*
14	9	5	
12	8	4	

d. Show the final contours labelled with corrected Daylight Factors on your plan, and graph them in section

From the Climate Exercise, section 3.5 (Criterion A Analysis tools, and summary questions), the daylight factors required in my climate for this space are: *10% on darkest days (winter)*

From Section 9.3, Criterion C, the recommended minimum D.F. for this space is: <u>5</u> %

Does your first trial design as evaluated by footprints meet these criteria?
Essentially, yes. Work station corrected D.F. seems fairly steady at (corrected) 8% minimum *well above the 5% minimum. At 9 am Dec 21, overcast 44°N latiitude, .08 x 380 = 30 foot-* *candles (MEEB p. 774), which meets the "work station" recommendation (MEEB p. 736) of 30 ± !*
If not, what will you change for your second trial?

Does your first trial design as evaluated by footprints meet criterion B?
Yes; 8% - 12% range within work station is well within 3 x 8 = 24% maximum (corrected).

If not, what will you change for your second trial?

Alternate Evaluation Tool for Criteria A & B - "Daylight Model"

The great advantages of this evaluation tool are its combination of accuracy and the variety of resulting design information. No other tool presents such a wide range of evaluation possibilities. It also allows quick evaluation of minor design changes, such as the substitution of one set of shading devices for another.

1. Build a model of the space you are investigating. Your model must be at least 1/2"=1'-0" to 3/4"=1'-0" scale to accurately use a standard light meter. Materials should reasonably match the opacity, surface texture and % reflectance of the floors, walls and ceilings of your space. For large spaces with repetitive bays, such as the factory, build just one bay and insert opaque "side walls" where adjacent bays would otherwise occur. The reflectance of these "walls" should be 50%.

The most important model details are those around the daylight openings; the size and depth of the mullions, the depth and reflectivity of the sill, louvres or other shading devices, and the reflectivity of surfaces just outside the daylight openings.

Louvres can be constructed at a larger scale than the windows, as long as they accurately show their own proportion of louvre-to-opening. Thus, a set of 1 1/2" scale louvres made to cover a 3/4" scale window will give reasonably accurate results - with only half as much louvre cutting/glueing as a 3/4" set of louvres would have required.

Glazing should be placed over openings; choose model materials that match the transmission characteristics of your actual glazing. Alternatively, you could omit model glazing and decrease your internal measurement to account for transmission loss through the glazing. Transmission loss factors can be found in MEEB, Table 19.5, p. 797 and Flynn and Segil, Table 2-18, p. 127.

Joints between walls/floors/ceilings/etc., should be taped to avoid cracks of daylight that would hurt the accuracy of the test. However, consider how the probe of the light meter will be inserted, and how it will accurately be positioned at various measurement points. You may need an openable flap of some sort.

Note: Refer to Benjamin Evans, <u>Daylight in Architecture</u>, Chapter 6 for more complete information on daylighting models.

2. Preparing your documentation
a. Set up a grid on your plan with intersections at which the daylight measurements will be taken. Unless a space is very small, at least 15 measurements (spaced equally throughout the floor area) should be taken or the grid should be a 5' or 10' grid. The surface of the light meter's probe should be at the approximate level at which DF are desired - the "workplane", which is usually about 30" above the floor (to model scale, of course!). This may require adding some material below the probe to bring it to the proper height.

If you know the size of the light meter probe in advance, it could be advantageous to put stops on your model floor at each measurement point, to quickly and accurately position the probe for each measurement.

b. Prepare several floor plans, clearly showing the grid, and allowing space to record daylight measurements. You may need to make more than one set of measurements; one set should be made for each design alternative you wish to explore.

3. Taking measurements.
a. The most critical factor is the sky condition, which should match as closely as possible the condition for which you are daylighting:
<u>Overcast skies</u> should be truly overcast - the actual position of the sun should not be discernable. Avoid "partly cloudy" skies.
<u>Clear skies</u> can be a problem, because the sun position should be at an altitude/azimuth similar to the time of day in the climate for which you are designing. Do <u>not</u> tilt the model a la "sun peg" approach to achieve this altitude/azimuth, because your building's openings will "see" a wildly different portion of sky vs. ground as your model leaves a horizontal position.

Another critical factor is the constancy of the outside light, against which interior measurements are compared to obtain daylight factor. If you are fortunate to have a "ratiometric" meter, which has at least two probes (one outside, one inside) and gives you the D.F. ratio directly, you can simultaneously measure exterior levels as well as interior ones, to assure that your DF will be based on accurate comparative measurements. This can also be accomplished by using two light meters simultaneously, one inside the model and one in an open area for the exterior measurement.

A somewhat less critical factor is the ground reflectance. Set your model on a surface with texture and reflectance matching the "real" condition. For the first 10' height of opening above grade, the 40' of horizontal surface in front of your model is particularly critical in this regard. Likewise, each 10' height increase should be matched by another 40' width of controlled ground reflectance.
b. For each point, measure and record the daylight both inside the model at that point and simultaneously outside the model in a large open area (be careful you don't obstruct the measurement by leaning over the meter).

4. When you have completed measurements, calculate your daylight factor for each point

DF = (interior fc/exterior fc)x(total reduction factor)

Plot the daylight factor contours on your plan, and graph them in section.

For criterion A (Analysis tools and summary questions), from the Climate Exercise section 3.5, the daylight factors required in my climate for this space are:
_____ %

From Section 9.3, Criterion C, the minimum recommended D.F. for this space is: _____ %

Does your first trial as evaluated by model measurements meet this criteria?

If not, what will you change for your second trial?

Does your first trial design meet criterion B?

If not, what will you change for your second trial?

10.1 INTRODUCTION

From the approximate window sizes of the Climate exercise and the more precise daylight factor predictions of Section 9.4, you know how daylighting serves the tasks in a particular space, and how that varies by the time of day and by season. Now you propose a scheme for supplementary electric lighting which responds to these changes. This enables you to find the energy requirements (watts/sq.ft.) of your overall lighting design and understand the contribution daylighting is making in reducing electric lighting's energy requirements.

Contents

10.2 DAYLIGHT PERFORMANCE

GOAL

Minimize energy use for electric lighting.

CRITERIA

A. Recommended minimum foot candles of illumination are provided for tasks using daytime electric lighting equal to or less than 1 watt per sq.ft.

DESIGN

Present full documentation of the design which you are evaluating in this exercise. Include at least the following:

A. Floor plans of your space (display/conference or factory), with the significant task areas labelled.

B. Sections of your space.

C. Partial roof plan and elevations (or paralines) clearly showing all sides and the top of your space.

D. Reflected ceiling plan(s) of your space showing positions of lighting fixtures and exposed structure (if any).

See Salem example Figures 10-2 through 10-4.

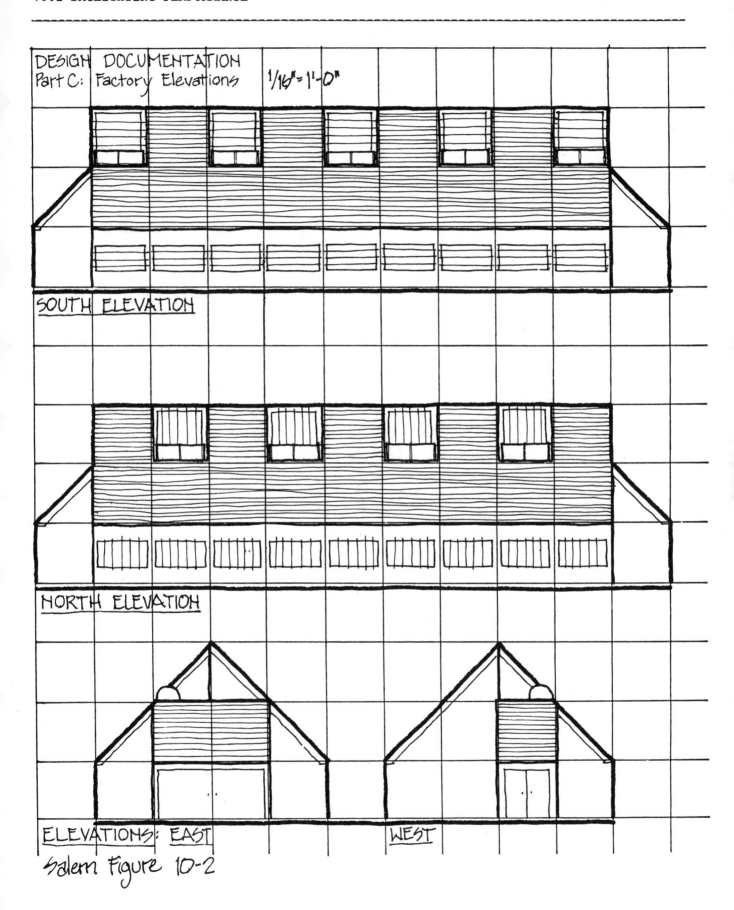

DESIGN DOCUMENTATION
Part C: Factory Elevations   1/16"=1'-0"

SOUTH ELEVATION

NORTH ELEVATION

ELEVATIONS: EAST   WEST

Salem Figure 10-2

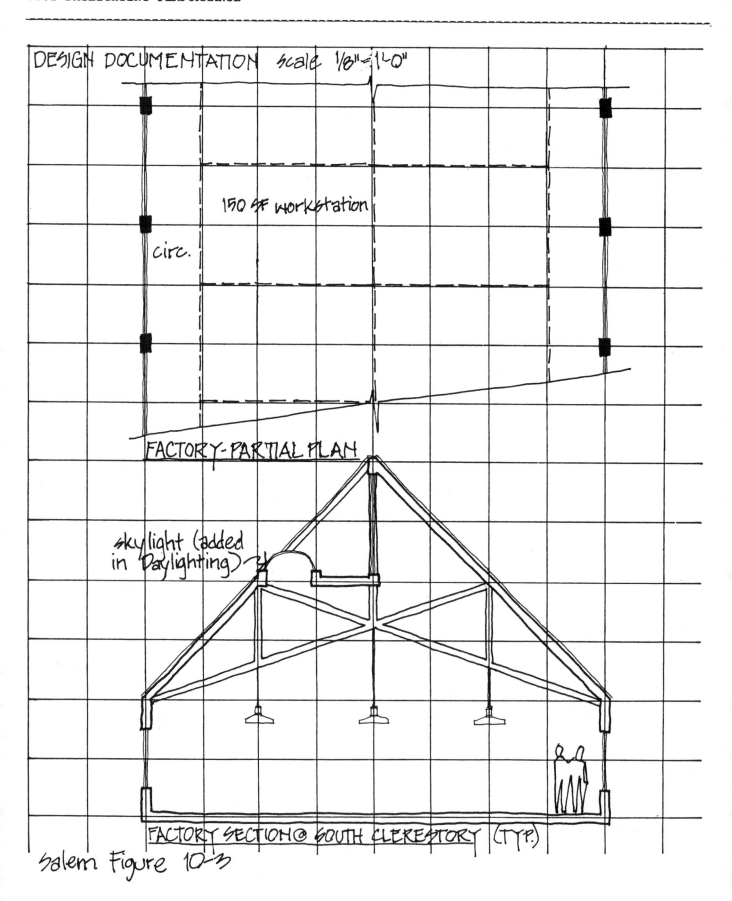

DESIGN DOCUMENTATION scale 1/8"=1'-0"

150 SF workstation

circ.

FACTORY-PARTIAL PLAN

skylight (added in Daylighting)

FACTORY SECTION @ SOUTH CLERESTORY (TYP.)

Salem Figure 10-3

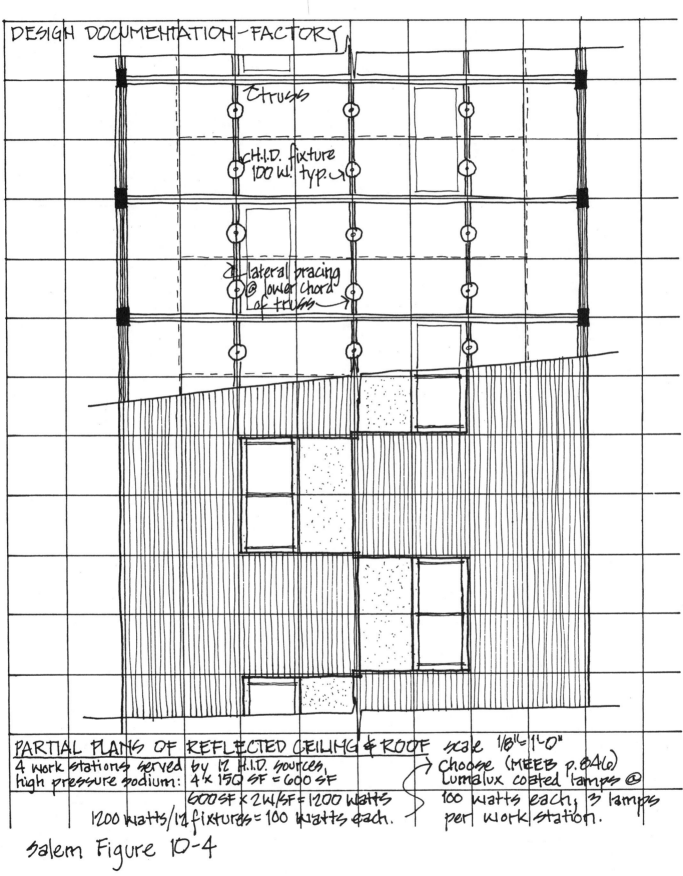

DESIGN DOCUMENTATION - FACTORY

truss

H.I.D. fixture
100 W. typ.

lateral bracing
@ lower chord
of truss

PARTIAL PLANS OF REFLECTED CEILING & ROOF   scale 1/8"=1'-0"
4 work stations served by 12 H.I.D. sources,   → choose (MEEB p.846)
high pressure sodium: 4 x 150 SF = 600 SF   Lumalux coated lamps @
                      600 SF x 2W/SF = 1200 watts   100 watts each, 3 lamps
    1200 watts/12 fixtures = 100 watts each.   per work station.

Salem Figure 10-4

for plans, sections, & elevations: if   1/16"=1'-0",   1 square = 10'.

for plans, sections, & elevations: if 1/16"=1'-0", 1 square = 10'.

SUPPLEMENTARY ELECTRIC LIGHTING
Criterion A, step 5: partial reflected ceiling plans @ 1/16"=1'0"

Condition A.   winter
                 7-8 am
                 4-5 pm

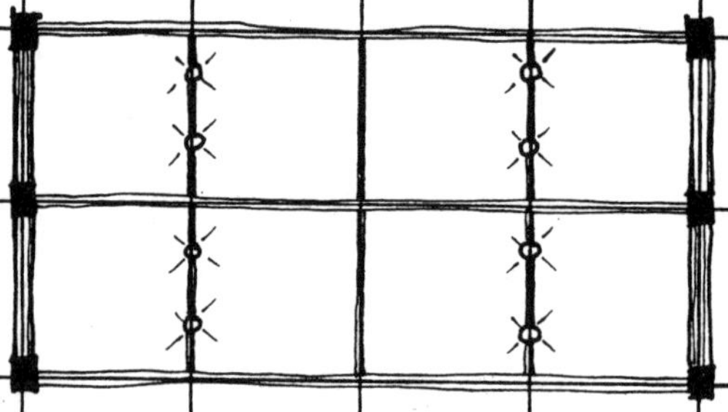

Condition B.   winter
                 8-9 am
                 3-4 pm
                 spring/fall
                 6-7 am
                 5-6 pm
                 summer
                 6-7 pm

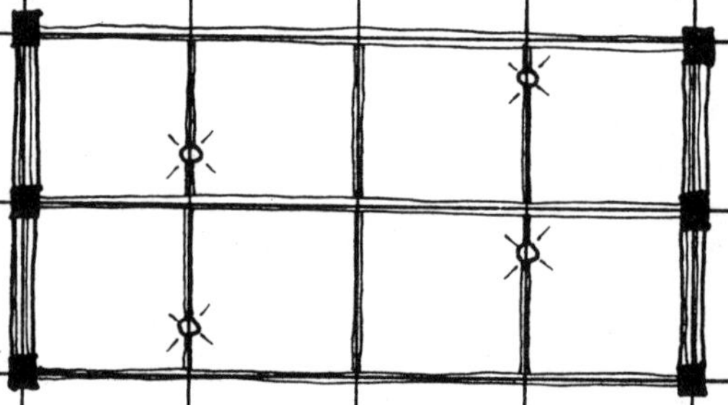

Condition C.   spring/fall
                 7-8 am
                 4-5 pm
                 summer
                 6-7 am
                 5-6 pm

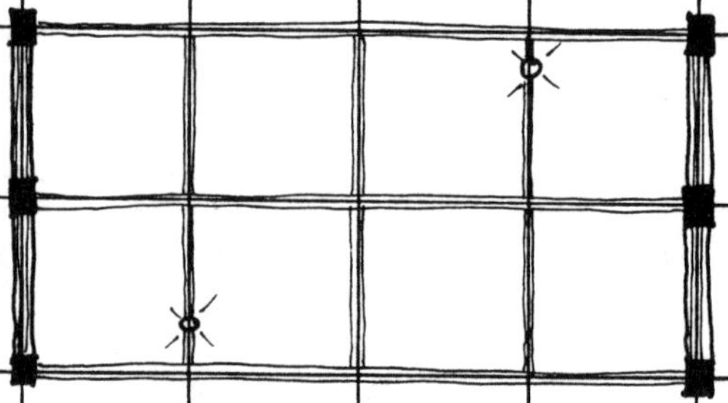

Salem Figure 10-5

_____

VALUATION

. **Criterion A**: Provide recommended minimum footcandles
f illumination for your tasks using daytime electric
ighting equal to or less than 1 watt per s.f.

**Evaluation Tool For Criterion A - "Specifying 100%
Electric System"**

. Show on your plans how you would employ a strategy of
task and ambient lighting for night conditions.
2. Complete this table to relate electric lighting
footcandle levels to watts/sq.ft. of installed lighting,
refer to MEEB Figure 20.2, p. 867.

Design Condition	Type of electric light source installed	Approx. fc to be provided by source	Task area to be lit, X sq.ft.	Watts/sq.ft. required for this fc	Total watts = of electric light
Night lighting	_HID_	50 WKsta. _30 circ_	150 WKsta. +94 circ	2 _1.5_	300 _+141_ watts
	High pressure sodium (30 is minimum pro-vided by daylight)			*MEEB, p. 867*	*Note:*

**Evaluation Tool for Criterion A - "Sizing the Daytime
Backup System"**

1. To size the supplementary daytime electric light
system, determine when and where daylight is less than
adequate by itself. Because the daylight factor is the
ratio between interior illumination and available light
outside, you can use it to determine actual illumination
levels (fc on working plane) from available daylight:

interior fc from daylight = DF x exterior fc available
daylight

Convert the daylight factors you found in Lighting
Exercise 9.4 p. 9-46 and 9-49 to footcandle levels for
the sky condition you are investigating. (Note that 1
footcandle = 1 foot lambert.)

Use the Equivalent Sky Brightness tables (MEEB Table
19.2a for overcast skies, 19.2c for clear skies where
direct sun is excluded from the window opening, p. 774-
776), to find the interior illumination of your space in
fc for minimum and maximum DF for seasons when sky
conditions you tested are applicable in your climate.

*in Section 2.0,
p. 2-1, we
allotted 500
watts of light per
work station.*

(See Salem Figure 9-25)

10.0 LIGHTING FINALIZATION

10.2 DAYLIGHTING PERFORMANCE

Note: For clear sky conditions, list the maximum and minimum fc for each hour's entry, for each window orientation (north, south, east, west).

Trial 1	Dec 21 Sky Condition *overcast* (Max *11*% DF)	(Min *8*% DF)	Mar/Sept 21 (1) Sky Condition *overcast* (Max *8*% DF)	(Min *6*% DF)	June 21 (2) Sky Condition *clear* (Max *6*% DF)	(Min *4*% D
8 am	11 fc	8 fc	59 fc	44 fc	N78, S81 fc	N52, S54 fc
10 am	66	48	107	80	75 126	50 84
noon	87	63	130	97	57 162	38 108
2 pm	66	48	107	80	75 126	50 84
4 pm	11	8	59	44	78 81	52 54

(1) Assume DF midway between Dec. and June, due to adjustable shading devices and internal-dominated load.

(2) Since DFs were calculated for overcast skies, the validity of these numbers is compromised.

Trial 2	Dec 21 Sky Condition (Max & DF)	(Min % DF)	Mar/Sept 21 Sky Condition (Max % DF)	(Min %DF)	June 21 Sky Condition (Max % DF)	Min % DF
8 am	___ fc	___ fc	___ fc	___ fc	___ fc	___ fc
10 am						
noon						
2 pm						
4 pm						

2. Compare these fc levels with the minimum recommended ones (See Climate Exercise section 3.4).

Task Area	Recommended Footcandles	Hours Occupied Begin:	Finish:	
Factory work station	50	6 am	10 pm	See p. S3-15 and S6-25.

3. Graph and label the following information on the chart below, to understand how the daylight you have compares to illumination required by the task.
a. Band of interior fc (min to max) for the chosen sky condition in all seasons applicable.
b. Line of recommended fc for each task area (for the hours required only)

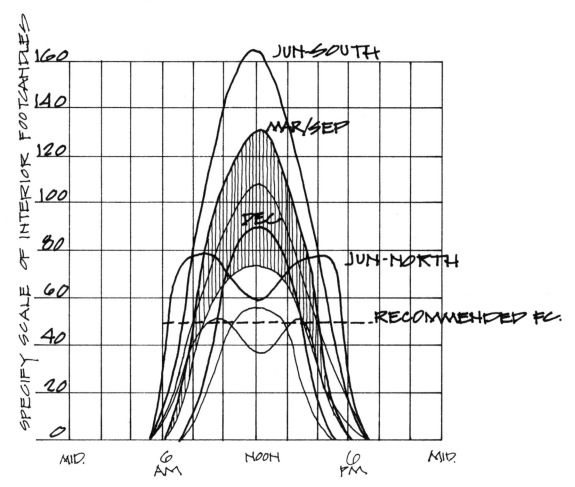

4. From the graph, determine <u>when and where</u> daylight, by itself, is less than adequate for each task. List the supplementary fc needed from electric lighting in order to meet the minimum light required for this task. Then compare this supplementary electric lighting level with the total electric lighting provided for night conditions.

Season & task	Time(s) that Daylight is available, but inadequate	Supplementary fc of electric light required	% of total installed electric lights needed to provide supplementary fc	Condition
Trial 1:			$(= \dfrac{25 \text{ ft. c. needed}}{50 \text{ ft. c. total electric provided}})$	
Winter	7am – 8am, 4pm – 5pm	25	50%	A
	8am – 9am, 3pm – 4pm	15	30%	B
Spring/ Fall	6am – 7am, 5pm – 6pm	20	40%	B
	7am – 8am, 4pm – 5pm	5	10%	C
Summer	6am – 7am, 5pm – 6pm	5	10%	C
	6pm – 7pm,	20	40%	B
Trial 2:				

Call Condition A = 2/3 of lamps
        B = 1/3 of lamps
        C = 1/6 of lamps

5. Finally, show on your plans which of your installed electric lights will be used under the various supplementary conditions. *(See Salem example Figure 10-5)*

Evaluation Tool for Criterion A - "% Savings from Daylighting"

1. From the information you listed in the two previous Evaluation Tools, complete this table:

Trial 1	Column 1 sf of elec light needed	X	Column 2 Watts/sq.ft. of electric light	X	Column 3 Hours per day of operation under these conditions	=	Column 4 Total watt hours electric light
Night: summer:	(whole factory) 2				3		23,400
winter:	3900		2		6		46,800
Least Daylight:	(Condition A; Salem Fig. 10-5)						
summer:	3900		2 x 2/3 = 1.3		0		
winter:	3900		1.3		2		10,140
Intermediate Daylight:	(Condition B; Salem Fig. 10-5)						
summer:	3900		2 x 1/3 = .7		1		2,730
winter:	3900		.7		2		5,460
Most Daylight:	(Condition C; Salem Fig. 10-5)						
summer:	3900		2 x 1/6 = .3		2		2,340
winter:	3900		.3		0		

Total
hrs/day operation   16

Total Watt hours   Summer: 28,470
per 24 hrs   Winter: 62,400

Trial 2	Column 1 sf of elec light needed X	Column 2 Watts/sq.ft. of electric X light	Column 3 Hours per day of operation under these conditions	Column 4 Total watt hours = electric light
Night: summer:				
winter:				
Least Daylight:				
summer:				
winter:				
Intermediate Daylight:				
summer:				
winter:				
Most Daylight:				
summer:				
winter:				

Total
hrs/day operation _____

Total Watt hours   summer:
per 24 hrs   winter:

2. Compare the actual internal gains that will result from electric lights with your estimate in the Climate exercise, section 3.6.

$$\frac{\text{total summer watt hours per 24 hours (Col. 4)}}{(\text{sq.ft.}) \ (24 \text{ hours})} = \text{watts per sq.ft.}$$

$$\frac{28,470}{3900 \times 24} = .3 \ watts/sq.ft.$$

Watts per sq. ft. = _0.3_

4.0 LIGHTING FINALIZATION

4.2 DAYLIGHTING PERFORMANCE

Same as Climate:

$$\frac{500\ watts}{150\ sq.ft.} = 3.33\ watts/sq.ft.$$

hat did you assume in the Cooling Exercise as the
atts/sq.ft. due to electric lighting?

(In Thermal Finalization, all electric lights assumed off during peak cooling hours; this exercise confirms that possibility.)

. To estimate the energy saved by daylighting compare
our total watts per 24 hours with what that would be if
o daylighting were admitted:

otal electric = Night sq.ft. x watts/sq.ft. x total hrs/day operation = _124,800_watt-hrs.
          (Col. 1)   (Col. 2)

$$3900 \times 2 \times 16 = 124,800$$

energy = 100-100 x (total watt hours per 24 hours, Col. 4) =
savings   total electric watt hours per 24 hours

ummer: $100-100 \times \dfrac{28,470}{124,800} = 77\%$ *savings*

inter: $100-100 \times \dfrac{62,400}{124,800} = 50\%$ *savings*

summer _77_ % Trial 1
winter _50_ %
summer ___ % Trial 2
winter ___ %

oes your first trial design meet Criterion A for
aylighting Performance? *Yes; well under 1 watt/sf for
lmost all daylight hours.*
ave you provided night time lighting not exceeding 2
watts/sq.ft.?   Have you provided daytime electric
ighting of less than 1 watt/sq.ft.?
es.
If not, what will you change for your second trial?

10.3 BACKUP SYSTEM INFORMATION

In finalizing you calculated the % electrical energy
savings due to daylighting for a given sky condition and
season.   This information can be expanded in several
ways.

You can model the predominant sky conditions for all
seasons in your climate, thereby being able to calculate
the electrical energy savings on an annual basis rather
than a single condition.

Reducing the use of electric lighting also means a
reduced cooling load (1 watt = 3.41 Btu's), which will
affect the sizing of your passive cooling design
components, as well as any mechanical cooling system.

Design of electric lighting systems involves many more
issues than watts per sq.ft.   We have included some

Electric light sources have widely varying color rendition characteristics. See MEEB Table 19.26, p. 853; and Fig. 18.56 p. 768 and Fig. 18.52 p. 764.

Lamp Type	Application
Incandescent	1. Decorative display lighting 2. Religious worship halls 3. Work closets or other very confined spaces 4. Stage spotlighting 5. Tasks which require a small light source
Fluorescent	1. Office and other relatively low-ceiling applications 2. Flashing advertising signs 3. Islands at service stations 4. Display cases in stores 5. Desk lamps 6. Classrooms or training centers 7. Cafeterias
High-intensity discharge	1. Stores and some office areas 2. Auditoriums 3. Outdoor area lighting 4. Outdoor floodlighting 5. Outdoor building security lighting 6. Marking of obstructions

Fig. 10-1   Examples of Energy-Conserving Lamp Applications from Energy Conservation Standards by Dubin and Long. Copyright (c) 1978. Used with the permission of McGraw-Hill Book Company.

See also William Lam, Perception and Lighting as Formgivers for Architecture, 1977.

Finally, a valuable graphic tool for electric lighting design is to render a night perspective of your design (interior and exterior).

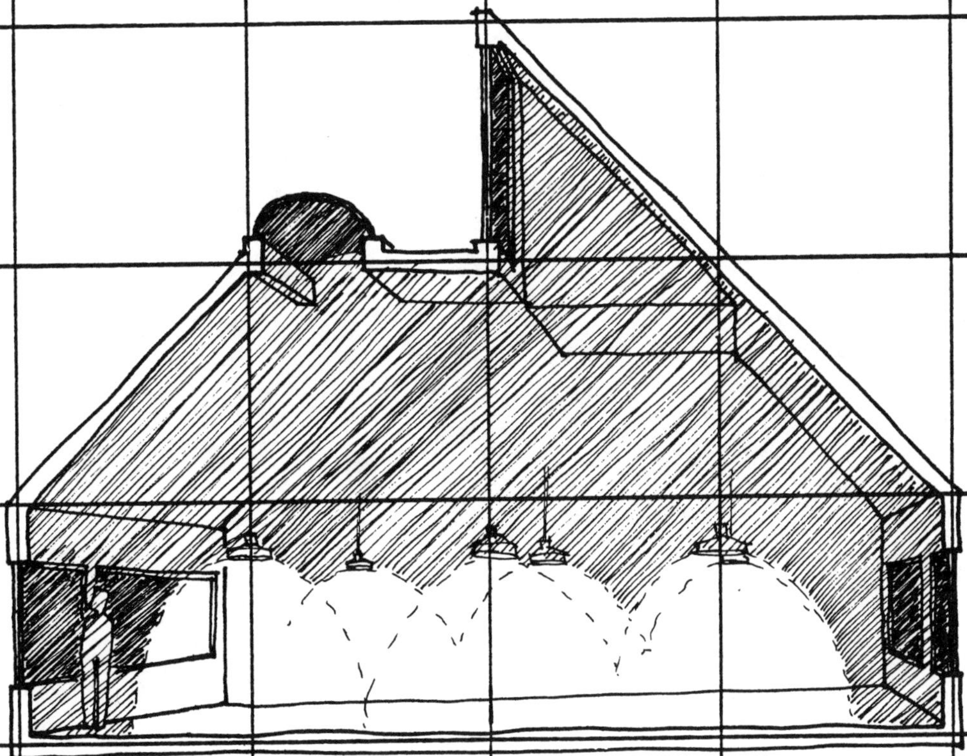

NIGHT LIGHTING (TYP.)

Salem Figure 10-6

12.1 INTRODUCTION _____

Just as you set the context for your design with climate analysis for the thermal exercises, some preliminary analysis of the water and waste context must be done.

You start by exploring the water supply and the soil conditions for your location, the hot water usage and minimum number of fixtures for your building program, and resulting total water use for conventional fixtures.

You then select the one most appropriate combination of fixtures for the conditions you have analyzed.

Contents

--$12-$

12.2 ANALYZING ---

GOAL

Determine the importance and feasibility of conserving
water in your site/building design.

A. Analysis Tool - "Annual Rainfall Comparisons"
Your annual rainfall (from climate data) is _41_ inches/ *Just a bit on the heavy side.*
year. From MEEB, Fig. 2.20, p. 51, determine whether
your annual rainfall seems unusually low, about average,
or unusually heavy.

B. Analysis Tool - "Soil Characteristics"
Soil conditions are a major influence on the required
size of drainage fields. Based on generalizations about
regional soil types assume these site soil
characteristics:

----

ocation	Soil Type	Time Required for Water Level to Drop 1" in a Test Hole:
adison odge City alem	Medium to good drainage; Sandy loam or sandy clay	5 minutes
noenix	Excellent drainage; Coarse sand or gravel	1 minute
narleston	Medium to poor drainage; Clay with large amounts of sand or gravel	10 minutes

Salem: About average suitability for septic tanks.

igure 12-1 Soil Characteristics

. Analysis Tool - "Required Minimum Plumbing Fixtures"
he population of your buildings, combined with their
unctions, determines the minimum number of plumbing
ixtures required in toilet rooms (see 4.6 Analysis Tool
. for population).

actory population ___16_

)ffice population ___37_ *(at most)*

otal   ___53_

len and women are equally likely to be hired as office
and factory workers here, so up to two-thirds of this
cotal should be able to be accomodated by either sex's
coilet room. Use MEEB, table 11.8, p. 459-461 to determine minimum number of fixtures.

53 x 2/3 = about 35
Assume a potential population of 35 males and 35 females.

	men	women	both
water closets	_2	_2	_4
urinals	*not required, but could be additional to WC's*		
lavatories	_2	_2	_4
drinking fountains	___	___	_1
vending machines	___	___	___
showers	___	___	___
other: slop sink	___	___	_1

Since most of these people are short-term (conference/display generates 33 people...), assume "public buildings, offices, etc." category.

Our employees are considered not exposed to the excessive heat or other occupational hazards listed...

D. Analysis Tool - "Conventional Fixture Water Usage Estimate"

Determine the average consumption per person assuming conventional fixtures. Use the population determined above, multiplied by the average gallons per person. If your building has use characteristics which indicate a greater need for fixtures than this procedure would lead you to believe then you should design for a larger number. Use MEEB, table 11.9, p. 466 to determine gallons per person.

	Gallons per person per day	Population	Total Gallons per day
Office	5	37	185
Factory	25	16x2 shifts	900
		TOTAL	1085

"Stores" is the closest logical category...

QUESTIONS

You now have an idea how much water is available in your location (input), how much water might be used, and how effectively waste may be processed on your site (output). Use this information to answer the following questions.

A. Viewed from the standpoint of availability, is water conservation important in your location?

Not yet (but California has its eyes on the Columbia, so...)

B. Viewed from the standpoint of conventional waste treatment on site (septic tank and drainfield), is water conservation important in your location?

Yes. The septic field will be fairly large, and might constrain site planning, especially considering possible expansion.

C. How does your need to conserve water compare to the other locations?

About average - neither critical, nor generous, water-waste relationship.

E. Analysis Tool - "Water and Waste Flow Diagrams"

The flows in a conventional system <u>using 1000 gal. per day</u> might look like this:

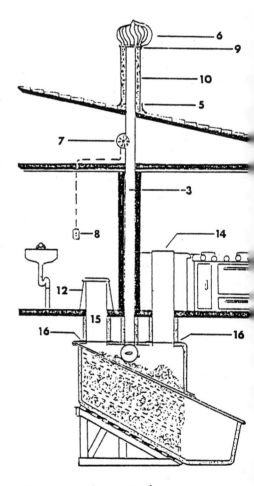

Compost toilet.

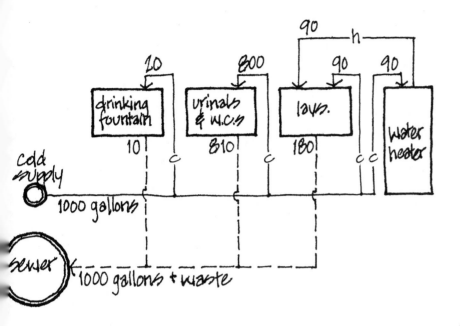

Compost toilets are a good idea, but require a basement, which isn't cheap. Conservation flush valve WC's are chosen, because they don't stop up as easily as tank types. If water availability were a greater issue in Salem, we'd go to tanks and keep a plunger handy.

Figure 12-2 Typical Conventional System Flows

A septic tank system should conserve water compared to a conventional system. Assemble a system which achieves what you feel is an appropriate unit of conservation for your area by selecting fixtures from the following list, and putting them together in a diagramatic system for your building:

Conventional:	Supply fixture units (MEEB p. 415)	Gals/day* For this building group
flush valve water closet	10	200
flush tank water closet	5	100
flush valve urinal	5	100
flush tank urinal	3	60
lavatory	2	40
shower	4	80
service sink	3	60

Conserving:		
conserving wc, valve	--	120
conserving wc, tank	--	60
conserving urinal, valve	--	40
conserving urinal, tank	--	24
compost toilet	--	0
conserving lavatory	--	16

		Gallons/Day	
#	Fixture	Convent.	Conserv.
4	WC's, valve	800	480
4	Lavator.	160	64
1	Service sink	60	60
1	Urinal, tank	60	24
	TOTAL	1080	628

Figure 12-3 Typical Plumbing Fixtures

For examples of alternative waste and water systems see:
MEEB Chapter 2(p. 50-52) and Chapter 9.
B. Vale, The Autonomous House
C. Stoner, ed., Goodbye to the Flush Toilet
M. Milne, Residential Water Conservation (this is a good
general review of water conservation components)
J. Lackie, et.al., Other Homes & Garbage, Chapter 5 -
Waste; Chapter 6 - Water.

*Note: The total gallons per day (1,000 in the above
diagram) was divided by the total supply fixture units to
get a general indication of the gallons per day per
fixture unit. This assumes that water supply fixture
units are a reasonable approximation of water demand
(they consider use patterns and flow rates) for each
fixture. Conserving fixture gallonages were determined
as follows:

WC: $\dfrac{\text{conserving model}\ 3\ \text{gal/flush}}{\text{conventional}\qquad 5\ \text{gal/flush}}$ = 0.6 conventional usage

Urinals: $\dfrac{\text{conserving model}\qquad 1\ \text{gal flush}}{\text{conventional}\qquad 2.5\ \text{gal flush}}$ = 0.4 conventional usage

Lavatories: $\dfrac{\text{conserving model}\ 2\ \text{gpm flow rate}}{\text{conventional}\qquad 4.5\ \text{gpm flow rate}}$ = 0.4 conventional usage

----

QUESTIONS

A. What percent reduction of water use did you achieve compared to a conventional system?

$\frac{628}{1080}$ = 58% of conventional system, or 100 - 58 = 42%—reduction.

B. Compared to the conventional system what percentage of your usage was the following:

conventional inputs		your system
rain	0	0
city main	100%	100
other	0	0

outputs		
grey water	0	0
black water	100%	100
city sewer	100%	septic tank
other	0	0

12.3 CONCEPTUALIZING --

GOAL

A. Use rainwater where possible.
B. Decrease the use of fresh (potable) water and the production of waste water.

DESIGN STRATEGIES

A. Site Scale
Minimize rain run off to encourage soil recharging.
B. Cluster Scale
Water moving downhill requires no pumping energy or pressurized piping; store water at higher elevations than it is used, and treat waste at still lower elevation.

--S12--

C. Building Scale

"Service" spaces, which include toilet and mechanical rooms, can help organize the clustering of buildings and the other spaces they serve.

Toilet spaces have special environmental characteristics, which usually suggest acoustic isolation from other spaces, and a ventilation pattern unshared by most other spaces.

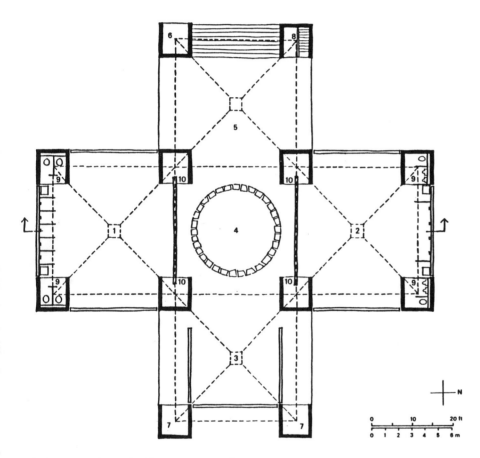

Kahn, Trenton Bath House. (Ronner, 1977, p. 94)

D. Component Scale

Plumbing fixtures are often grouped together within buildings for cost savings in plumbing, and for convenience in clustering environments containing water usage.

DESIGN

Choose an existing building/site which has a clear conceptual approach, at any scale, to achieving these goals. Document your choice as follows:

A. identify the location, program, architect (if any) and the source of your information.
B. Include xerox copies or drawings (whatever is quick and easy for you) to explain the design.

EVALUATION

Evaluation Tool - "Building Response Diagrams"
Diagram how this design is organized to achieve these goals.

This more general part is not included in the Salem, Oregon example.

12.4 SCHEMING

GOALS

A. Use the site and buildings to the degree possible to heat water, collect rainwater and dispose of wastes.
B. Arrange the elements of the water/waste system within the cluster and buildings to minimize plumbing.

CRITERIA

A. Provide enough solar collectors to heat at least 50% of the hot water used in your buildings.
B. Provide adequate collection area and water storage for your buildings' needs. (If you elect to use this method)
C. Size the septic tank and drainfield to adequately treat the waste generated in your building considering your soil conditions.
D. Provide enough composting toilets to meet your buildings' needs.

DESIGN STRATEGIES

A. Site Scale
Locate the drainfield and septic tank at a lower elevation than the building.

Utilize the naturally irrigated area over the drainfield as a flat unpaved open area for view or active use (excluding cars, etc.).

Provide adequate clearances to other site elements.

Minimum Horizontal Distance	Building Sewer	Septic Tank	Disposal Field	Seepage Pit or Cesspool
Buildings or Structures	2 ft (.6 m)	5 ft (1.5 m)	8 ft (2.4 m)	8 ft (2.4 m)
Property line adjoining private property	Clear	5 ft (1.5 m)	5 ft (1.5 m)	8 ft (2.4 m)
Water supply wells	50 ft (15.2 m)	50 ft (15.2 m)	100 ft (30.5 m)	150 ft (45.7 m
Streams	50 ft (15.2 m)	50 ft (15.2 m)	50 ft (15.2 m)	100 ft (30.5 m
Large trees	---	10 ft (3 m)	---	10 ft (3 m)
Seepage pits or cesspools	---	5 ft (1.5 m)	5 ft (1.5 m)	12 ft (3.7 m)
Disposal field	---	5 ft (1.5 m)	4 ft (1.2 m)	5 ft (1.5 m)
Domestic water line	1 foot (.3 m)	5 ft (1.5 m)	5 ft (1.5 m)	5 ft (1.5 m)
Distribution box	---	---	5 ft (1.5 m)	5 ft (1.5 m)

Figure 12-4 Location of Sewage Disposal System, table
I-1, p. 186, Uniform Plumbing Code.

Drainfields may not be used for parking, and they should
be free from trees and shrubs because roots may block the
lines.

B. Cluster Scale
Where rainfall is to be collected and stored for use,
consider arranging roof slopes so that the rain water
falling on the roof readily converges upon a single
location, above the cistern inlet.

Celebrate the flow of water from collection area to
storage area.

Group the things that use water in the same area.

C. Building Scale
If the domestic hot water solar collectors aren't integrated with the roof, consider using them to provide
the roof with an exterior shaded area.

D. Component Scale
Provide the collectors with an unshaded south facing
area. Set them at an angle above horizontal equal to or
less than your latitude.

DESIGN

Document your design as follows:

A. Select and circle the water and waste system you will
use in your buildings/site.

INSIDEOUT

----

supply < $\begin{matrix} \text{city} \\ \text{site} \end{matrix}$ <well

cistern & collection area

waste < $\begin{matrix} \text{city} \\ \text{site} \end{matrix}$ < septic tank & drainfield

composting toilets

Supply: city
(but size cistern just for curiosity)
Waste: septic tank and drainfield

B. Site or cluster plan showing parking, drives and depending on what system you are using septic tank, drainfield, cistern, well, supply and waste line from city lines, rain collection areas.
C. Floor plans showing location of hot water storage toilet rooms, and location of plumbing lines.
D. Building section(s) showing solar collectors, hot water storage, location of plumbing lines, and if you have them, cisterns and composting toilets.

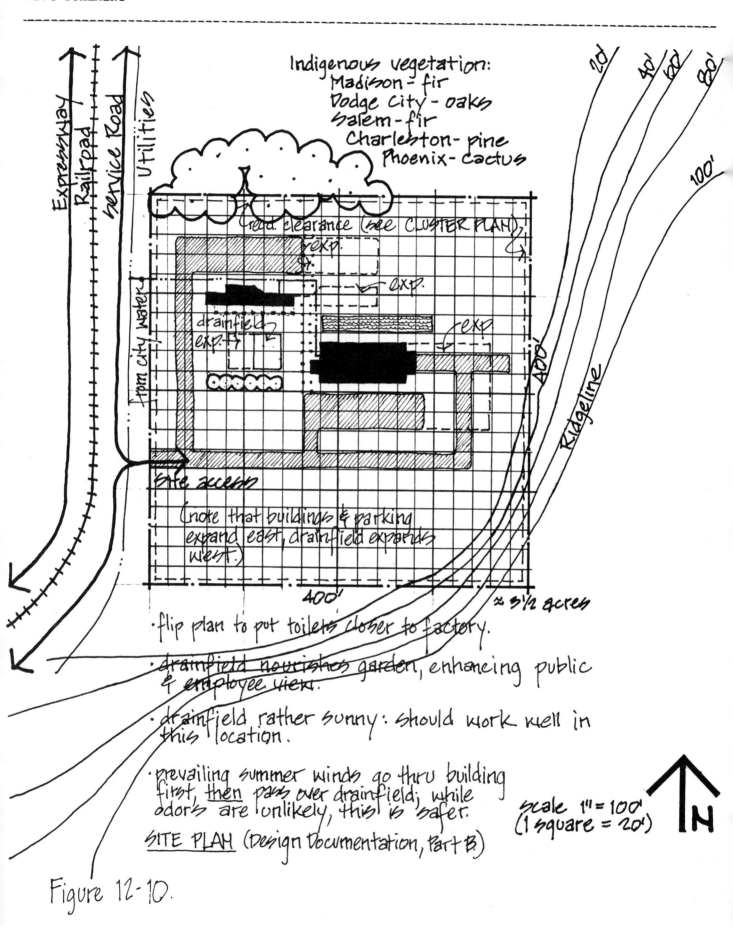

Indigenous vegetation:
Madison - fir
Dodge City - oaks
Salem - fir
Charleston - pine
Phoenix - cactus

road clearance (see CLUSTER PLAN)

exp.

exp.

exp.

drainfield exp.

Site access

(note that buildings & parking expand east, drainfield expands west.)

400'

≈ 3½ acres

· flip plan to put toilets closer to factory.

· drainfield nourishes garden, enhancing public & employee views.

· drainfield rather sunny: should work well in this location.

· prevailing summer winds go thru building first, then pass over drainfield; while odors are unlikely, this is safer.

SITE PLAN (Design Documentation, Part B)

scale 1" = 100'
(1 square = 20')

Figure 12-10.

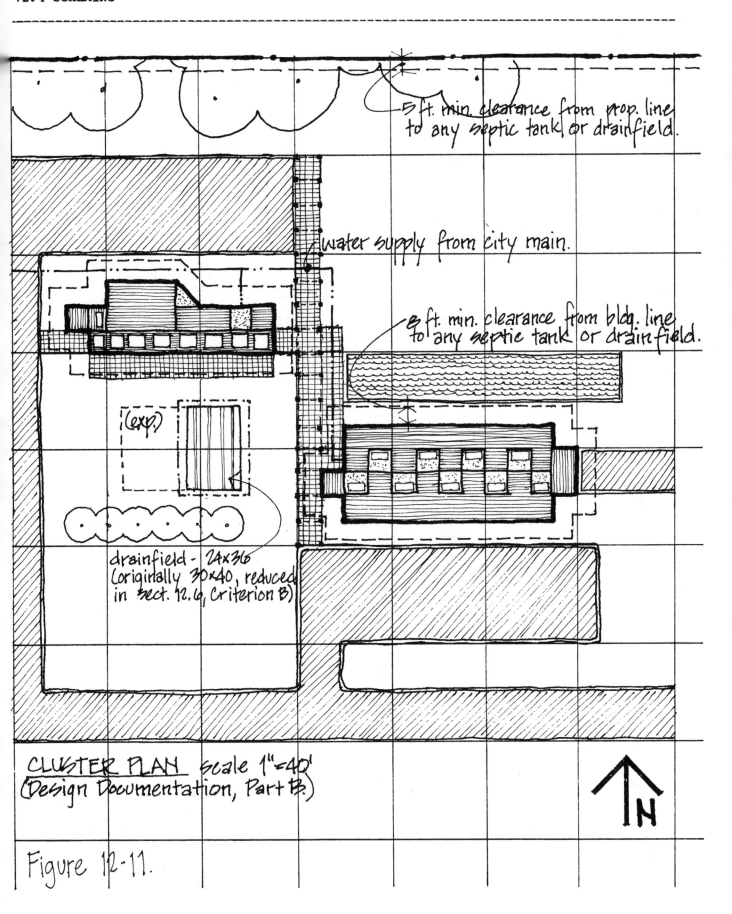

5 ft. min. clearance from prop. line to any septic tank or drainfield.

water supply from city main.

5 ft. min. clearance from bldg. line to any septic tank or drainfield.

(exp.)

drainfield - 24x36 (originally 30x40, reduced in sect. 12.6, Criterion B)

CLUSTER PLAN scale 1"=40' (Design Documentation, Part B.)

N

Figure 12-11.

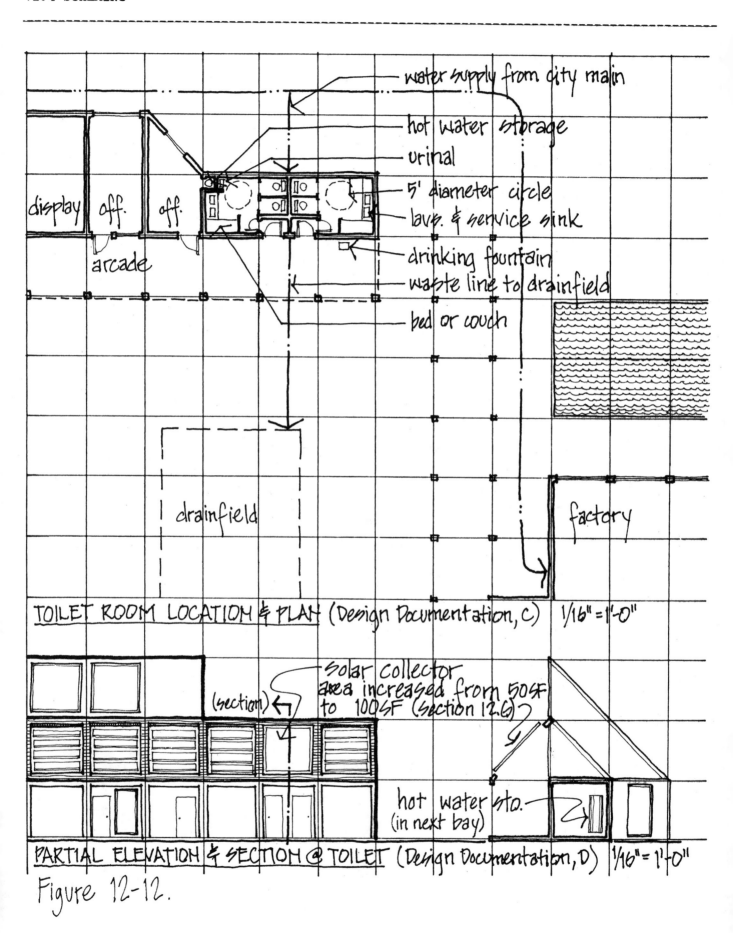

water supply from city main

hot water storage
urinal
5' diameter circle
lavs. & service sink
drinking fountain
waste line to drainfield
bed or couch

display off. off.

arcade

drainfield

factory

TOILET ROOM LOCATION & PLAN (Design Documentation, C)   1/16"=1'-0"

(section)

Solar collector area increased from 50SF to 100SF (section 12.6)

hot water sto. (in next bay)

PARTIAL ELEVATION & SECTION @ TOILET (Design Documentation, D)   1/16"=1'-0"

Figure 12-12.

for plans, sections, & elevations :   if   1/16"=1'-0",   1 square = 10'.

----

for plans, sections, & elevations: if 1/16"=1'-0",   1 square = 10'.

for plans, sections, & elevations: if   1/16"=1'-0",   1 square = 10'.

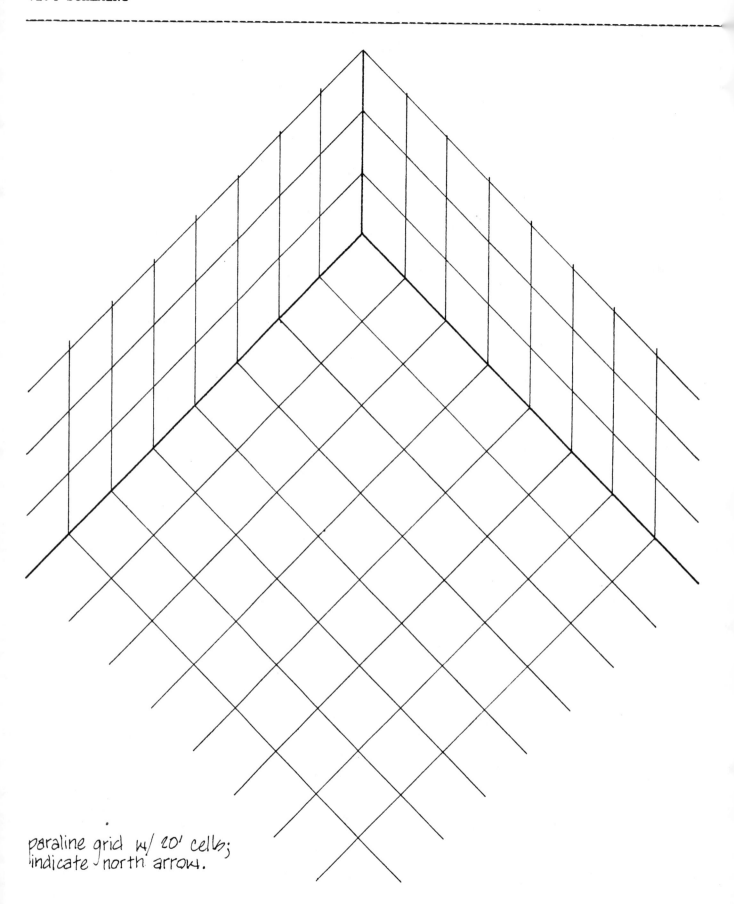

paraline grid w/ 20' cells;
indicate north arrow.

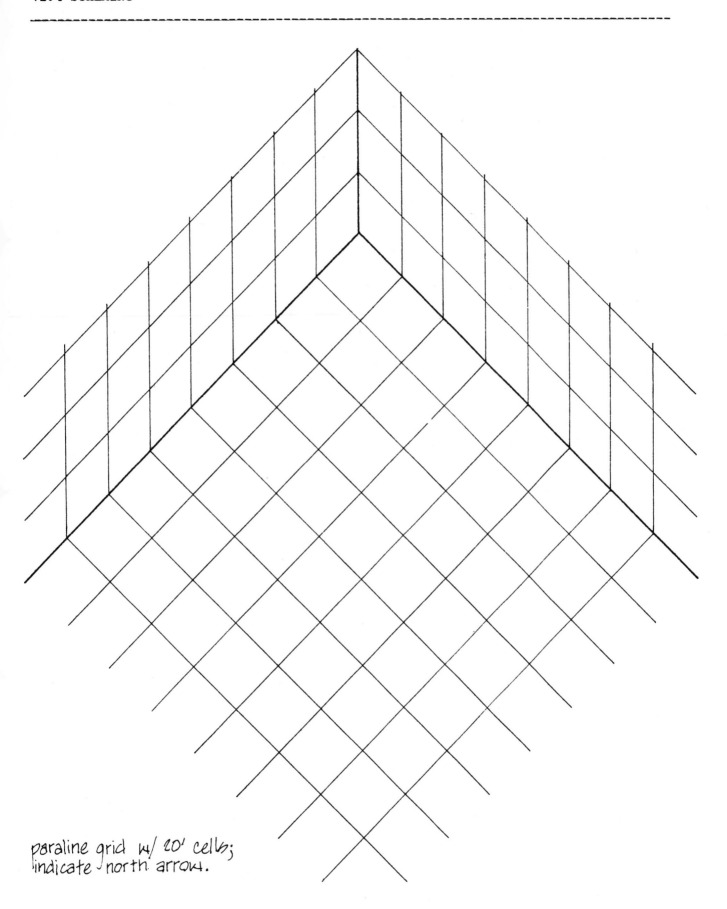

paraline grid w/ 20' cells;
indicate north arrow.

EVALUATION

A. Criterion A: Provide enough solar collectors to heat at least 50% of the hot water used in your buildings in the month with the most solar radiation.

Evaluation Tool for Criterion A - "Solar Collector Sizing"

sf of collector per 100 employees*

	sf of collector per 100 employees*	
Phoenix	30 to 45	
Charleston	42 to 63	
Salem	50 to 75	*...50 s.f. in our building.*
Madison	50 to 75	
Dodge City	36 to 54	

Figure 12-5 Estimated Solar Collector Sizing
*Based on 50 to 75 sf for Salem; adjusted collector area by a percent equal to the increase in the annual radiation ona horizontal surface for the other locations. This neglects ambient temperature and actual collector tilt. Thus, a very rough rule of thumb!

Does your first trial meet the design criteria?   *Yes, minimally.*

If not, what will you change for your second trial?

B. Criterion B: Provide adequate collection area and water storage for your buildings' needs.

Evaluation Tool for Criteria B - "Cistern Sizing"
Monthly usage = __630__ gallons/day x 31 days = ___19,500___ gallons/month.

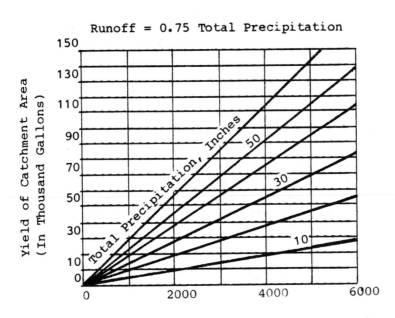

Runoff = 0.75 Total Precipitation

Horizontal Area of Catchment (In Square Feet)

Figure 12-6   Rainfall Catchment Area Yield (Lehr, et al
Domestic Water Treatment, p. 46)

Determine the average yield of the catchment area
(= approximate ground floor area) for each month.

Aver. inches rain		Gallons collected
Jan	6.9	15,900
Feb	4.8	11,000
March	4.3	9,900
April	2.3	5,300
May	2.1	4,800
June	1.4	3,200
July	.4	900
August	.6	1,200
Sept	1.5	3,400
Oct	4.0	9,200
Nov	6.1	14,000
Dec	6.9	15,900

*Assume entire roof of both
buildings (excluding arcade) is
available for catchment.
Horizontal area of these roofs:
Office: 1250
Factory: 3900
Total = 5150*

Salem, Oregon annual = 41 in.

*Total annual, about 102,000
gallons.
(Salem climate data indicates
maximum monthly rain is 15
inches, which yield 34,500
gallons.)*

---

cistern size = (sum of average consumption for months
(in gallons) in which consumption exceeds catchment yield) − (yield from catchment area in those months)

When the collectable rainfall from the roof <u>exceeds</u> the consumption on a monthly basis store water in a cistern from the excess capacity months for the below capacity months.

Doesn't happen in our average months.

When collectable rainfall from the roofs doesn't usually exceed consumption for most months, store water in a cistern which is typical of a maximum one month rainfall. Maximum one month rainfall catchment yield = cistern size in gal.

20,000 gallon cistern would seem fine. In a really rainy month, 34,500 collected − 19,500 used = 15,000 left in storage.

Does your first trial meet the criteria? *Not if we wanted full water needs to be met from rain alone.*
If not, what will you change for your second trial?
If rain were to be used, we'd need much greater catchment area, and a larger cistern.

C. Criterion C: Considering your soil conditions, size the septic tank drainfield to adequately treat the waste generated in your building.

Evaluation Tool for Criterion C – "Septic Tank and Drain Field Size"
To determine the square footage of drainfield required, multiply the number of fixtures determined in analysis by:

34 for coarse sand and gravel
43 fine sand
68 sandy loam or sandy clay
102 clay with considerable sand or gravel
153 clay with small amount of sand or gravel

Salem, sandy loam: 68 x 10 fixtures – 680 s.f.

(In poor soils, a stand-by drainage field "reserve" is sometimes required, equal in size to the original field.)

The number of fixtures multiplied by 170 equals the minimum septic tank capacity in gallons.

10 fixtures x 170 = 1,700 gal.

Does your first trial design meet the criteria?

If not, what will you change for your second trial?

Drainfield approx. 30 x 40 = 1200 s.f., more than adequate.

Note: Derivation of the Septic Field Rule of Thumb is based on the Uniform Plumbing Code 1976 edition, table I-2, p. 187, table I-4, p. 179, and table 4-1, p. 38. Fixture units were determined by multiplying the number of fixtures by 4, which is an average number between lavs at 2 and valve WC's at 6. Codes seem to require about an equal number of lavs and WC's for most buildings.

Fixture units were converted to minimum septic tank capacity in gallons by multiplying by 42.5. 42.5 is the average of 50 (which is the minimum septic tank capacity in gallons, 750, divided by the maximum fixture units served, 15, from table I-2) and 35 (which is the minimum septic tank capacity in gallons, 3500, divided by the maximum fixture units served 100).

.425 was divided by 100 to convert septic tank capacity into sf of leaching area per 100 gallons of septic capacity. See table I-4.

20 sf of leaching area per 100 gallons of tank capacity was used for porous soils and 90 sf of leaching area per 100 gallons of tank capacity for impervious soils. See table I-4.

Therefore:

Fixtures x 4 (av. fixture units) x 42.5 (conversion to septic tank capacity) ÷ 100 (conversion to leaching area in S.F.) x 20 for poor soils or 90 for good soils.

    4 x 42.5 ÷ 100 x 20 = 34
    4 x 42.5 ÷ 100 x 90 = 153

fixtures x 34 in good soil = sf leaching field
         x 153 in poor soil = sf leaching field

D. Criterion D: Provide enough composting toilets to meet your buildings' need.
Evaluation Tool for Criteria D - "Compost Toilet Capacity."
3 to 10 persons per unit in a residential application.
(Stoner, Goodbye to the Flush Toilet, p. 132)

If we were using composting toilets, this would suggest that the longer term employees (32 in 2 shifts at the factory, and 4 in the office building) be treated as equivalent to a "residential" application. Thus, four composting toilets should serve the 36 long term employees, with some capacity left for short-term visitors.

Evaluation Tool for Criteria D - "Removing Compost"
Draw a diagramatic section through the toilet room
showing how compost is removed and how access to the
removal area is achieved:

Did your first trial meet the design criteria?

If not what will you change for the second trial?

(Not using composting toilets)

12.5 DEVELOPING

GOAL

Design toilet facilities which are both adequately sized
and comfortable for all users, including the handicapped.

CRITERIA

A. Your design meets handicapped requirements of the 1979
Uniform Building Code.
B. Vision into either toilet room is blocked beyond the
entry way.

---S12-

DESIGN STRATEGIES

A. Component Scale
Clear space within a toilet room should be sufficient to inscribe a circle of 5' diameter. Put fixtures back to back to reduce the length of pipes and increase ease of installation.

DESIGN

Develop the design of your toilet rooms to achieve the goal. Document your design as follows:

Floor plan of toilet rooms @ 1/4" = 1'-0"

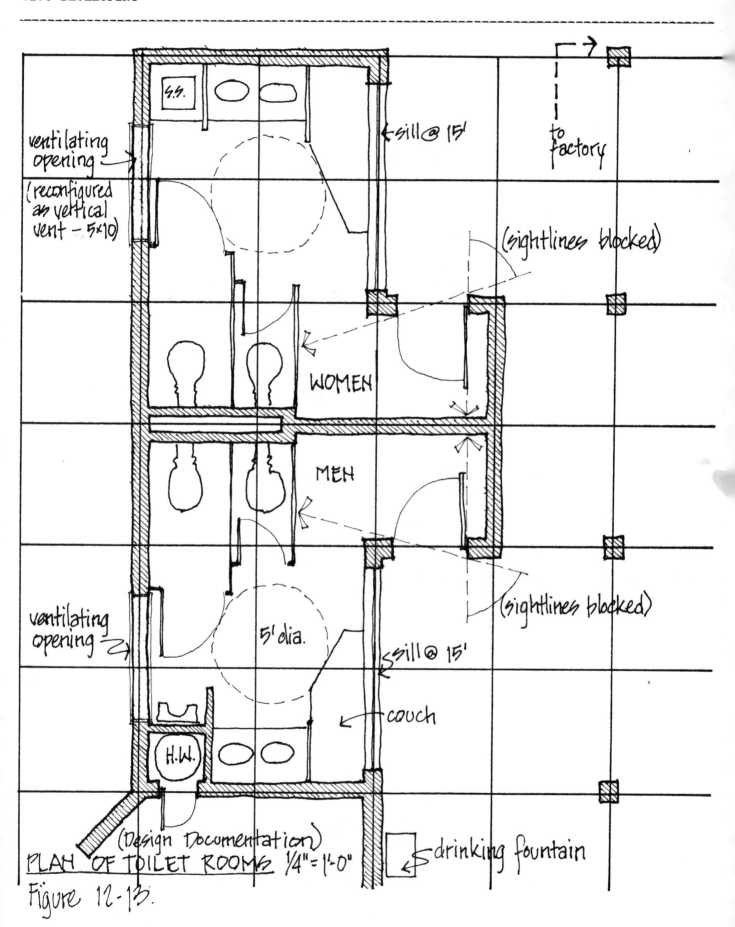

ventilating opening

(reconfigured as vertical vent - 5×10)

← sill @ 15'

to factory

(sightlines blocked)

WOMEN

MEN

ventilating opening →

5' dia.

sill @ 15'

couch

(sightlines blocked)

(Design Documentation)
PLAN OF TOILET ROOMS 1/4"=1'-0"

Figure 12-13.

drinking fountain

EVALUATION

A. **Criterion A:** Meets handicapped requirements for the 1979 Uniform Building Codes.

Evaluation Tool for Criteria A - "Code Review"
The 1979 Uniform Building Code specifies (Section 1811):
Doorways to toilet rooms: 30" door width, with 44" clear beyond each side of door.
Clear space within toilet room: sufficient to inscribe a circle of 5' diameter. Doors may encroach this circle by a maximum of 1' when open.
WC stall: one for each sex must have 42" wide stall, with 48" clear in front of stool within stall. (Stall door cannot encroach on this space, so it usually swings outward.)
A minimum of 44" clear access is needed in front of the stall (into which door may swing).
Lavatory: under one lavatory, a space 26" wide, 27" high and 12" deep.

(Section 12.4's plan was fine-tuned to provide specific stall dimensions.)

Does your first trial design meet the criteria?

If not, what will you change for your second trial?

B. **Criterion B:** Vision into either toilet room is blocked behind the entry way.

Evaluation Tool for Criterion B - "Sight Lines"
Draw sight lines into your toilet rooms as if the doors were not in place. Show the result on the 1/4"=1'-0 plan.

Does your first trial design meet criterion B?

If not, what will you change for your second trial?

(Section 12.4's plan did not meet criteria)
Section 12.4's plan expanded to block vision through doorways.

12.6 FINALIZING

GOAL

Insure that the major parts of your systems are accurately sized so that they perform well and conserve water.

CRITERIA

A. Supply and waste pipes are adequately sized to carry their expected loads.
B. Drainfield and septic tank are adequately sized to carry the expected loads of the conserving system.
C. Solar collectors should be sized tp supply 100% of the hot water requirement in the month with the most solar radiation.
D. The cistern is sized to meet the needs of the conserving system.

DESIGN

Draw a schematic diagram of your water and waste system showing the major elements and the flow of water through the system in gallons.

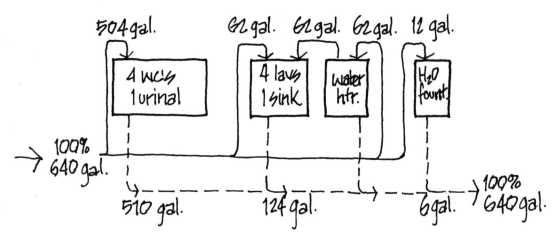

SYSTEM SCHEMATIC
Figure 12-14.

`VALUATIONS

   Criterion A: Supply and waste pipes are adequately
.zed to carry their expected loads.

`valuation Tool for Criterion A - "Supply Pipe Sizing"
`or the water/waste system you have chosen, size the
`table water supply system and the waste lines to the
`eptic tank (or filter).

`upply: (MEEB, Example 10.4, p.413 may be helpful)

`iven: main (or pump) pressure 40 psi
       main location: at street 6' below grade
       (pump location: submersible, in cistern)
       total equivalent length of pipe = 150% actual
       length

`. Estimate the diameter of the supply pipe __1"_ (to   *About 150 feet.*
`evelop your ability to guesstimate)
                                                              4wc x 10 = 40
`. Total actual length of pipe, main (or cistern) to   *1 urinal (tank) = 3*
`urthest fixture: *4 lav x 2 = 8*
                                                              1 service sink = 3
                                    Trial 1   Trial 2   *1 drink. fount. = 1*
`. Determine the supply fixture units   *55 F.U.*
`MEEB, Table 10.10)   _55_   _55_
`. Determine flow required gal/min   *In a valve-dominated*
`MEEB, Fig. 10.27)   _53_   _53_   *system, 55 F.U. con-*
`. Pressure available to overcome friction   *verts to 53 gallons*
`n piping: *per minute.*
`ain or pump pressure   _40_   _40_ psi   *given*

`ater meter friction (MEEB, Fig. 10.28) _25_   _8_ psi

`emainder   _15_   _32_ psi

`ixture min. pessure (MEEB, Tab. 10.9) _15_   _15_ psi

`emainder   *oops!* _0_   _17_ psi

`tatic head (height x .434 constant) _____   _4_ psi   *(6' below grade + 3'*
                                                              to valve) x .434
`emainder, available to overcome friction _____ _13_ psi

`nit friction loss, psi/100' of pipe _____   5.77 psi   *13 psi x 100'*
                                                              ―――――――――――――――――
`= (pressure available to   100' constant   *150 actual length x 1.5*
   overcome friction)   x
   ―――――――――――――――――――――――――――――――
      total equivalent length of piping

---

6. Size (cast iron) (MEEB, Fig. 10.26) _2_ "diam.

7. Velocity is _6_ ft/sec

Does your first trial design meet the criteria?
If not, what will you change for your second trial?
No; 1" pipe was too small!

Try 1 1/2" size.

Evaluation Tool for Criterion A - "Waste Pipe Sizing"
(MEEB Examples 11.1 and 11.2 may be helpful.)
1. Locate and diagram waste stacks and list the number of
fixtures on each branch; size stacks and branches from
MEEB, Table 11.5.

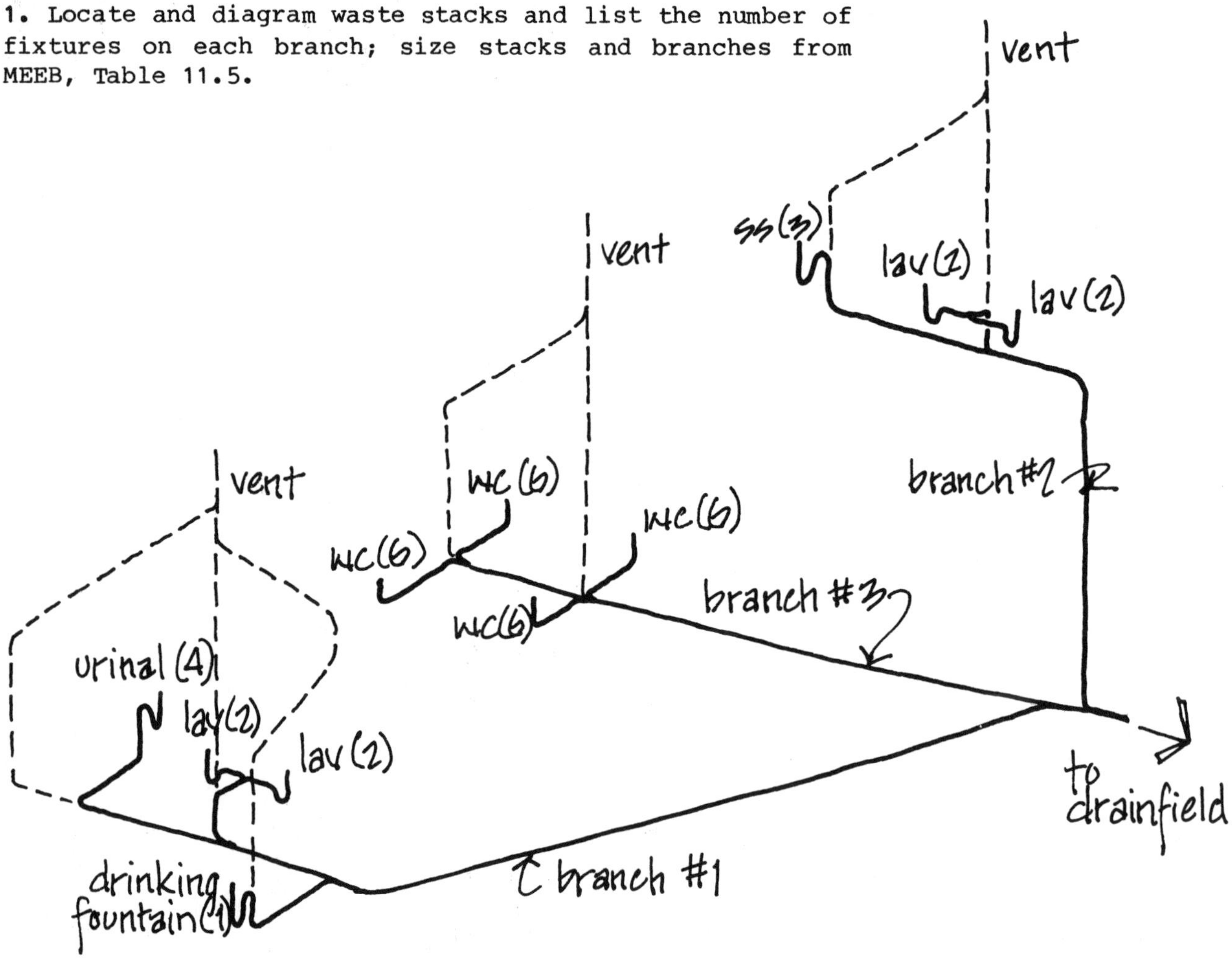

WASTE STACK DIAGRAM
(criterion A; waste pipe sizing, part 1.)

Figure 12-15.

Branch #1 Men's lavatories				Branch #2 Women's lavatories				Branch #3 Common wall			
Fixt. type	#	Waste FU each	Total FU	Fixt. type	#	Waste FU each	Total FU	Fixt. type	#	Waste FU each	Total FU
Lav	2	2	4	Lav	2	2	4	WC	4	6	24
Urinal	1	4	4	Sink	1	3	3				
dr.fnt.	1	1	1								

Total			9	Total			7	Total			24
Size horiz branch			2 1/2"	Size horiz branch			2 1/2"	Size horiz branch			4"
Capacity			12	Capacity			12	Capacity			160

Note: Capacity is the <u>total possible</u> FU that this size line can handle (MEEB, 11.5).

2. Minimum size <u>vertical stack</u> to serve all facilities (MEEB, Tab. 11.5) is determined by the sum of waste FU:

_____total F.U.   _____inch stack   capacity _____FU

(Our design does not involve a vertical stack.)

3. Minimum size of <u>vent</u> required to serve total toilet room fixtures (MEEB, Table 11.6): 1 1/2 in

       30 ft max length

9 + 7 + 24 = 40 waste F.U.

(serves 42 F.U. max.)

----

4. Minimum size building drain, using a <u>minimum</u> 1/8" fall/ft. (MEEB, Table 11.4): <u>_4_</u> in

capacity <u>_180_</u> FU

5. Given the horizontal distance between the toilet room and the septic tank, what **vertical drop** is required to maintain the minimum slope of this drain line?

<u>_30_</u> ft horizontal distance x 1/8" fall/ft = <u>_4___</u> inch   *No problem!*
vertical drop

Is this compatible with elevations on your site?   *Yes.*
Remember, trenches in drainage field will be even lower, to maintain 1/8" fall per foot.

B. Criterion B: Drainfield and septic tank are adequately sized to carry their expected loads.

Evaluation Tool for Criterion B – "Sizing Drainfield and Septic Tank"
The total gallons/day to be handled by your septic tank and drainfield combination, from your conservation potential analysis, or schematic:

<u>_640_</u> gallons/day total   <u>_40_</u> waste FU total

<u>Septic tank size:</u> MEEB, Table 11.10, based on number of fixture units:

<u>_40_</u> drainage FU, needs <u>_2,000_</u> gallons septic tank

<u>Drainage field size</u> MEEB, Table 11.12 lists the length of tiles needed for various trench sizes, in various soil conditions. By assuming either 24" or 36" wide trenches, the length and spacing of the tile lines and their trenches can be determined from MEEB, Tables 11.12 and 11.13.

Determine the overall size of your drainage field:

Assume 2 ft. wide trenches. Salem has 5 minutes for 1" drop; lineal feet of 2' trench = 21' per 100 gallons (MEEB 11.12).

21 x $\frac{640}{100}$ = 134 lineal feet: this could be 4 lines at 34' long
6 lines at 22' long

2 ft. trench spacing = 6 feet (MEEB 11.13)
Sizes of drainfield:
4 lines x 6' apart = 24'; 34' long = 816 s.f.
6 lines x 6' apart = 36'; 22' long = 792 s.f.

----

nsider the possibility of building (and drainage field)
pansion. Show on the site plan how your drainage field
ets the restrictions placed on its location (see MEEB,
g. 9.7, p. 358).

w does this final size compare with the estimate you
de earlier?

Drawn at 30 x 40
Earlier estimate (section 13.4
Criterion C): 680 s.f.
Final size: about 800 s.f.

Criterion C: Solar collectors are sized to supply 100%
the hot water requirement in the month with the most
lar radiation.

**aluation Tool for Criterion C - "Estimate Hot Water
age"**
this building group, hot water is only used in the
vatories. Hot water for washing hands and face becomes
comfortably hot at about 110°F. Assuming that
ergy-conscious management has resulted in the water
ater thermostat set no higher than 120°F, that an
erage water usage temperature is about 100°F, and that
ld water enters the buildings at GWT (see below for
ur climate's GWT, or ground temperature),

Salem: 55°

gallons @ 120° + Y gallons @ GWT = Z gallons at 100°
X+Y=Z=100%)

hot = x = $\frac{100-GWT}{120-GWT}$

% hot = $\frac{100-55}{120-55}$ = $\frac{45}{65}$ = 69%

cold = y = 100% - x

OTAL hot water = __124__ gallons/day in lavatories x __69__ % hot = __86__ gallons/day, hot (120°)

OTAL Btu/day: _120_ gallons/day hot x 8.33 lb/gal x(120° - _55_^{O}GWT = _65,000_ Btu/day.

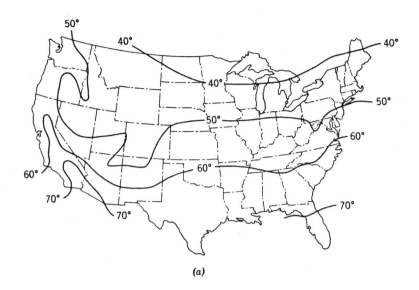

(a)

Figure 12-7   Ground Water Temperature, from MEEB, Fifth Edition 1971.   Used by permission.

Evaluation Tool for Criterion C - "Sizing DHW Solar Collectors"

Listed below are each climate's average insolation values for the worst and best solar months, at several tilt angles (from SOLMET data, distributed by U. S. Government)

	Worst Month				Best Month			
	Horiz.	$30°$	$60°$	Vert.	Horiz.	$30°$	$60°$	Vert.
Madison	389	429	525	477	1933	1364	1010	505
Dodge City	731	857	1047	952	2294	1565	1188	447
Salem	277	281 X	343	312	2141	1499 X	1166	555
Phoenix	931	1104	1214	1104	2737	1765	1073	346
Charleston	744	732	789	665	1859	1251	876	417

Figure 12-8   Solar Radiation Data

You need to choose a collector tilt angle that will:
 - be close to optimum in the best month (high insolation and collector efficiency), and
 - be somewhat better than the lowest insolation in the worst month, since hot water is also desirable in winter.

Since the $45°$ grid is s pervasive, let's stick wit $45°$.

Best month, $\frac{1499 + 1166}{2} = 1333$

Worst month, $\frac{281 + 343}{2} = 312$

_____

o do this, find the collector size required to provide
00% solar heating in the <u>best</u> month; then check its
erformance in the <u>worst</u> month (the one with the least
adiation, that's the coldest).   <u>Trial 1</u>   <u>Trial 2</u>

ollector tilt angle chosen: _____   _____

o size the collector:
. Collector efficiency: this graph is a typical
fficiency curve for selective surface, single glazed
lat plate collectors of moderately low cost.

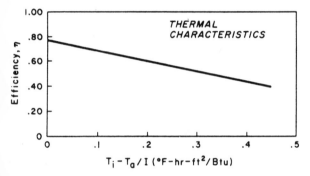

Figure 12-9 Collector Efficiency Curve

T_i = input temperature, a little lower than the
thermostat setting in your water heater.
(summer: higher, since a hot sunny day will heat your
tank well above this setting)
(winter: somewhat lower, since the setting will be the
maximum temperature, and you are feeding your maximum
collectors with colder water from the bottom of the
tank.)

T_a = outdoor daytime temperature (a little below the
maximum temperature for the month, from climate analysis)

I = insolation, in Btuh. Interpolate if necessary from
the insolation table in this exercise, to find the
Btu/day; the <u>hourly</u> insolation, during collection hours,
is about 18% of the daily total.

	Trial 1	Trial 2
Best month collector efficiency	60 %	_____ %
Worst month collector efficiency	20 %	_____ %

Assume:
Summer: 120°
Winter: 100°

Salem: Best month 78°
       Worst month 43°

Salem: 1333 x 18% = 240 Btuh/sf
       312 x 18% = 56 Btuh/sf

Best:
$\frac{120-78}{240}$ =.175; Efficiency = 60%

Worst:
$\frac{100-43}{56}$ = 1.0; off chart!
        Efficiency about 20%?

---12---

2. System Efficiency: Your collector will not usually operate during every hour that the sun is above the horizon; it will only collect during hours when enough sun enters the collector to heat water above the temperature at the bottom of the water heater tank. Therefore, reduce your collector efficiency by about 20% to account for hours of too-low-insolation, as well as heat losses in the lines to and from your collector.

System efficiency = (0.8) x (collector efficiency) =

.8 x 60% = 48% best
.8 x 20% = 16% worst

<u>Trial 1</u>:

_48_ % Best month   _16_ % Worst month

<u>Trial 2</u>:

____% Best month   ____% Worst month

3. Collector size, <u>Trial 1</u>:   Best month

$$\frac{\underline{65,000}\ Btu/day\ needed\ for\ hot\ water}{\underline{1333}\ \ Btu/sf\text{-}day\ (summer)\ x\ summer\ system\ effic.\ \underline{.48}} = \underline{102}sf\ (summer\ optimum)\ collector\ a\imath$$

Now <u>check</u> to see what % solar water heating this will produce in your worst month:

_312_ Btu/sf-day (winter) x _.16_ winter system effic. x _100_ sf collector area = _5,000_ Btu/day

What % of total daily Btu need is this? _8___ % $\frac{5,000}{65,000}$

<u>Trial 2</u>: Best month,

_____ Btu/day needed for hot water

$$\frac{}{\underline{\ \ \ \ }\ Btu/sf\text{-}day\ (summer)\ x\ summer\ system\ effic.\underline{\ \ \ }} = \underline{\ \ \ }sf\ (summer\ optimum)\ collector\ are\epsilon$$

Now <u>check</u> to see what % solar water heating this will produce in your worst month:

____Btu/sf-day (winter) x ___winter system effic. x ____sf collector area = _____Btu/day

What % of total daily Btu need is this ____%

For your collector size and tilt, what average yearly percent solar water heating might you expect, based on simple interpolation between the best and worst months?
About 54%

How does this collector area compare to your earlier estimate?
Earlier: 50 sf. This calculation: 100 sf.

D. Criterion D: The cistern is sized to supply the needs of your conserving system design.

Evaluation Tool for Criterion D - "Cistern Sizing"
(See MEEB, p. 50-52)
1. Calculate catchment (roof) area: _____sf Trial 1
_____sf Trial 2

2. Trial run on sizing - add monthly rainfall in gallons and identify your 'rainy season'. 1 cf = 7.48 gal

We are not using a cistern. If we were, we'd consider adding the parking lots to the catchment area, although oil could become quite a problem! Our roof area is not adequate for catchment, if we are to achieve any storage during the rainy months.

			Trial 1		Trial 2
Avg. Monthly Rainfall (in) ÷ 12 = (ft.)	Roof Area (sf)	Rain Collected (cf) x 7.48 = (Gal)		Roof Area (sf)	Rain Collected (cf) x 7.48 = (Gal)
Jan					
Feb					
Mar					
Apr					
May					
Jun					
Jul					
Aug					
Sep					
Oct					
Nov					
Dec					
		TOTAL _____(Gal)		TOTAL _____(Gal)	

What is your rainy season?_____

12.6 FINALIZING

----

3. Cistern size should accomodate at least minimum capacity (See MEEB, p. 52) plus the average rainy season rainfall. (Use and bypass should mean this size will accomodate above average seasons.)

Minimum capacity _____gal _____cf

Average rainy season _____gal _____cf

TOTAL _____gal _____cf

What are dimensions of your cistern? ___ x ___ x ___

4. Compare yearly supply to demand

Supply _____gal   Demand _____ gal

Is this ok? If supply is small, how can you increase supply, or reduce demand?

5. Compare use to supply on a month to month basis. Start with the month after the "rainy season" and assume a 90% full cistern at that time.

			Trial 1		Trial 2	
Month	Use (Gal)	Rainfall (Gal)	Balance* (Gal)	Backup Source (Gal)	Balance* (Gal)	Backup Source (Gal)

*Must not go below minimum capacity.

Did you run dry in your first trial design?   Why?

If so, what will you change for your second trial?

14.0 ACOUSTICS

14.1 INTRODUCTION

14.1 INTRODUCTION

Your design cycles through heating, cooling and lighting
have produced a combination of shapes and surfaces that
are optimized for energy conservation and the use of
passive solar heating and natural cooling sources, as
well as other design objectives you have brought to this
project. Specifically, combinations of non-porous surfaces (for thermal mass and for the admission of daylight) and highly porous materials (insulation) have been
distributed for heat, cold and light utilization.

The acoustics of space is highly influenced by the shape,
the surface materials, and enclosing wall and floor
construction.

Natural ventilation is a further major influence on
acoustics. Openings that freely admit cooling breezes
admit exterior noises just as freely. They also freely
allow interior noises to escape. The open plans that
facilitate natural ventilation also allow noises from one
area to move freely to other, perhaps quieter, spaces.

This exercise examines the often-conflicting goals of
acoustics as compared with your decisions in the preceeding energy-conscious exercises. Since acoustic
analysis is so often postponed until after most other
decisions are made, this conflict is both frequent and
visible.

This exercise will use dB at 500 Hertz (Hz), which is one
of the important frequencies in human speech. See MEEB,
figures 26.14 and 26.19.

Contents

14.0 ACOUSTICS

14.1 INTRODUCTION

14.2 ANALYZING

GOAL

A. Determine the the acoustical conflicts and compatibilities in the site and program.

A. Analysis Tool - "Site Noise Contours"

Three major sources of sound are present here: the freeway, the railroad, and the factory itself. All three may be considered "linear" sources rather than "point" sources: their sound is generated along a line (such as windows along the side of the factory) rather than at a point (such as a siren would produce).

INSIDEOUT

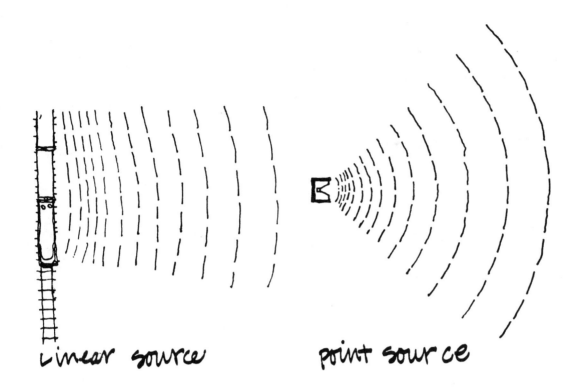

linear source   point source

From Environmental Acoustics by Doelle. Copyright (c)
1972. Used with the permission of McGraw-Hill Book
Company.

For linear sources, the sound level outdoors falls off
3 dB with each doubling of the distance from the line
source. (Egan, 1972, p. 91). Given:

Freeway (constant) average noise level is 60dB (500 Hz)
at 50 feet away.
Train (occasional) average noise level is 70 dB (500 Hz)
at 100 feet away.
Factory's noise level will be more accurately determined
later. For now, assume: factory (constant) average noise
level is 60 dB (500 Hz) at 1 foot away.

1. Plot the noise contours for each source separately on tracing paper.
2. Combine the contours for the fixed sources (freeway and train). Decibels (dB) are not added arithmetically. Use the following chart to add the sources.

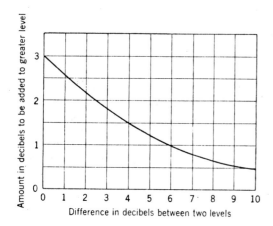

Figure 14.1 Chart for adding two noise levels or sound pressure levels (SPL). From MEEB, Fifth Edition 1971. Used by permission.

3. Plot the resulting combined "noise contours" at 10 dB intervals on the site plan, from the fixed noise sources (freeway and train).
4. On a tracing paper overlay, plot the noise contours at 10 dB intervals from all sides of the factory. (Since this is a "moveable" source in the design process, the overlay can be used to assess the acoustical merits of various factory locations on the site.)

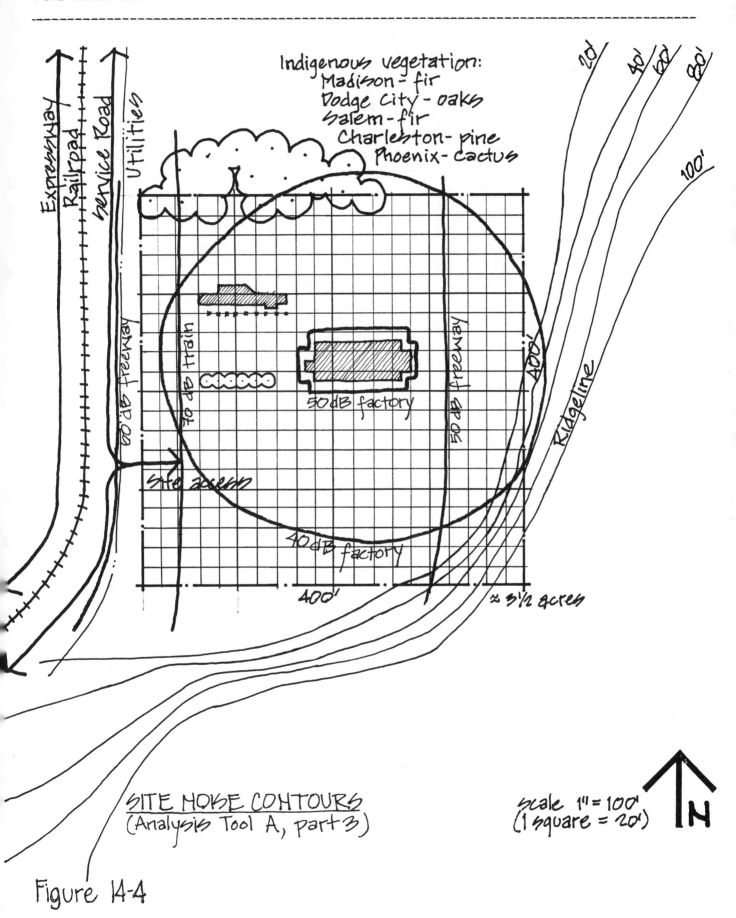

Indigenous vegetation:
Madison - fir
Dodge City - oaks
Salem - fir
Charleston - pine
Phoenix - cactus

Expressway
Railroad
Service Road
Utilities

60 dB Freeway
70 dB train
50 dB factory
50 dB freeway
400'
Ridgeline

Site access
50 dB factory
40 dB factory
400'
≈ 3½ acres

SITE NOISE CONTOURS
(Analysis Tool A, part 3)

scale 1" = 100'
(1 square = 20')
N

Figure 14-4

B. Analysis Tool - "Noise Criteria-NC"

Noise Criteria are used to establish a maximum sound level for background noise within a space, based on the activity within the space. As such they begin to quantify which spaces need more quiet than others and give you a preliminary means for assessing your design decisions.

Spaces with low NC requirements should be placed away from noisy (high NC) spaces. Where adjacent spaces have differing NC, an acoustical barrier between them is necessary. The greater the difference in adjacent NC, the more impenetrable an acoustical barrier is needed.

Silence-goal spaces are those in which attention is focused on one sound source, such as a lecturer. Background sound is excluded from silence-goal spaces. Background sound is admitted to quiet-goal spaces, so that no one sound source becomes predominant.

Find the NC appropriate for each space. (See MEEB, Table 27.11, p. 1255, for Noise Criteria (NC) data.)

Factory (speech or telephone communication not required, yet desirable) NC = _60_

Circulation and toilet rooms NC = _45-60_

Display and conference space NC = _30-35_

Private office spaces   NC = _30-35_

Questions

A. Which spaces are silence goal spaces and which are quiet goal spaces? *All are acceptable as quiet goals, although a fairly large "audience" in the display/conference room could approach a "silence" situation.*

B. Which are the acoustically similar program spaces? *Office and conference.*

C. Which are the acoustically incompatible spaces? *Toilet rooms and offices; toilet rooms and conference; factory and offices; factory and conference; except modest background sound from factory could be useful to acoustic privacy in offiices, and in toilet rooms.*

D. Are there potential acoustical conflicts between inside and outside? *Yes; factory noise; if sufficiently loud, could interfere with a conference usage.*

4.3 CONCEPTUALIZING _____

GOAL

A. Utilize wanted sounds on site, while minimizing noise (unwanted sound).

DESIGN STRATEGIES

A. Cluster Scale
Use buildings or other barriers to create quieter spaces, sheltered from noise sources.

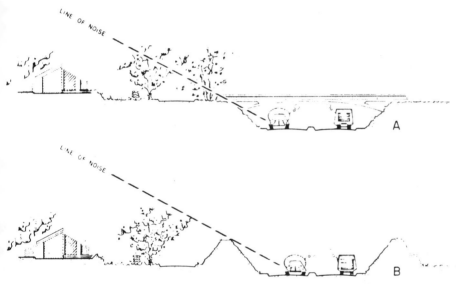

From Environmental Acoustics by Doelle. Copyright (c) 1972. Used with the permission of McGraw-Hill Book Company.

Quiet spaces for quiet uses.

B. Building Scale
Separate the quiet from the noisy spaces.
C. Component Scale
Noise sources are usually most economically quieted <u>at source</u>.

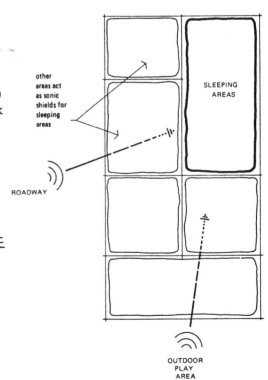

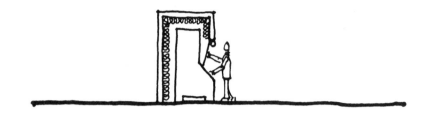

From Concepts in Architectural Acoustics by Egan.
Copyright (c) 1972. Used with the permission of
McGraw-Hill Book Company.

DESIGN

Choose an existing building/site which has a clear
conceptual approach, at any scale, to achieving these
goals. Document your choice as follows:
A. Identify the location, program, architect (if any) of
the design and the source of your information.
B. Include photocopies or drawings (whatever is quick and
easy for you) to explain the design.

EVALUATION

A. Evaluation Tool - "Building Response Diagrams"
Diagram how this design is organized to achieve these
goals.
*This more general part of the exercises is not included
in the Salem, Oregon example.*

14.4 SCHEMING ─────────────────────────────────────

GOALS

A. Separate or isolate acoustically incompatible spaces.
B. Exclude background sound from silence-goal spaces and admit background sound to quiet-goal spaces.

CRITERIA

A. Potential acoustical conflicts have been avoided.
B. Heating, Cooling, Daylighting performance has not been impaired in meeting the acoustic criteria.

DESIGN STRATEGIES

A. Cluster Scale
Consider introducing a sound source to make noisy sources (such as a courtyard fountain to mask traffic). When sound sources generated within buildings are preferable to noise sources offsite, these sounds can be similarly used for masking.

B. Building Scale
When ventilating winds bring noise into the building with them, consider placing masking sound sources near air vents, or placing the most noisy spaces nearest to noisy ventilating edges of building.

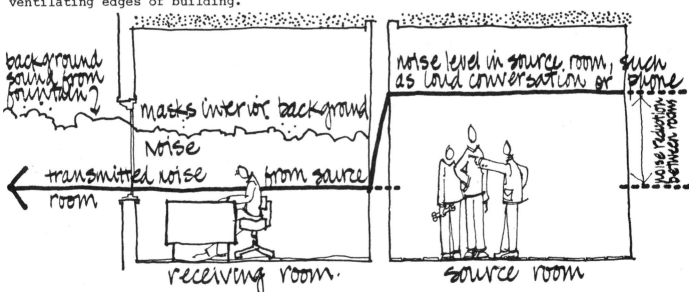

From Concepts in Architectural Acoustics by Egan. Copyright (c) 1972. Used with the permission of McGraw-Hill Book Company.

C. Component Scale
Thermally massive walls can be excellent sound barriers,
such as between noisy and quiet spaces.

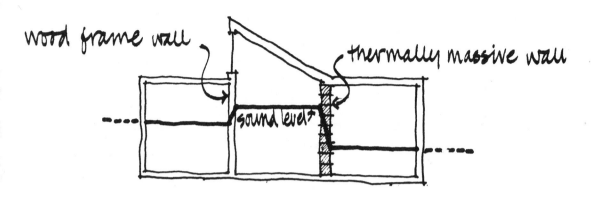

Sound barriers should have a minimum of openings, and
these openings must be fully closable when sound isolation is desired. When a "weaker" element, such as a
window or door, is used in a construction the composite
TL for the combination is usually closer to the TL of the
weaker element.

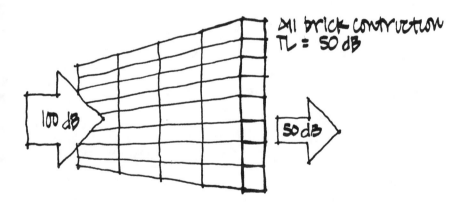

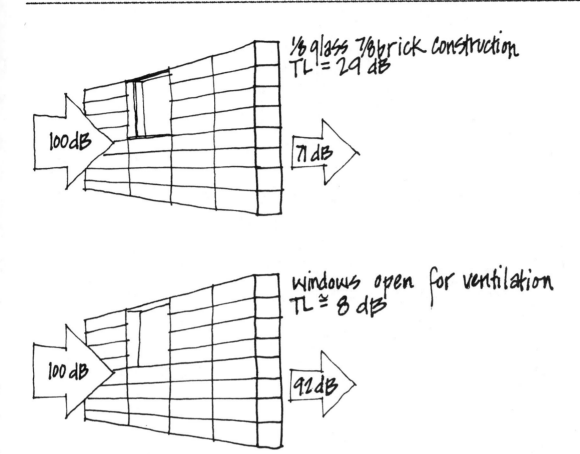

Acoustically absorptive surfaces are rarely compatible
with thermal storage, so such surfaces should either
receive no winter sun, or be in a position to reflect it
toward thermal mass surfaces. Similarly, such surfaces
need no exposure to cooling winds.

DESIGN

Design your office building, factory and site to achieve
the Scheming goals. Document your design:

A. Site Plan
B. Cluster Plan
C. Floor Plans
D. Sections
E. Roof plan and elevations or paralines showing all
sides and top of your buildings.

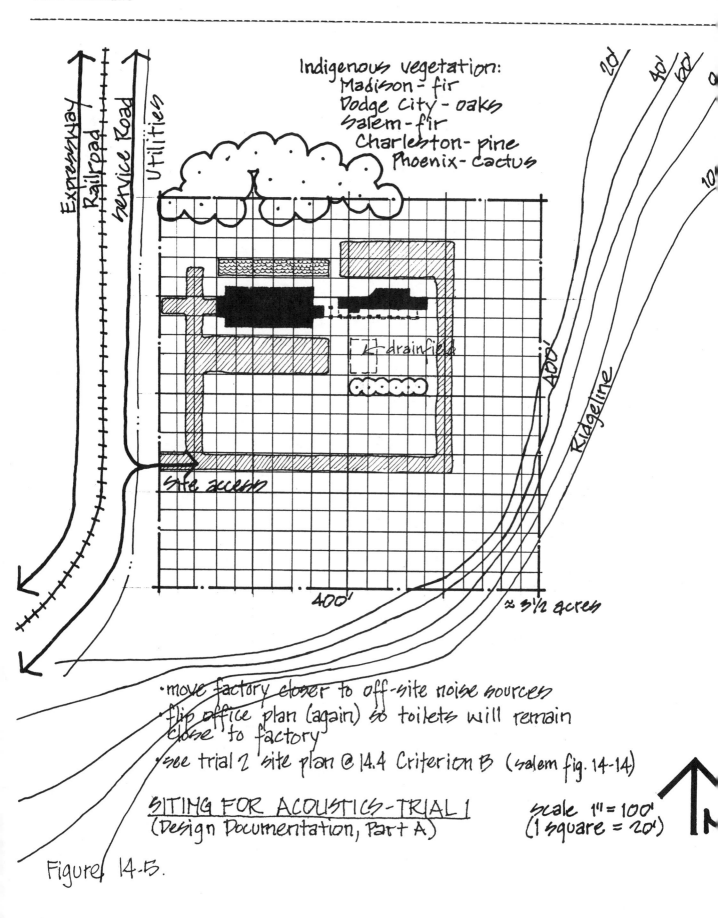

Indigenous vegetation:
Madison - fir
Dodge City - oaks
Salem - fir
Charleston - pine
Phoenix - cactus

Expressway
Railroad
Service Road
Utilities

drainfield

site access

Ridgeline

400'

≈ 3½ acres

400'

20'
40'
60'

10

400'

• move factory closer to off-site noise sources
• flip office plan (again) so toilets will remain close to factory
• see trial 2 site plan @ 14.4 Criterion B (Salem fig. 14-14)

SITING FOR ACOUSTICS-TRIAL 1
(Design Documentation, Part A)

scale 1" = 100'
(1 square = 20')

Figure 14-5.

N

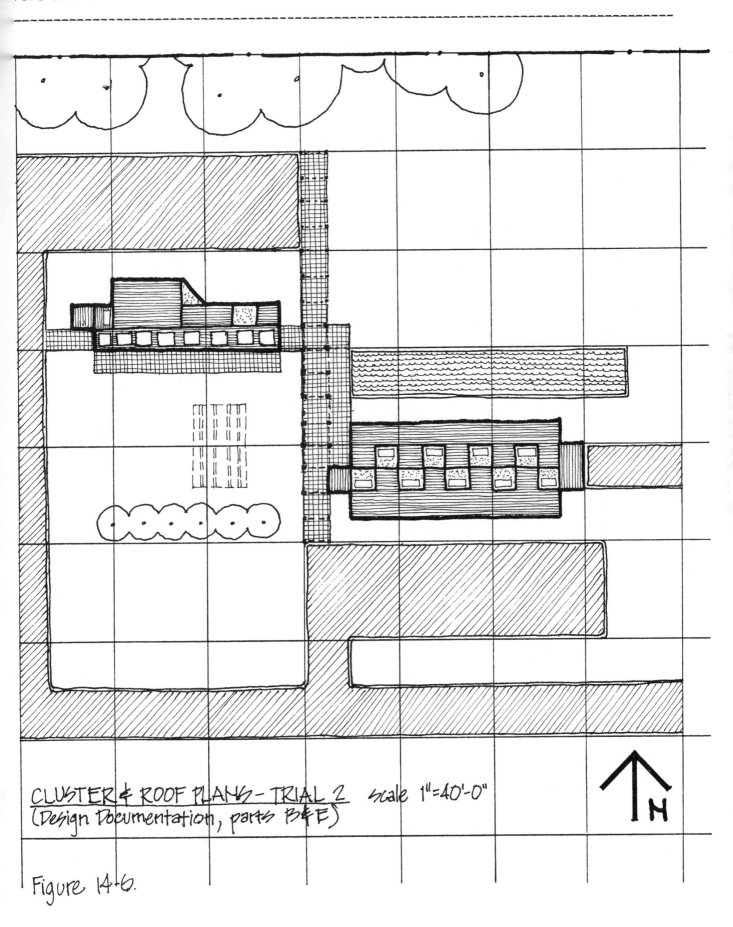

CLUSTER & ROOF PLANS - TRIAL 2   scale 1"=40'-0"
(Design Documentation, parts B&E)

Figure 14-6.

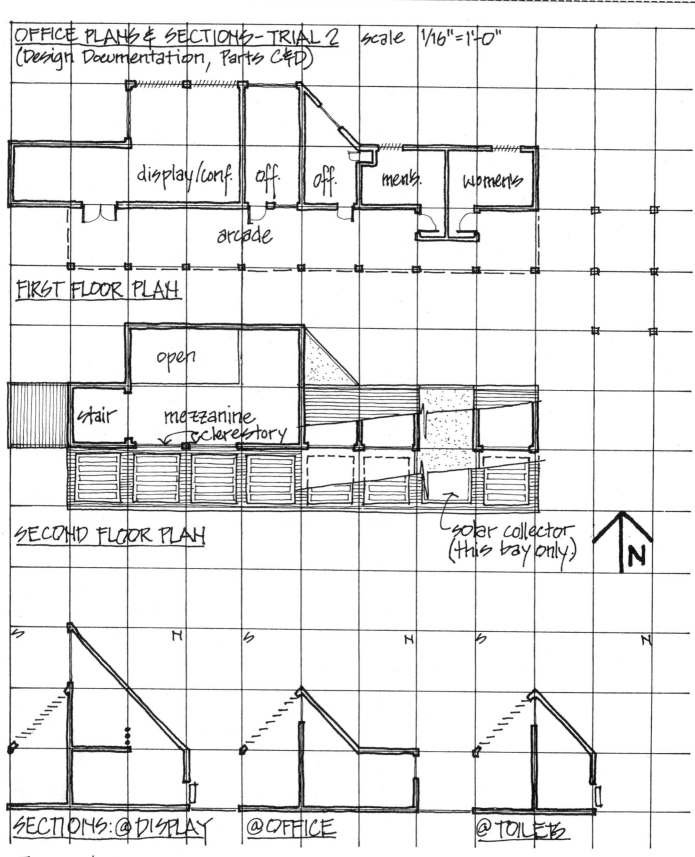

OFFICE PLANS & SECTIONS—TRIAL 2   scale 1/16"=1'-0"
(Design Documentation, Parts C&D)

display/conf.   off.   off.   mens.   womens

arcade

FIRST FLOOR PLAN

open

stair   mezzanine   clerestory

SECOND FLOOR PLAN

solar collector
(this bay only)

N

SECTIONS: @DISPLAY   @OFFICE   @TOILETS

Figure 14-7.

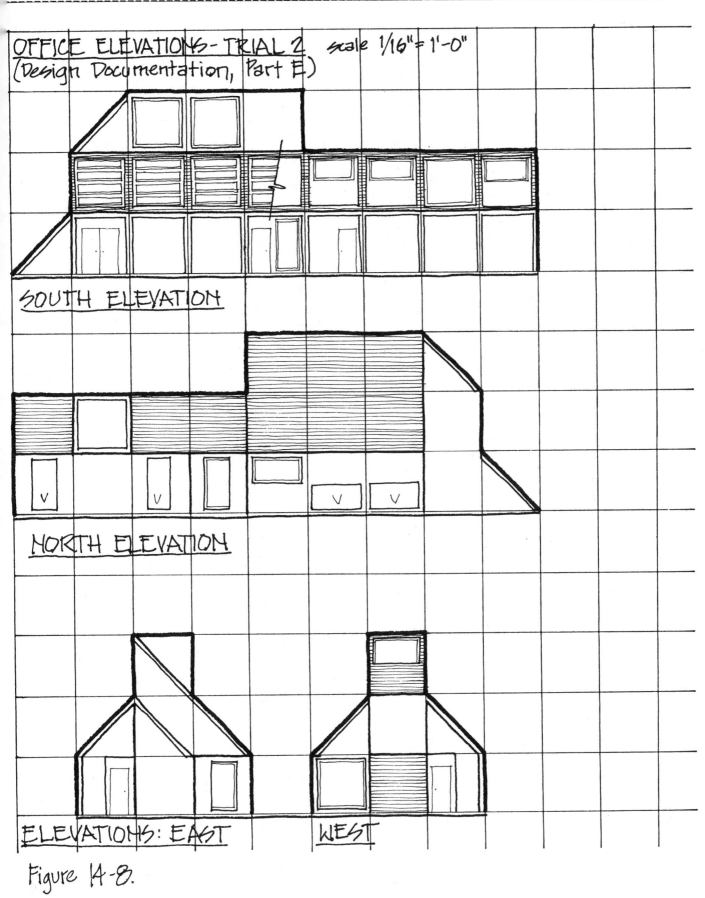

OFFICE ELEVATIONS – TRIAL 2   scale 1/16" = 1'-0"
(Design Documentation, Part E)

SOUTH ELEVATION

NORTH ELEVATION

ELEVATIONS: EAST   WEST

Figure 14-8.

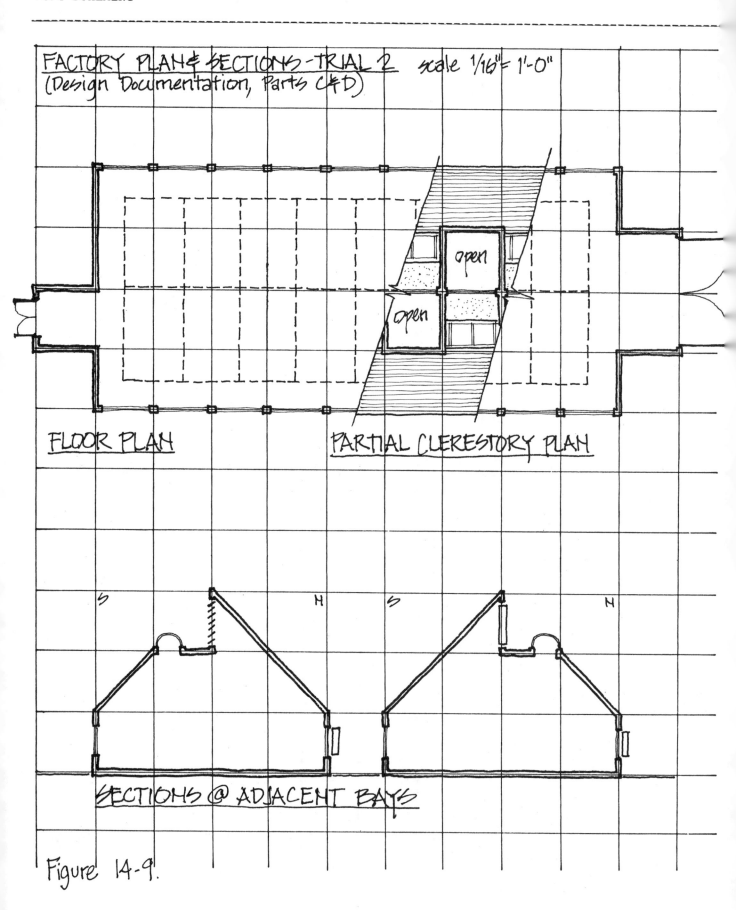

FACTORY PLAN & SECTIONS - TRIAL 2   scale 1/16"= 1'-0"
(Design Documentation, Parts C&D)

FLOOR PLAN   PARTIAL CLERESTORY PLAN

SECTIONS @ ADJACENT BAYS

Figure 14-9.

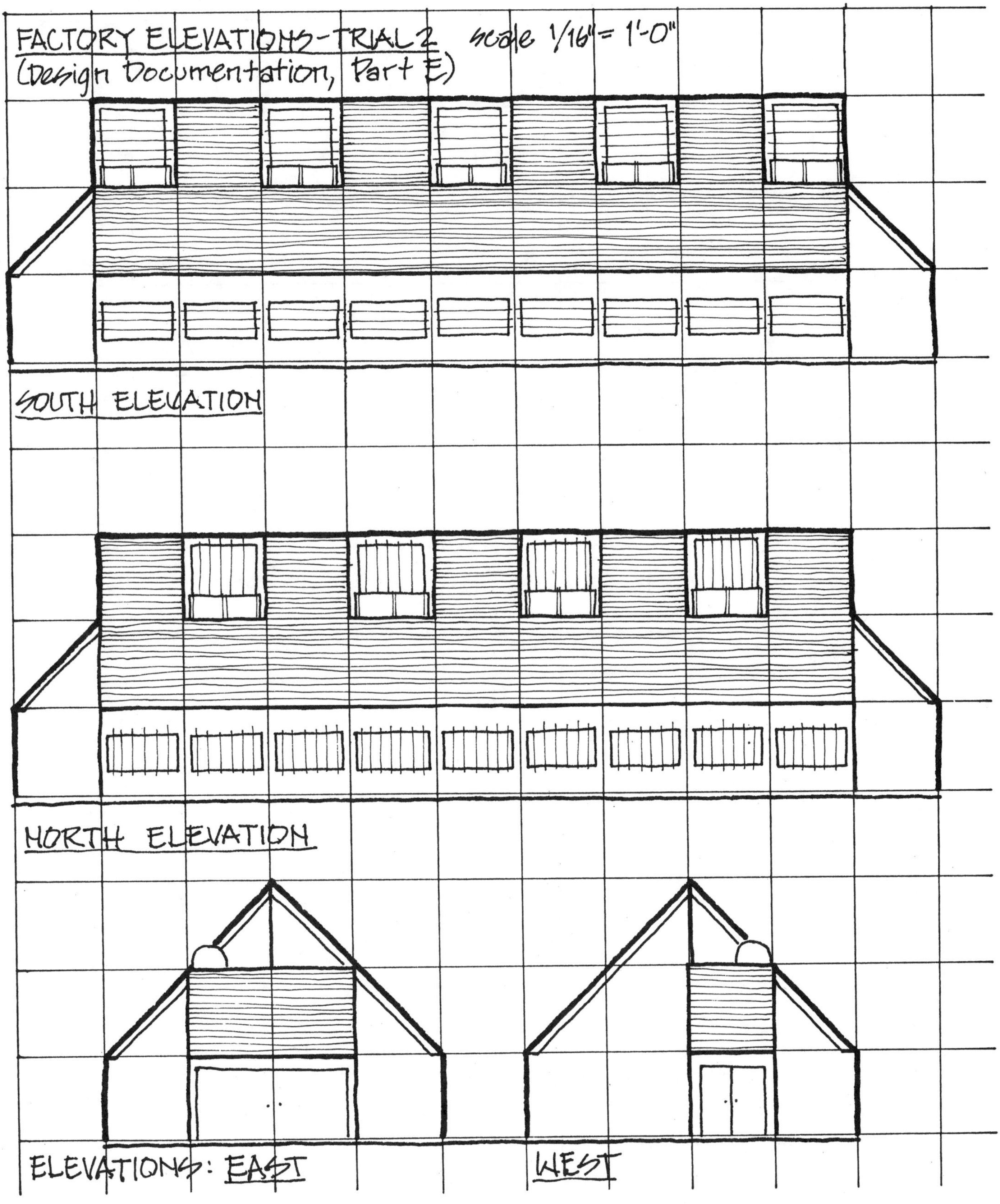

FACTORY ELEVATIONS – TRIAL 2   scale 1/16" = 1'-0"
(Design Documentation, Part E)

SOUTH ELEVATION

NORTH ELEVATION

ELEVATIONS: EAST   WEST

Figure 14-1Q.

for plans, sections, & elevations: if   1/16"=1'-0",   1 square = 10'.

for plans, sections, & elevations: if   1/16"=1'-0",   1 square = 10'.

paraline grid w/ 20' cells;
indicate north arrow.

paraline grid w/ 20' cells;
indicate north arrow.

EVALUATION

A. Criterion A: The potential acoustical conflicts identified in 14.2 Analyzing Summary, Question D, have been
avoided on the site and within the buildings.

Evaluation Tool for Criterion A - "Site Noise Contours"
1. Locate your buildings on a site plan.
2. Use MEEB, Fig. 26.12, p. 1171 to add the dB (at 500
Hz) from the three noise sources (freeway, train and
factory) to achieve overall noise contours for the site.
(The total dB from more than one source is not the sum of
the individual dB measurements.) We are assuming that
our factory equipment has a peak noise level at 500 Hz;
remember that this is only one frequency, that noise
levels vary with frequency, and that factory equipment in
particular might peak at other frequencies.

**Evaluation Tool for Criterion A - "Acoustical Renderings
or Noise Drawings"**
1. For each space identify the NC and whether the goal is
silence or quiet (from the Analysis Section.)

*Factory: speech or telephone communication not required,
yet desirable: NC = 60, "quiet"*
Circulation and toilet rooms: NC = 45 to 55, "quiet"
*Display and conference: NC = 30 to 40, "quiet"
(occasionally approaching "silence", when a small
audience listens to one person speaking)*
Office: NC = 30 to 40, "quiet"

2. In your cluster plan clearly show all barriers and all
openings.
3. Draw an acoustical picture (in plan, section, perspective, model, whatever) which shows sounds at their
sources and how they are used as background sound or
controlled by isolation or separation.

Does your first trial design meet Criterion A?
Yes.
If not, what will you change for your second trial?

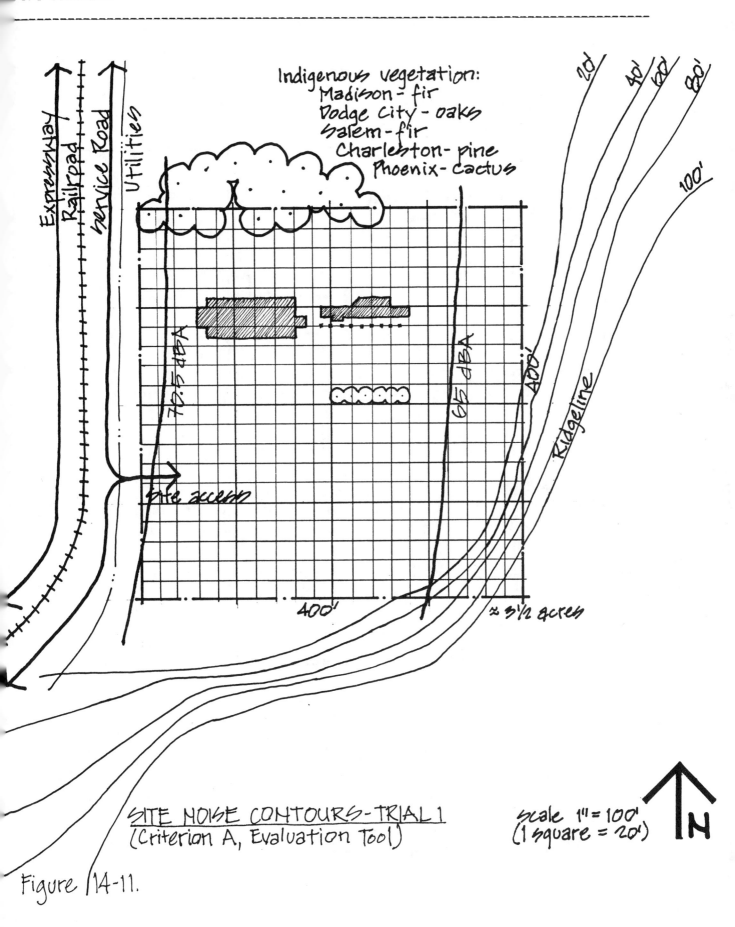

Indigenous vegetation:
Madison - fir
Dodge City - oaks
Salem - fir
Charleston - pine
Phoenix - cactus

SITE NOISE CONTOURS - TRIAL 1
(Criterion A, Evaluation Tool)

scale 1" = 100'
(1 square = 20')

Figure 14-11.

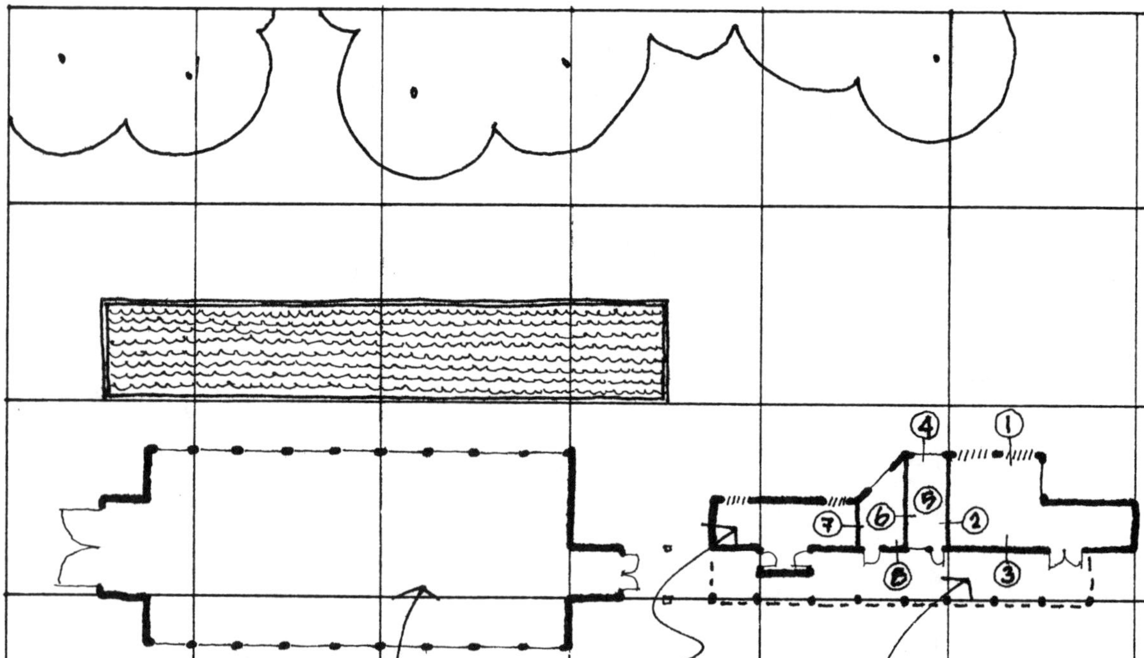

noise
sources: a. freeway &   b. factory   c. toilets   d. circulation corridor/
              train   arcade
              (off-site)

ACOUSTICAL CONFLICTS
1. factory as major noise source/office building needs quiet:
     SEPARATION by toilets - turn office building around.

2. toilet noise → office   ISOLATION - barrier #7
3. outside noise → offices   "   - barrier 4
4. office → office   "   - barrier 6
5. office → arcade   "   - barrier 8   (see 14.5, Crit. B)
6. office → mezzanine (ceiling)   "   - barrier 5
7. display → arcade   "   - barrier 3
8. display → office   "   - barrier 2
9. display → outside   "   - barrier 1

10. freeway & train noise sources: block noise from
    office with factory.

CLUSTER PLAN   scale 1"=40'-0"
(Criterion A; Acoustical Renderings, part 2)

Figure 14-12.

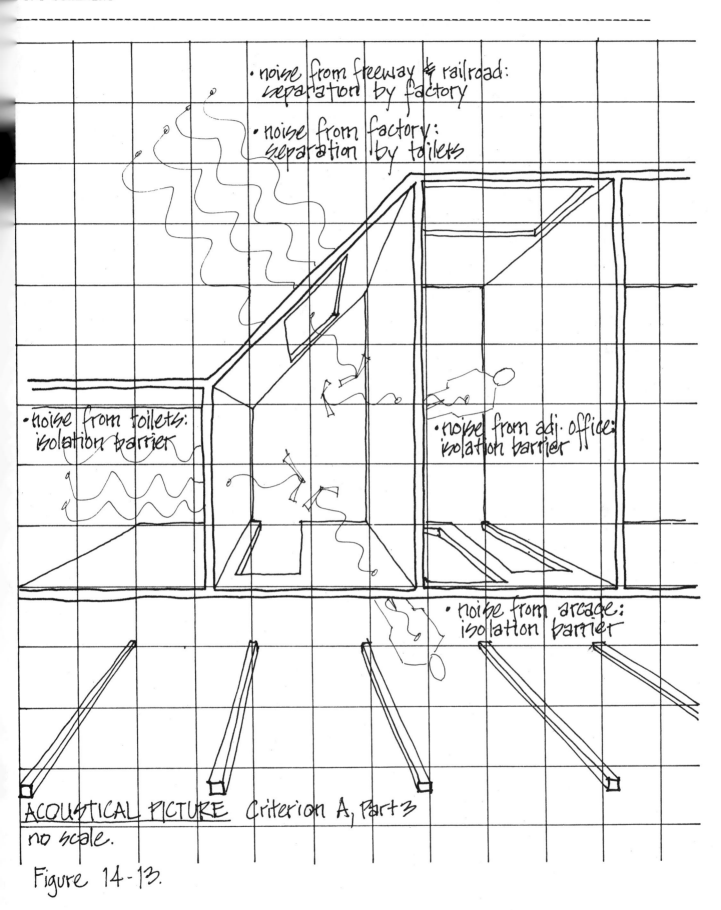

- noise from freeway & railroad: separation by factory
- noise from factory: separation by toilets
- noise from toilets: isolation barrier
- noise from adj. office: isolation barrier
- noise from arcade: isolation barrier

ACOUSTICAL PICTURE Criterion A, Part 3
no scale.

Figure 14-13.

B. Criterion B: Heating, Cooling, Daylighting Check:
Re-check the criteria that produced your site and building organization in the energy-conscious exercises of
heating, cooling and daylighting. If acoustic goals are
incompatible with other considerations, decide on your
priorities and explain your decisions.

Does your first trial design preserve the heating, cooling and daylighting performance achieved in those exercises? *No, factory is blocked from summer wind, and
office building loses some early morning winter sun.*
If not, what will you change for your second trial?

*Switch the buildings <u>back</u> to where they were on the site,
and live with the slight increase in noise in the office
building; let the factory noise help drown out the
freeway noise, because at least the factory noise is
work-related.*

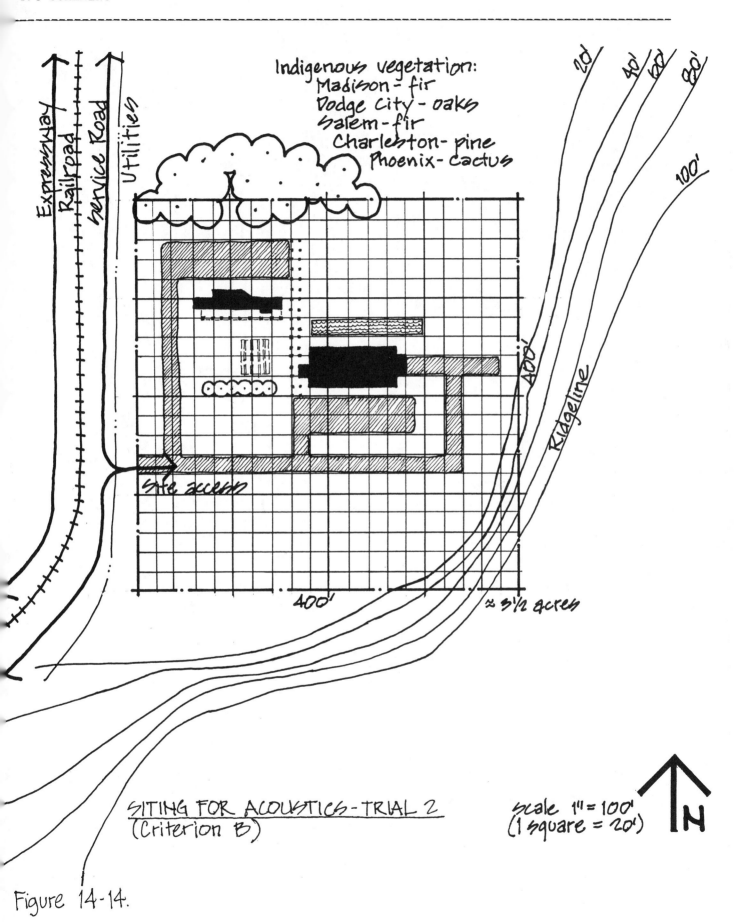

Indigenous vegetation:
Madison - fir
Dodge City - oaks
Salem - fir
Charleston - pine
Phoenix - cactus

SITING FOR ACOUSTICS - TRIAL 2
(Criterion B)

scale 1" = 100'
(1 square = 20')

Figure 14-14.

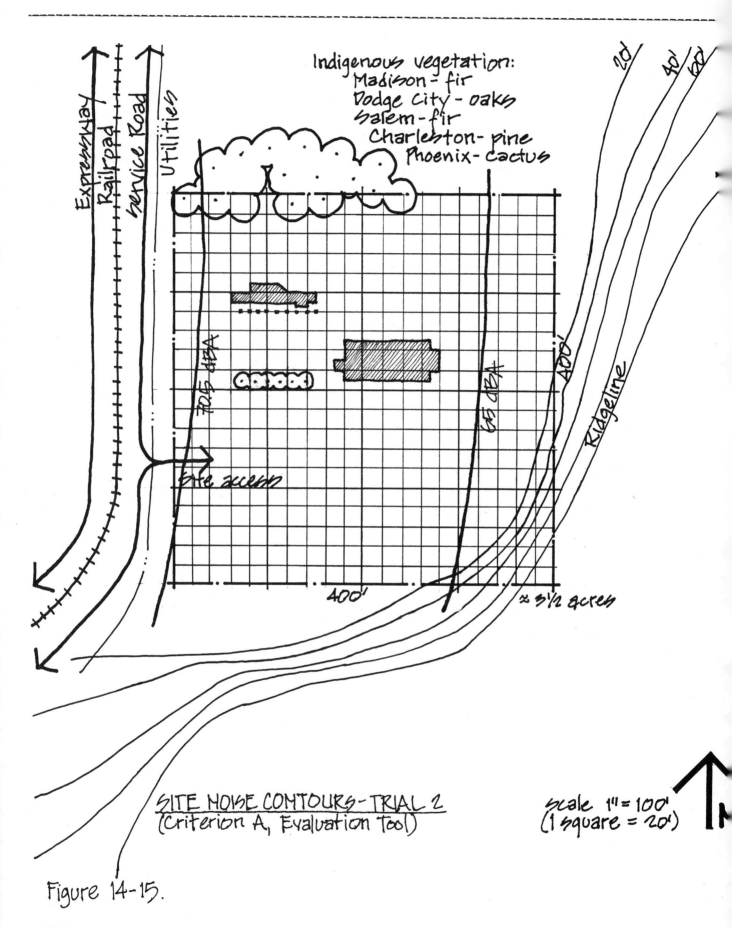

Indigenous vegetation:
Madison - fir
Dodge City - oaks
Salem - fir
Charleston - pine
Phoenix - cactus

Expressway
Railroad
Service Road
Utilities

70.5 dBA

65 dBA

20'
40'
60'
400'

Ridgeline

Site access

400'

≈ 3½ acres

SITE NOISE CONTOURS - TRIAL 2
(Criterion A, Evaluation Tool)

scale 1" = 100'
(1 square = 20')

N

Figure 14-15.

_____---__

14.5 DEVELOPING _____

GOALS

A. Make spaces acoustically comfortable with your choice of interior materials and finishes.
B. Achieve necessary sound isolation characteristics with your choice of wall construction and finish materials.

CRITERIA

A. **Criterion A:** Spaces are live, neutral or dead as appropriate for their function.
B. **Criterion B:** All barrier walls meet the recommended STC rating.

DESIGN STRATEGIES

A. **Component Scale**
Spaces with predominantly hard surface materials (including wood, masonry, plaster, glass and concrete) are reverberant; a sound stays "alive" longer within these spaces. Spaces with predominantly soft surface materials (rugs, drapes, acoustic tile, exposed insulation boards or batts) do not reinforce sounds, so they "die" away very soon, as they would outdoors.

"Live" spaces encourage sound creation, such as whistling in concrete stairwells, or singing in the shower. They also blend sounds together, and can become acoustically exhausting with prolonged exposure to a high level of mixed and largely unwanted sounds.

"Dead" spaces discourage the creation of sound, and tend to keep sounds distinct from one another. A space may be chosen to be alive or dead simply to provide for acoustical variety in a collection of spaces. Contrasts between spaces can be enhanced by liveness or deadness; a quiet space that is "dead" seems especially quiet if entered from a noisy, "live" one.

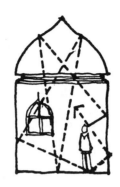

Dead Space   Live Space

Adjacent and very different NC spaces will require
acoustic barriers, as will adjacent spaces with low
NC's.

Privacy is a factor in providing acoustic barriers - you
may want to provide a barrier to overhearing speech from
an adjacent space, even if it has the same NC.

Noise transmission can be reduced by increasing the mass
of the barrier up to the point beyond which diminishing
returns may be expected.

Noise transmission can be reduced by structural discon-
tinuity within the barrier.

A very stiff partition will not resist transmission as
well as might be expected from its mass alone.

Materials chosen for high sound absorption will generally
have little value as sound barriers, due to their porous
and lightweight character.

Acoustical barriers can be rated by their Sound Transmission Class (STC), with more effective barriers having
higher STC's. As a guide:

STC 35 Normal speech can be heard and understood through
barrier
STC 40   Normal speech is heard as a murmur through
barrier
STC 50   Very loud sounds can be heard only faintly
through barrier

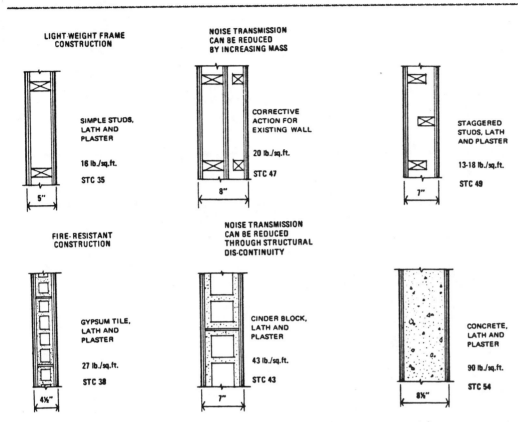

Figure 14-2   STC Ratings for Typical Wall Constructions
from <u>Architectural Interior Systems</u> by J. E. Flynn and A.
W. Segil (c) 1970 by Litton Educational Publishing, Inc.
Reprinted by permission of Van Nostrand Reinhold Co.
For other partition alternatives, check tables A.1 and
A.2 in the back of MEEB.

DESIGN

Develop your design for the factory, one 200 sq.ft.
office and the display conference room to achieve the
acoustical goals above. Document your design:

A. A sectional perspective of each room indicating finish
materials.

B. A section at 3/4"=1'-0" of each acoustical barrier and
key these to the cluster plan.

C. Indicate on the following chart your design intentions
for each of these three spaces:

Space	Live or Dead	Reason
Factory	*Medium Live*	*As long as overall noise level stays well within "safe" levels, a relatively quiet space is expected. Communication is of secondary importance here.*
Office	*Relatively dead*	*A small, private space is reinforced by being a bit on the "dead" side.*
Display/ Conference	*Relatively live*	*This needs to feel like a rather large, spacious room; a bit of liveness will reinforce this impression.*

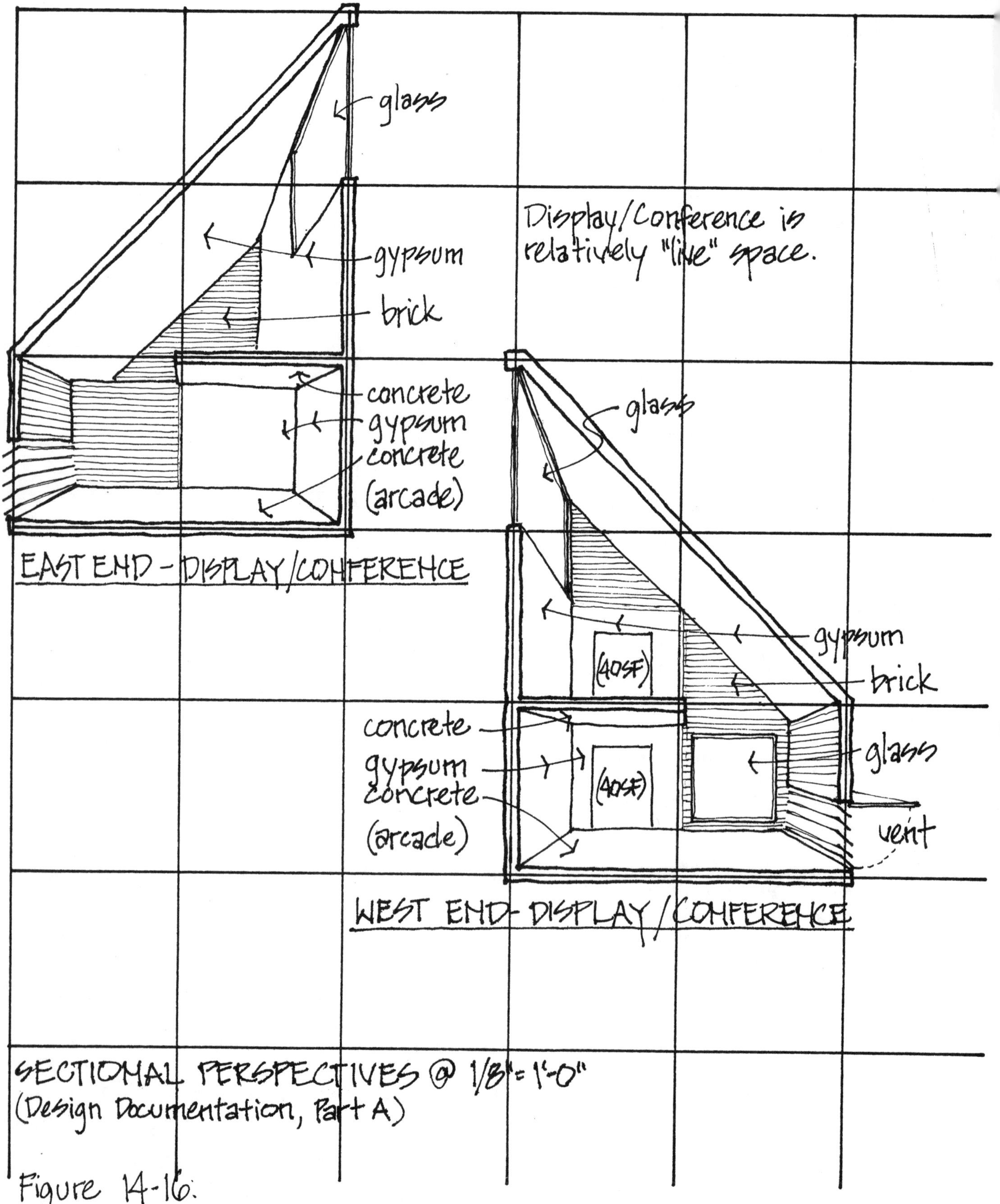

glass

gypsum

brick

concrete
gypsum
concrete
(arcade)

Display/Conference is
relatively "live" space.

EAST END - DISPLAY/CONFERENCE

glass

(40SF)

gypsum

brick

glass

concrete
gypsum
concrete
(arcade)

(40SF)

vent

WEST END - DISPLAY/CONFERENCE

SECTIONAL PERSPECTIVES @ 1/8"=1'-0"
(Design Documentation, Part A)

Figure 14-16.

SECTIONAL PERSPECTIVES @ 1/8" = 1'-0"
(Design Documentation, Part A; cont.)

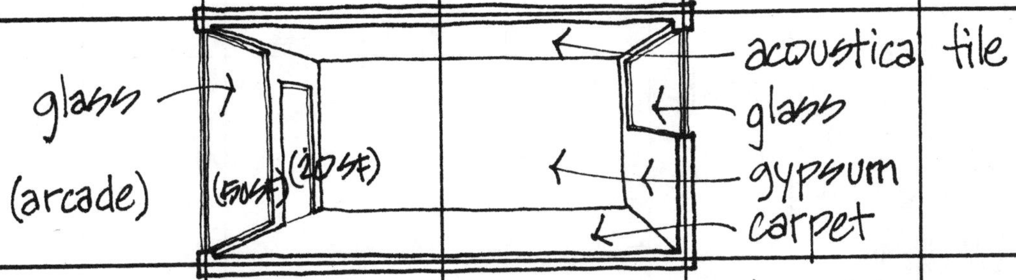

glass →

(arcade)

(50SF) (20SF)

acoustical tile
glass
gypsum
carpet

OFFICE (UNDER MEZZANINE) - relatively "dead" space.

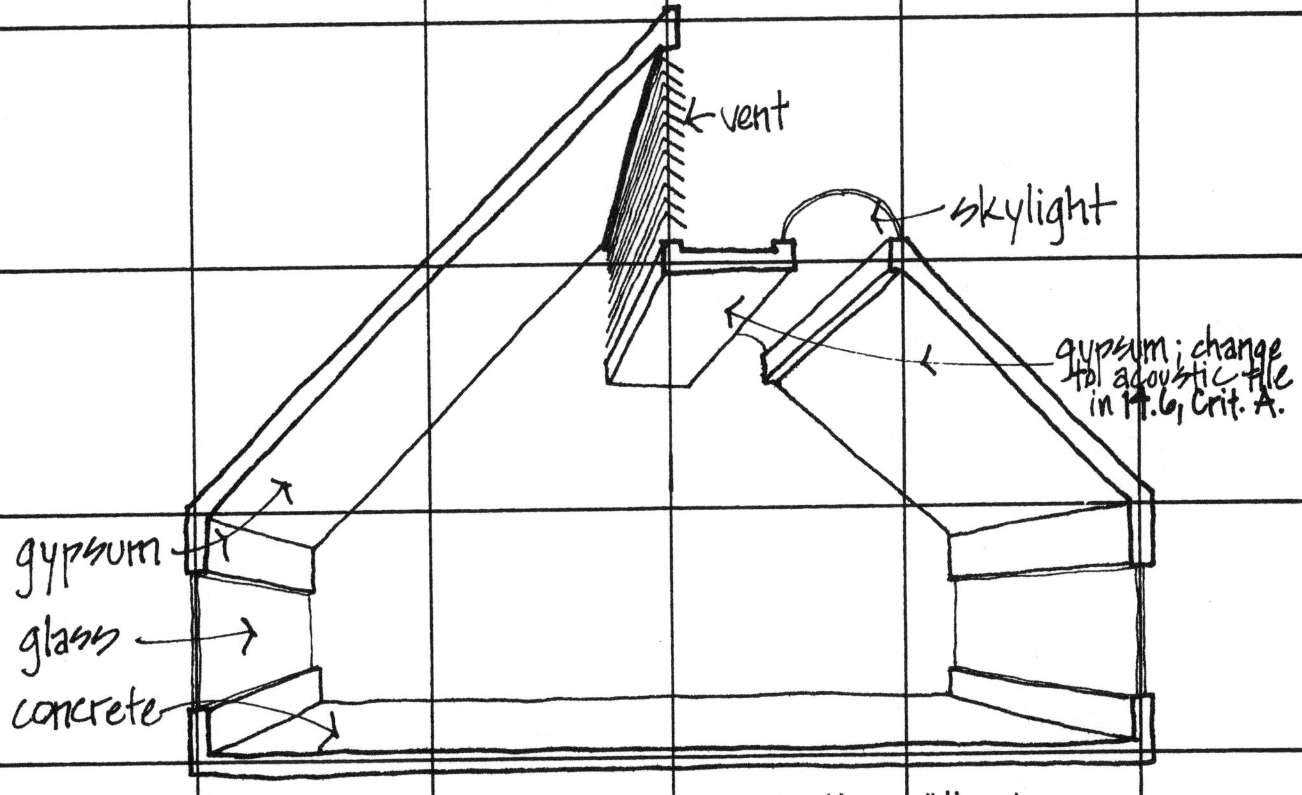

← vent

skylight

gypsum; change
to acoustic tile
in 14.6, Crit. A.

gypsum →

glass →

concrete

FACTORY - TYPICAL BAY - medium "live" space.

Figure 14-17

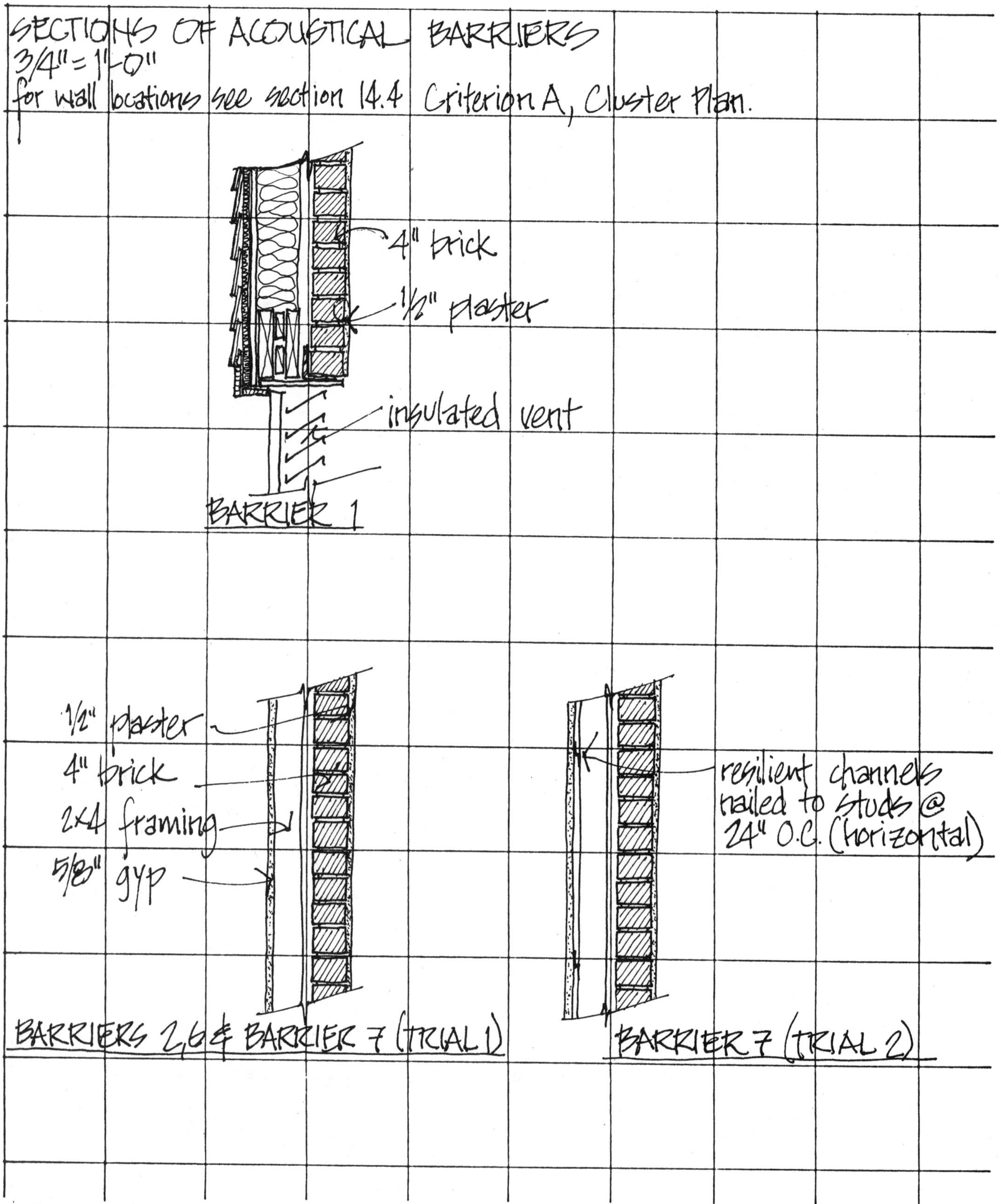

SECTIONS OF ACOUSTICAL BARRIERS
3/4" = 1'-0"
for wall locations see section 14.4 Criterion A, Cluster Plan.

4" brick

1/2" plaster

insulated vent

BARRIER 1

1/2" plaster
4" brick
2x4 framing
5/8" gyp

resilient channels
nailed to studs @
24" O.C. (horizontal)

BARRIERS 2,6 & BARRIER 7 (TRIAL 1)

BARRIER 7 (TRIAL 2)

Figure 14-18.

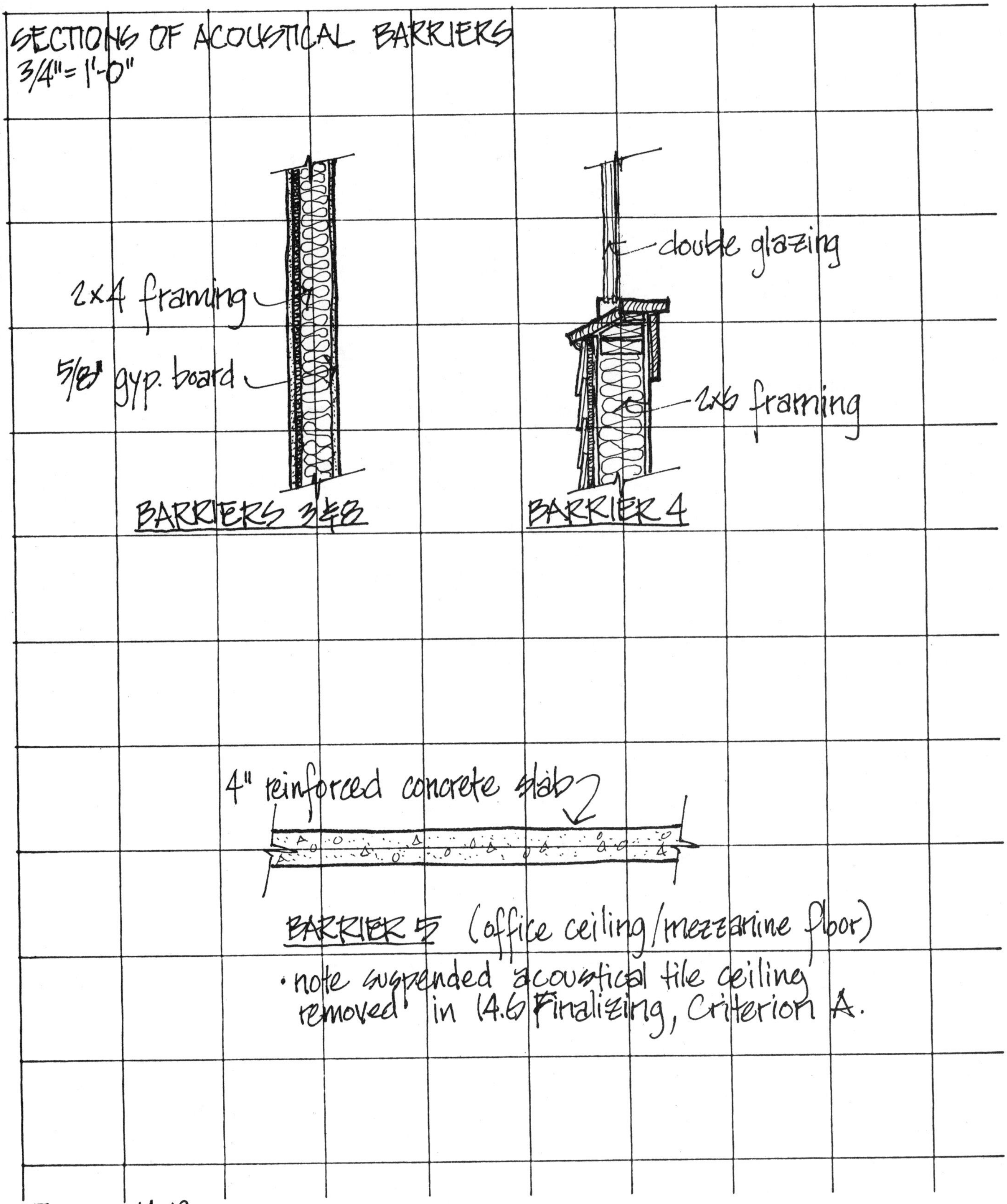

SECTIONS OF ACOUSTICAL BARRIERS
3/4"= 1'-0"

2x4 framing

5/8" gyp. board

BARRIERS 3&8

double glazing

2x6 framing

BARRIER 4

4" reinforced concrete slab

BARRIER 5 (office ceiling/mezzanine floor)

• note suspended acoustical tile ceiling
  removed in 14.6 Finalizing, Criterion A.

Figure 14-19.

EVALUATION

A. Criterion A: Spaces are live, neutral or dead as appropriate for their function.

Evaluation Tool for Criterion A - "Room Absorbency"

An early step in nearly all acoustics calculations is to determine the total absorbency in each space in which acoustics are important.

Absorbency is measured in SABINS, and is equal to the coefficient of absorption times the surface area (essentially, the % of sound which will be absorbed times the area absorbing it):

absorbency = (α) x (sf)

Total room absorbency is simply the sum of the absorbencies of all surfaces. For buildings which are naturally ventilated, the openable window areas present a potentially substantial increase in absorbency since an open window's co-efficient of absorption is 1.0, far greater than most building materials. (Of course, $\alpha = 1.0$ also means TL=0. Thus no acoustic barrier value.) Therefore, for ventilated buildings, two absorbency calculations will be performed. Although such absorbency calculations can be done at a variety of frequencies, we will continue to concentrate on the mid-frequency range of 500 Hz.

Use the coefficients of absorption MEEB Table 27.1, p. 1206 - 1208 to calculate the total absorbencies of your rooms on the tables below.

Trial 1:

Factory	Area (sf)	Coefficient of Absorption at 500Hz		Absorbency: Area x α (Sabins)		
		"closed"	"open"	"closed"	"open"	
floor	3900	.015	.015	58	58	
walls	3100	.05	.05	155	155	
openings:						
closed	2100	.04	NA	84	NA	(includes operable windows, vents and doors)
open (such as windows)	2100	NA	1.0	NA	2100	
ceiling	4680	.05	.05	234	234	(includes skylight area which is assumed non-operable)
people, # occupants	16	.50	.50	8	8	
			TOTAL =	539	2555	Sabins

Trial 2:

Factory	Area (sf)	Coefficient of Absorption at 500 hz		Absorbency: Area x α (Sabins)		
		"closed"	"open"	"closed"	"open"	
floor	3900	.015	.015	58	58	
walls	3100	.05	.05	155	155	
openings:						
closed	2100	.04	NA	84	NA	
open (such as windows)	2100	NA	1.0	NA	2100	
ceiling						
acoustic tile	1710	.70	.70	1197	1197	*Acoustic tile added in*
gypsum	2970	.05	.05	149	149	*next evaluation tool.*
people, # occupants	16	.50	.50	8	8	
			TOTAL =	1651	3667	Sabins

Trial 1:

Display/Conf	Area (sf)	Coefficient of Absorption at 500Hz		Absorbency Area x α (Sabines)		
		"closed"	"open"	"closed"	"open"	
floor	800	.015	.015	12	12	
walls *brick*	350	.03	.03	11	11	
gypsum	820	.05	.05	41	41	
openings:						
closed	480	.04	NA	19	NA	*(includes 80 sf doors to*
open (such as windows)	480	NA	1.0	NA	480	*stairwell, upstairs and down)*
ceiling	1040	.05	.05	52	52	*(sloped and flat)*
people, # occupants	33	.5	.5	17	17	
			TOTAL =	152	613	Sabins

Trial 2:

Display/Conf	Area (sf)	Coefficient of Absorption at 500 Hz		Absorbency Area x α (Sabins)	
		"closed"	"open"	"closed"	"open"
	400	.015	.015	6	6
floor	400	.57	.57	228	228
walls brick	350	.03	.03	11	11
gypsum	820	.05	.05	41	41
openings:					
closed	480	.04	NA	19	NA
open (such as windows)	480	NA	1.0	NA	480
ceiling	1040	.05	.05	52	52
people, # occupants	33	.5	.5	17	17
			TOTAL =	374	835 Sabins

Trial 1:

Office	Area (sf)	Coefficient of Absorption at 500Hz		Absorbency Area x α (Sabins)		
		"closed"	"open"	"closed"	"open"	
floor	200	.57	.57	114	114	
walls brick	100	.03	.03	3	3	
gypsum	380	.05	.05	19	19	
openings:						
closed	120	.04	NA	5	NA	(includes 20 sf door)
open (such as windows)	120	NA	1.0	NA	120	
ceiling	200	.70	.70	140	140	
people, # of occupants	2	.5	.5	1	1	
			TOTAL =	282	397 Sabins	

Trial 2:

Office	Area (sf)	Coefficient of Absorption at 500Hz		Absorbency Area x α (Sabins)	
		"closed"	"open"	"closed"	"open"
floor	200	.57	.57	114	114
walls *brick*	100	.03	.03	3	3
gypsum	380	.05	.05	19	19
openings: closed	120	.04	NA	5	NA
open (such as windows)	120	NA	1.0	NA	120
ceiling	200	.05	.05	10	10
people, # occupants	2	.5	.5	1	1
			TOTAL =	152	267 Sabins

Evaluation Tool for Criterion A - "Room Liveness Graph"
1. Complete this table for your rooms:

	Volume ft^3	Total room absorption in sabins			
		1st Trial		2nd Trial	
		"closed"	"open"	"closed"	"open"
Factory	71,500	539	2555	1651	3667
Display/conference	10,000	152	613	374	835
Office	2,000	282	397	152	267

2. Use this information with the graph below (Flynn and Segil, 1970, p. 54) to evaluate your rooms.

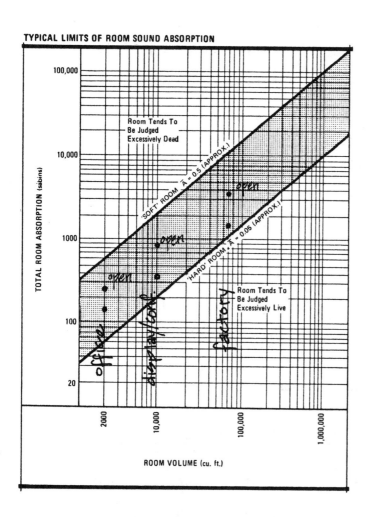

TYPICAL LIMITS OF ROOM SOUND ABSORPTION

(Trial 1)
Factory: neutral open, live
when closed. Looks ok for now.
Conference: neutral open, o
the live side when closed, ok.
Office: neutral-to-dead closed,
quite dead when open, ok for
now.

Figure 14-3 Room Liveness Graph from <u>Architectural Interior Systems</u> by J. E. Flynn and A. W. Segil (c) 1970 by Litton Educational Publishing, Inc. Reprinted by permission of Van Nostrand Reinhold Co.

Does your first trial design meet Criterion A?
yes
If not, what will you change in your second trial?

B. Criterion B: All barrier walls have STC ratings greater than or equal to the ratings in the MEEB Table 27.12, p. 1260-61.

Evaluation Tool for Criterion B - "STC Rating of Barrier Walls"
1. List all your barriers as keyed to the cluster plan.

---

2. Enter the recommended STC for each barrier from MEEB Table 27.12.

3. Find the actual STC and transmission loss (TL) for each barrier using MEEB Appendix A (use index A, then table A.1, then table A.2). See also Table 27.3, p. 1220, 27.5, p. 1230, and 27.6, p. 1231 and the discussion of walls with windows p. 1229 - 1231.

Some TL values at 500 Hz for glass are:

1/8" single plate glass	26
1/8" with rubber gasket	30
Laminated glass (viscoelastic between glass)	33
1/4" & 1/8" double glazing;	
with 2" air space	35
with 4" air space	42

(From Egan, 1972, p. 75)

Barrier	Recommended STC		1st Trial Actual STC	TL @ 500 Hz	2nd Trial Actual STC	TL @ 500 Hz
1	38	W5	42	41		
2	42	W5/W29	42/36	41/37		
3	40	W29	42	37	double glazed glass STC:	
4	38	glass	35	35	MEEB p. 1230, 1231	
5	42	Fla	44	75	glass TL: MEEB p. 1220	
6	38	W5/W29	42/36	41/37		
7	47	W5/W29	42/36	41/37	W5/W39	47/37
8	40	W29	36	37		

Does your first trial design meet Criterion B?
Yes, except for #4, #7 and #8.
If not, what will you change in your second trial?
#4, office exterior walls: live with STC=35, because 4" air space is more expensive and a bit thermally disadvantageous. #7, office to toilet wall: this one is rather critical, so go to resilient clips between studio and wall board. #8, office to arcade: live with this, since glass within these walls has similar STC, and office contact with circulation may be desirable for security reasons.
14.6 FINALIZING

GOALS

A. The offices and factory are acoustically comfortable places to work.

CRITERIA

A. Criterion A: Noise levels in the factory are within allowable dB levels.
B. Criterion B: The office can be naturally ventilated and stay within the NC goal.

DESIGN

Present full documentation of the design being evaluated in this section if it has changed from the design as documented in 14.4 Scheming and 14.5 Developing.

EVALUATION

A. Criterion A: Noise levels in the factory are within allowable dB levels.

Evaluation Tool for Criterion A – "Reverberation Time T_R"
The time during which sounds reverberate within a space is another useful characteristic for acoustic analysis. In "live" spaces, reverberation lasts longer than in "dead" ones. Although none of our spaces is critical for stage/audience situations for which T_R is usually calculated, T_R is also used as a guide in some noise control calculations.

$$T_R = \frac{.049 \times \text{Volume (cf)}}{\text{Absorbency (SABINS)}}$$

1. Determine recommended T_R for your spaces using MEEB, Fig. 26.26, p. 1188.
2. Calculate the T_R for your spaces using the formula above.

Space	Recommended T_R	Actual T_R 1st Trial closed	open	2nd Trial closed	open
Factory	.5 to 2	6.5	1.4	2.1	.96
Office	.6 to .8	.35	.25	.64	.37
Display/ Conference	.8 to 1.2	3.3	.8	1.3	.6

Do your spaces meet Criterion A in the first trial?
No.
If not, what will you change for your second trial?
Office: Trial 1 results are really dead! So, remove that ceiling tile.
Display/conference: too live. Add carpet to mezzanine floor.

Second trials: See 14.. Developing, Criterion A.

Factory, Trial 1:
closed:
$$\frac{.049 \times 71,500}{534} = 6.5 \text{ seconds}$$
- what a place to practice organ!

open:
$$\frac{.049 \times 71,500}{2555} = 1.4 \text{ seconds}$$
ok.

So: add acoustic tile to factory ceiling.

Factory, Trial 2:
closed:
$$\frac{.049 \times 71,500}{1651} = 2.1 \text{ seconds}$$
ok.

open:
$$\frac{.049 \times 71,500}{3667} = .96 \text{ seconds}$$
ok.

Evaluation Tool for Criterion A - "Room Noise Level-Sound Source in Room"

The factory contains 16 machines, each producing sound as well as manufactured goods and heat. The inherent noisiness of a given machine is called its Sound Power level, PWL, and is typically referenced to 10^{-12} watt (see Egan, 1972, p. 50, 51 for a more thorough discussion).

Once this PWL is known, the resulting noise level, the SPL (Sound Pressure Level), made by this machine in any given space can be determined. The graph below shows how the PWL of a noise source is changed by placing that source in a space of given volume and T_R. (In highly reverberant, smaller rooms the resulting noise level can actually be greater than PWL.)

1. Determine allowable dB level at 500 Hz, using MEEB Fig. 26.23, p. 1184 or Table 27.11, p. 1255.

factory NC = _____60_____

allowable dB at 500 Hz = __63__

2. Given: PWL of one machine 500 Hz = 70dB. Use graph
below and the T_R from above, to find the resulting SPL of
one machine.
closed _-15;_ SPL = 70-15=55dB, one machine
open _-18;_ SPL = 70-18=52 dB, one machine

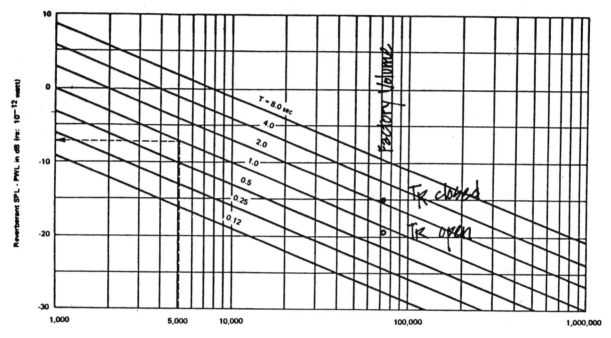

Figure 14-4 Sound Absorption reference, from
Concepts in Architectural Acoustics, p. 51 by Egan.
Copyright 1972. Used by permission of McGraw-Hill Book
Company.

3. Once the SPL in the factory from one machine is known,
the other 15 machines' noise can be added. For the
closed building condition, use McGuinness, Fig. 26.12 (p.
1171) to add dB sources. Each machine adds progressively
less dB to the previous subtotal.

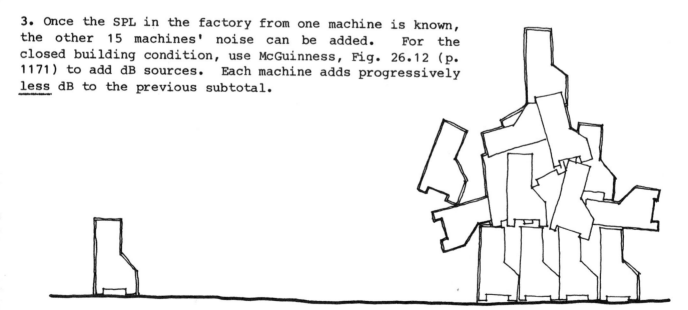

	1st trial	2nd trial	
Machine #1	55		dB
add #2	3		dB added
	58		SUBTOTAL
add #3	1.8		dB added
	59.8		SUBTOTAL
add #4	1.3		dB added
	61.1		SUBTOTAL
add #5	.9		dB added
	62.0		SUBTOTAL
add #6	.8		dB added
	62.8		SUBTOTAL
add #7	.6		dB added
	63.4		SUBTOTAL
add #8	.6		dB added
	64.0		SUBTOTAL
add #9	.6		dB added
	64.6		SUBTOTAL
add #10	.5		dB added
	65.1		SUBTOTAL
add #11	.5		dB added
	65.6		SUBTOTAL
add #12	.4		dB added
	66.0		SUBTOTAL
add #13	.4		dB added
	66.4		SUBTOTAL

---

add #14	.4	_____	dB added
	66.8	_____	SUBTOTAL
add #15	.3	_____	dB added
	67.1	_____	SUBTOTAL
add #16	.3	_____	dB added
TOTAL	67.4	_____	dB

Note: A much less tedious, though more mathematically complicated, approach is illustrated in Egan, 1972, p. 15.

IL one machine = 55dB

$$55 = 10 \log (I/10^{-16})$$

$$5.5 = \log (I/10^{-16})$$

$$3.15 \times 10^5 = I/10^{-16}$$

for one machine:

$$I = 3.15 \times (10^5 \times 10^{-16}) = 3.15 \times 10^{-11}$$

for 16 machines:

$$I = 16(3.15 \times 10^{-11})$$

$$I = 50.4 \times 10^{-11} = 5.04 \times 10^{-10}$$

$$IL = 10 \log (5.04 \times 10^{-10}/10^{-16})$$

$$IL = 10 \log (5.04 \times 10^6)$$

IL 16 machines = 67dB

$$IL = 10(6.70) = 67.0 \ dB$$

#14

4. What is the <u>difference</u> between the dB level of one machine and the total created by 16 machines?

1st trial _12.0 dB_   2nd trial _____

5. To estimate the SPL of the open building, as a safe estimate, add the above difference to the SPL of one machine in the <u>open</u> building:

_64.0dB_1st Trial _____ 2nd Trial   52 + 12 (p. S14-43, 44)

Is your first trial design within allowable dB levels determined in step 1 for both open and closed buildings?
allowable 63; 64 is close.
If not, what will you change in your second trial design?
Since we are only 1 dB over allowable SPL, we can add absorption at the machines when they are installed.

INSIDEOUT

_____

. Criterion B: The office can be naturally ventilated
and stay within the NC goal.

Evaluation Tool for Criterion B - "Final Site Noise Contours"
In the scheming phase it was assumed that 60 dB came
through the exterior walls from the factory interior.
Now that you know the actual SPL in the factory as well
as the characteristics of the wall, you can finalize the
amount of sound that escapes.

1. SPL in open factory: _64_ dB at 500 Hz

2. Use MEEB Fig. 27.23, p. 1229, to determine how the
open windows affect the barrier characteristics of the
exterior factory wall:

a. Which wall are you evaluating? *Assume factory walls
are same as office exteriors, W29 from MEEB.*
b. From 14.5 Developing Criterion B, p. 14-41, TL of this
wall:
 _37_ dB (500 Hz)

c. % of wall in ventilating openings	1st trial	2nd trial	
	_50_ %	___ %	
d. Resulting TLc	_3_ dB	___ dB	*(For TL of 35 or more)*
e. Resulting noise level at factory exterior wall			
SPL - TLc =	_61_ dB	___ dB	*64 - 3 = 61*

**Evaluation Tool for Criterion B - "Room Noise Sound
Level-Sound Source Outside"**
1. Establish the noise levels outside the office window.

a. The factory is _60_ feet away from the office.

About 60. Distance to actual private offices, here.

b. Add the constant sources using MEEB Fig. 26.12, p.
1171:
factory noise at office _41_ dB (assume a 3dB drop 1 foot
away from the factory as before in 14.2 Question A.)

freeway at office _55_ dB

total _55_ dB constant   *55 + .4 = 55.4, or 55*

*61 - 3 = 58 dB at 1 foot for
linear source, -3 dB for each
doubling of distance:*
55 dB at 2 ft.
52 dB at 4 ft.
49 dB at 8 ft.
46 dB at 16 ft.
43 dB at 32 ft.
40 dB at 64 ft.

----

c. Add intermittent sources

train at office _67_ dB

total _67_ dB intermittent & constant $67 + .4 = 67.4$, or 67

2. Use MEEB Fig. 27.23, p. 1229 to find the resulting "predicted noise level" <u>inside</u> the office.

a. Which office wall are you evaluating?

W29 from MEEB, opens onto arcade

b. PNC from 14.2 Analysis: _30 to 40_

c. Allowable dB at 500 Hz from MEEB Fig. 26.23, p. 1184
_35 to 45_ dB

d. TL of this wall from 14.5 Developing Criterion B
_37_ dB (500 Hz)

	<u>1st trial</u>	<u>2nd trial</u>
e. % wall in ventilating openings	_50_ %	_____ %
f. Resulting TLc	_3_ dB	_____ dB
g. Predicted constant noise indoors; constant exterior dB-TLc =	_52_ dB	_____ dB $55 - 3 = 52$
h. Predicted intermittent noise indoors; intermittent exterior-TLc =	_64_ dB	_____ dB $67 - 3 = 64$

Is the indoor noise level of your first trial design within allowable dB for Criterion B?
No; about 12 dB above, at constant noise level. It's a noisy site!
If not, what will you change for your second trial?
Ventilation is important, so find revised NC, when windows are open;
52 dB at 500 Hz conforms to NC = 47; (MEEB p. 1255). This is somewhat above "large offices"; moderately good to fair listening conditions.

Note: when windows can be shut (as could be temporarily the case): from MEEB Fig. 27.22 p. 1129
TL_1 = 37 dB, wall (14.6 finalizing)
TL_2 = 35 dB, glass; 50% of wall area
$TL_1 - TL_c$ = about 2 dB, from graph in MEEB
TL_c = 37 - 2 = about 35 dB
Therefore 55 dB constant outside - 35 dB = 20 dB; well within NC for private offices; and 67 dB intermittant outside - 35 dB = 32 dB, also well within NC.

BIBLIOGRAPHY

ASHRAE (American Society of Heating, Refrigerating and Air-Conditioning Engineers): Handbook of Fundamentals 1972 & 1977, and Standard 90-75: Energy Conservation in New Building Design 1975. ASHRAE, 345 East 47 St., New York City, 10017.

Balcomb, J. D. et al, Passive Solar Design Handbook Vol. 2, 1980, U. S. Department of Energy, "Passive Solar Design Analysis".

Crowther, Karen, "Night Ventilation Cooling of Mass" in Passive Cooling Applications Handbook, 1980, U. S. Department of Energy and Lawrence Berkeley Laboratory.

Evans, Benjamin, Measuring and Reporting Daylight in Buildings, and The Use of Models for Evaluation of Daylighting Design Alternatives, both unpublished. (Lawrence Berkeley Laboratory Materials)

Flynn, J. E. and Segil, A. W., Architectural Interior Systems, 1970. Van Nostrand Reinhold Company, 135 W. 50 St., New York City, 10020.

Hopkinson, R. G. et al, Daylighting, 1966. William Heinemann Ltd., Kingswood, Todworth Surrey, England.

Knowles, Ralph, Energy and Form, 1978. The M.I.T. Press, 28 Carleton Street, Cambridge, Massachusetts 02142.

Koenigsberger, Ingersall, Mayhew, Szokolay, Manual of Tropic Housing and Building, 1974. Longman Group Limited, Suite 1012, 19 W. 44th St., New York City, 10036.

Libbey-Owens-Ford, Sun Angle Calculator, 1974, 811 Madison Avenue, Toledo, Ohio 43695.

McGuinness, W., and Stein, B., Mechanical and Electrical Equipment for Buildings, 5th edition, 1971. John Wiley and Sons, Inc., 605 Third Avenue, New York City, 10016.

McGuinness, W., Stein, B., and Reynolds, J., Mechanical and Electrical Equipment for Buildings, 6th edition, 1980. John Wiley and Sons, Inc., 605 Third Avenue, New York City, 10016.

Mazria, Edward, The Passive Solar Energy Book, 1979. Rodale Press, Emmaus, Pa.

Millet, Marietta and Bedrick, James, Manual: Graphic Daylighting Design Method, 1980. Lawrence Berkeley Laboratory, 1 Cyclotron Rd., Berkeley, CA 94720. Also Architectural Daylighting Design Procedure, June 1978. Unpublished (charts as adapted from Hopkinson, Daylighting).

NOAA, (National Oceanic and Atmospheric Administration), Local Climatological Data. U. S. Department of Commerce, National Climatic Center, Asheville, N. C.

Olgyay, Victor, Design With Climate, 1963. Princeton University Press, Princeton, N. J. 08540.

Olgyay, Aladar and Victor, Solar Control and Shading Devices, 1957. Princeton University Press, Princeton, N. J. 08540.

Shurcliff, William A., Thermal Shutters and Shades, 1977. Brickhouse Publishing Co., Andover, Ma. 01810.

Stenhouse, D. S. & Wolf, C. M., "Daylighting Design for the Sacramento State Office Building Competition", Proceedings of the Second National Passive Solar Conference, 1978. American Section International Solar Energy Society, Killeen, Texas 76541.

Uniform Building Code, 1979. International Conference of Building Officials, 5360 South Workman Mill Road, Whittier, California 90601.

U. S. Department of Commerce, Climatic Atlas of the United States, 1968. Environmental Services Administration.

Villecco, M., Selkowitz, S., Griffith, W., "Strategies of Daylight Design", AIA Journal September 1979.

Watson, Donald, Energy Conservation Through Building Design, 1979. McGraw Hill.

Wright, David, "National Solar Cooling" Proceedings of the 3rd National Passive Conference 1979. American Section of the International Solar Energy Society, Killeen, Texas 76541.

NOTES

NOTES

NOTES

NOTES